The Springer Series on Demographic Methods and Population Analysis

Volume 39

In recent decades, there has been a rapid development of demographic models and methods and an explosive growth in the range of applications of population analysis. This series seeks to provide a publication outlet both for high-quality textual and expository books on modern techniques of demographic analysis and for works that present exemplary applications of such techniques to various aspects of population analysis.

Topics appropriate for the series include:

- General demographic methods
- Techniques of standardization
- Life table models and methods
- Multistate and multiregional life tables, analyses and projections
- Demographic aspects of biostatistics and epidemiology
- Stable population theory and its extensions
- Methods of indirect estimation
- Stochastic population models
- Event history analysis, duration analysis, and hazard regression models
- Demographic projection methods and population forecasts
- Techniques of applied demographic analysis, regional and local population estimates and projections
- Methods of estimation and projection for business and health care applications
- Methods and estimates for unique populations such as schools and students

Volumes in the series are of interest to researchers, professionals, and students in demography, sociology, economics, statistics, geography and regional science, public health and health care management, epidemiology, biostatistics, actuarial science, business, and related fields.

More information about this series at http://www.springer.com/series/6449

Robert Schoen
Editor

Dynamic Demographic Analysis

Springer

Editor
Robert Schoen
Population Research Institute
Pennsylvania State University
University Park, PA, USA

ISSN 1389-6784 ISSN 2215-1990 (electronic)
The Springer Series on Demographic Methods and Population Analysis
ISBN 978-3-319-26601-5 ISBN 978-3-319-26603-9 (eBook)
DOI 10.1007/978-3-319-26603-9

Library of Congress Control Number: 2016930170

Springer Cham Heidelberg New York Dordrecht London

Printed on acid-free paper

Springer International Publishing AG Switzerland is part of Springer Science+Business Media (www.springer.com)

Preface

A central strength of the discipline of demography is its mathematical core, which builds on the logically closed system in which the demographic processes of fertility and mortality shape a population's size and composition. Mathematical demography is now in transition, moving from a focus on the fixed rate models of the past to dynamic models, where the underlying vital rates change over time. That increased level of analytical complexity brings new and difficult challenges.

This volume presents state-of-the-art work on how current research is enlarging the scope of dynamic mathematical demography. It ranges across the field, including studies of fertility, mortality, population heterogeneity, the dynamics of population size and structure, and the thorny problem of age-period-cohort analysis. The primary audience of this book is academic demographers, but it will also be of interest to demographers in government and business and to some actuaries and statisticians.

As editor, I want to express my appreciation to Springer, for its many efforts in promoting demographic research, and especially to its editors Evelien Bakker and Bernadette Deelen-Mans. Vladimir Canudas-Romo, Kenneth C. Land, Adrian Raftery, and Roland Rau provided me with helpful comments. Let me also acknowledge the positive, facilitating role of the Population Association of America, which has greatly strengthened the field by bringing together mathematically oriented demographers from around the world. My greatest thanks go to the chapter authors, whose diligent work has made this volume possible.

University Park, PA, USA
August 2015

Robert Schoen

Contents

Chapter 1
Introduction

Robert Schoen

The twentieth century saw profound changes in human demography. The "demographic transition" saw much of the world change from high rates of birth and death to relatively low rates of birth and death, a shift accompanied by unprecedented longevity, population aging, and the emergence of sustained below replacement fertility in many developed nations. Those dramatic demographic changes created a population landscape in the year 2000 that was profoundly different from what existed in the year 1900.

A distinguishing feature of demography is its logical coherence: the size and structure of a closed population are fully determined, over time, by the population's rates of birth and death. Mathematical demography exploits that logical closure. The basic equation underlying the field is quite simple in form, and can be written

$$d\ P/dt = r\ P \tag{1.1}$$

where P denotes population, r the rate of change, and d/dt indicates differentiation with respect to time. Looking at total population size and constant growth rate, r, that differential equation is readily solved, yielding.

$$P(t) = P(0)\ e^{rt} \tag{1.2}$$

where P(t) is total population size at time t. The insight that constant demographic rates yield exponential growth was a prominent feature of Malthus' early thinking on population dynamics.

R. Schoen (✉)
Population Research Institute, Pennsylvania State University, University Park, PA 16802, USA
e-mail: rschoen309@att.net

R. Schoen (ed.), *Dynamic Demographic Analysis*,
The Springer Series on Demographic Methods and Population Analysis 39,
DOI 10.1007/978-3-319-26603-9_1

An exponential solution to Eq. (1.1) continues to result when the population is subdivided by age, and when the rates of mortality and fertility vary over age. The dominant paradigm of twentieth century mathematical demography was Lotka's stable population, where constant age-specific rates of birth and death were shown to imply exponential population growth and a time invariant age composition (Lotka 1907, 1939). Exponential growth and a constant composition are preserved even when the population is divided into multiple states, with persons free to move among them (Rogers 1975).

Unfortunately, the exponential solution to Eq. (1.1) breaks down in models recognizing age or state when the vital rates vary over time. Because the temporal ordering of vital events does matter, the products of matrices of vital rates do not commute, and cannot be represented by an exponentiated sum of matrices (cf. Gantmacher 1959). No analytical solution is generally possible, and population evolution over time must be described by a product integral that needs to be evaluated term by term (Volterra and Hostinsky 1938). Only a small number of special cases can be solved analytically (see Schoen 2006, Chap. 7). The shift from fixed rate to dynamic models thus represents a qualitative change in analytical complexity.

Mathematical demography is now a field in flux, moving past the fixed rate stable paradigm to confront the challenges of dynamic rates. New discoveries have upset well established beliefs, and entrenched conventions have had to be reconsidered. The twin demographic perspectives of period (a year or several years in a cross-section) and cohort (a closed group followed over time) have proven insufficient, and a new perspective, CAL or the wedge period, has been advanced (Brouard 1986; Guillot 2003). The Bongaarts-Feeney dynamic mortality model has led to new relationships between changing mortality, survivorship, and the lags between period and cohort behavior (Bongaarts and Feeney 2002; cf. Schoen 2006, Chap. 5). The Bongaarts-Feeney dynamic fertility model has led to the finding that a population can decline in size even when every birth cohort more than replaces itself (Bongaarts and Feeney 1998; Schoen and Jonsson 2003).

This volume on *Dynamic Demographic Analysis* presents work that advances dynamic analyses across a broad range of demographic topics. While lacking the overarching model that the stable population provided for fixed rate analyses, research on dynamic demography is making significant progress in a number of areas.

Chapter 2, in Part I of the volume, provides the paper by Joshua R. Goldstein and Thomas Cassidy on "Amplified changes: An analysis of four dynamic fertility models." Fertility analysis has long recognized the distinction between period and cohort perspectives, and examined the relationships between them. Ryder's classic work on "demographic translation" (Ryder 1964), further highlighted the concepts of "quantum" (fertility level) and "tempo" (fertility timing). The current pattern of below replacement fertility in much of the more developed world has heightened interest in the connections between period quantum and cohort timing. Beginning with Ryder's conceptual framework, Goldstein and Cassidy explicate four models of dynamic fertility, analyze them, and apply them to contemporary data. Their work leads to new insights into the issues encountered by such models, and the continuing tension between the heuristic appeal of the cohort perspective's focus on the behavior of a real group of people versus the reality that fertility behavior unfolds period by period.

Part II, Dynamic Mortality and Morbidity, considers several topics brought to the fore by qualitative changes in survivorship. The age pattern of mortality rates has not changed appreciably over recent decades, but the age pattern of deaths has. Infant mortality now affects a tiny proportion of births in low mortality countries, where the great majority of persons survive to the age of retirement. The focus of interest has accordingly shifted to mortality at older ages, where the modal age at death has emerged as a new index of longevity. In Chap. 3, "Am I halfway? Life lived = expected life," Vladimir Canudas-Romo and Virginia Zarulli explore the intriguing concept of the age where the number of years lived equals the average future life expectancy. Given alternative models, the chapter looks at how the "prime" of life has evolved and is likely to continue evolving.

Chapter 4, "Revisiting life expectancy rankings in countries that have experienced fast mortality decline" by Michel Guillot and Vladimir Canudas-Romo, exploits the wedge period perspective by examining the cohort truncated life expectancy at birth. Their empirical and theoretical work reveals a new index, the maximum truncated cohort life expectancy, and shows how there is a momentum of mortality disadvantage.

Chapter 5, "Changing mortality patterns and their predictability: The case of the United States," by Christina Bohk and Roland Rau, takes on the daunting task of forecasting future mortality. Bohk and Rau seek to synthesize previous approaches by focusing on rates of mortality improvement and by extrapolating American experience in combination with selected reference countries. Their median point estimates indicate that in 2050, life expectancy at birth for U.S. women is likely to reach 88.8 years, and for U.S. men, 85 years.

Chapter 6, "Modeling the dynamics of an HIV epidemic" by Jason R. Thomas and Le Bao, extends the methodology underlying models of epidemics, and applies the approach to the case of Tanzania. In the face of inadequate data, sophisticated mathematical models can track the spread of an epidemic, determine its peak, and provide insights into its future course.

In Part III, Analyzing Heterogeneity, we turn to analyses of mortality in non-homogeneous populations. The complex effects of differences in susceptibility to death have long intrigued demographers, and a considerable amount of work has been done in the area since the pioneering paper by Vaupel et al. (1979). Chapter 7 offers "Revisiting mortality deceleration patterns in a gamma-Gompertz-Makeham framework," by Felipe Ribeiro and Trifon I. Missov. A slowing of the pace of mortality decline is viewed as a consequence of population heterogeneity. The rationale is that since mortality declines produce more frail populations at older ages, that increase in frailty exerts an upward pressure on death rates at high ages. Overall mortality is estimated by a gamma-Gompertz-Makeham model, with the robustness of that parameterization carefully examined.

Chapter 8 is "Demographic consequences of Barker frailty," by Alberto Palloni and Hiram Beltran-Sanchez. Barker frailty is a heightened susceptibility to mortality at adult ages that stems from experiences around birth and early childhood. Palloni and Beltran-Sanchez show mathematically and via simulations that when mortality has been declining, Barker frailty can accentuate the mortality deceleration that naturally arises in frailty models (cf. Chapter 7). Analyses of Latin American countries suggest that, empirically, the Barker effect can substantially reduce gains in life expectancy at older ages.

Chapter 9, "Dynamic subpopulation ordering and mortality crossovers", by Elizabeth Wrigley-Field and Felix Elwert, takes a different tack toward analyzing heterogeneity. A mortality crossover occurs when population A, which has had lower age-specific mortality than population B, begins to exhibit higher age-specific mortality. In a standard frailty model, that can happen when one part of population B has lower mortality than a part of population A, and the subpopulation frailty composition shifts. For example, in the United States, a Black-White crossover has been reported, and explained by the argument that higher mortality among Blacks at younger ages leads to a more robust Black population at the high ages. Wrigley-Field and Elwert push the analysis further, and expose one of heterogeneity's unexpected tricks. When the heterogeneity of frailty occurs along multiple dimensions, the order of subpopulation mortality is dynamic, and multiple crossovers are possible.

Part IV presents analyses that extend the scope of stationary and stable population models. In Chap. 10, "The continuing retreat of marriage: Figures from marital status life tables for United States females, 2000–05 and 2005–10," I present empirical results from a new approach called Rate Estimation from Adjacent Populations (REAP). In multistate models with N states, the population composition by state at the beginning and the end of an interval constrain overall mortality over the interval, as well as N−1 of the rates of transfer between the N living states. In the marital status model considered, there are four living states and five rates of transfer between them. The REAP approach is able to use population data by age and state, plus two exogenous sets of transfer rates, to produce marital status life tables that reflect recent American marriage and divorce behavior. Marriage is continuing to retreat, though at a modest pace, while the probability of divorce is holding steady at about 43–46 %.

Chapter 11, "Emigration and the stable population model: Migration effects on the demographic structure of the sending country" by Cristina Bradatan, breaks new ground by examining the effects of outmigration on a population with below replacement fertility. The results, using data on Romanian migration to Spain, show that outmigration can have only a modest effect on further reducing the long term decline in population size, and can actually lead to a small reduction in the population's dependency burden.

Chapter 12 is "Exploring stable population concepts from the perspective of cohort change ratios: Estimating the time to stability and intrinsic r from initial information and components of change" by David A. Swanson, Lucky M. Tedrow, and Jack Baker. Cohort change ratios, sometimes referred to census survivorship (or progression) ratios, are used to examine the transition to stability in the presence of migration. Ergodicity, the "forgetting" of initial conditions, is examined through multiple indices, and the analysis provides an evaluation of the role of migration in the convergence to stability.

Part V considers analyses of the dynamics of population size and structure. Chapter 13, "Estimating the demographic dynamic of small areas with the Kalman Filter" by Manuel Ordorica-Mellado and Victor M. Garcia-Guerrero, offers a new approach to the estimation of small area population size. Data are drawn from censuses, remote sensing data (satellite images) of the subject area, and knowledge of the type of homes in the area. Estimations over time are then made using the Kalman Filter, an unbiased linear estimator frequently used in engineering applications but rarely seen in the social sciences. An application to Mexican data suggests that the Kalman Filter can produce improved estimates of small area populations.

In Chap. 14, Nan Li takes up the question "Are the pension systems of low fertility populations sustainable?" Below replacement fertility heightens the contrast between "pay-as-you-go" pension systems, which are funded on a period (year to year) basis, and "funded" pension systems, in which each cohort essentially funds its own pensions. The chapter compares the two systems using a time-based cohort old-age dependency ratio, and assesses the feasibility of funded systems.

Chapter 15, by Hana Sevcikova, Nan Li, Vladimira Kantorova, Patrick Gerland, and Adrian E. Raftery, discusses procedures used to create "Age-specific mortality and fertility rates for probabilistic population projections." In July 2014, the United Nations released its first probabilistic population projections for all world countries, based on the Lee-Carter model (Lee and Carter 1992). The chapter describes the methods used to convert projected total fertility and life expectancy values to time series of age-specific vital rates. Those methods include the new Coherent Kannisto Method to avoid sex mortality crossovers, new approaches to avoid "jump-off" (or initial value) bias, and the Rotated Lee-Carter method to adjust for mortality deceleration.

Part VI takes up the venerable Age-Period-Cohort problem, which has once again become a focus of interest. Since, by definition, Age equals Period minus Cohort, the three variables are collinear and not amenable to conventional statistical analyses. The challenge is to avoid the problems inherent in collinearity while recognizing the separate influences of factors related to age, period, and cohort. In Chap. 16, Sam Schulhofer-Wohl and Y. Claire Yang tackle the issue by "Modeling the evolution of age and period effects." They avoid collinearity by expressing cohort effects in terms of age-period interactions. Conceptually, that approach avoids the problems that follow from the linear model's assumption that the effect of age is the same at all times, that the effect of period is the same for all ages, and that cohort effects do not change over time. Empirically, the chapter shows that their age-period interaction model provides a clearly superior fit to that of the linear model when applied to 150 years of age-specific mortality rates for Sweden.

Chapter 17, "Bayesian ridge estimation of age-period-cohort models" by Minle Xu and Daniel A. Powers, takes a different approach. It employs ridge regression where the ridge estimator is viewed from a Bayesian perspective, i.e. it is estimated in the light of available prior information. As is (and should be) the case in Bayesian models, the final estimates depend on the prior values chosen; better information yields better results.

In sum, the work presented here expands and extends our understanding of demographic behavior. In a rapidly changing world, dynamic analyses bring us closer to contemporary reality.

References

Bongaarts, J., & Feeney, G. (1998). On the quantum and tempo of fertility. *Population and Development Review, 24*, 271–91.

Bongaarts, J., & Feeney, G. (2002). How long do we live? *Population and Development Review, 28*, 13–29.

Brouard, N. (1986). Structure et dynamique des populations: La pyramide des années à vivre, aspects nationaux et examples régionaux. *Espaces, Populations, Sociétés, 2*, 157–68.

Gantmacher, F. R. (1959). *Matrix theory*. New York: Chelsea.

Guillot, M. (2003). The cross-sectional average length of life (CAL): A cross-sectional mortality measure that reflects the experience of cohorts. *Population Studies, 57*, 41–54.

Lee, R. D., & Carter, L. (1992). Modeling and forecasting the time series of U.S. mortality. *Journal of the American Statistical Association, 87*, 659–671.

Lotka, A. J. (1907). Relation between birth rates and death rates. *Science, 26*(N.S.), 21–22.

Lotka, A. J. (1939). (orig.). *Analytical theory of biological populations*. Translated and with an Introduction by D.P. Smith & H. Rossert. New York: Plenum Press, 1998.

Rogers, A. (1975). *Introduction to multiregional mathematical demography*. New York: Wiley.

Ryder, N. B. (1964). The process of demographic translation. *Demography, 1*, 74–82.

Schoen, R. (2006). *Dynamic population models*. Dordrecht: Springer.

Schoen, R., & Jonsson, S. H. (2003). A diminishing population whose every cohort more than replaces itself. *Demographic Research, 9*, 111–18.

Vaupel, J. W., Manton, K. G., & Stallard, E. (1979). Impact of heterogeneity in individual frailty on the dynamics of mortality. *Demography, 16*, 439–54.

Volterra, V., & Hostinsky, B. (1938). *Operations infinitesimals lineares*. Paris: Gauthier-Villars.

Part I
Analyzing Dynamic Fertility

Chapter 2
Amplified Changes: An Analysis of Four Dynamic Fertility Models

Joshua R. Goldstein and Thomas Cassidy

2.1 Introduction

In this chapter, we provide overviews of four models for the dynamics of fertility change. These models are applicable to populations with low and moderate birth rates, populations that have already completed the demographic transition and are now subject to the ups-and-downs of period fertility.

Implicit in each model is its own story of the behavior driving fertility. Each model offers its own lessons about how relatively small changes in one aspect of fertility (e.g., timing or targets) can produce large changes in another aspect such as total fertility.

Demographers use these models to try to understand why fertility change is occurring and to make informed predictions about the future level of births. Ideally, such models would allow a better understanding of the determinants of fertility change. But, as we will see, the stylized assumptions behind each model are better suited for providing insights into the variety of forces that influence fertility rather than the creation of measures that are free of "distortions" or "bias."

In this chapter, we present the models in the order that they were developed. We begin with Norman Ryder's (1964) classic formulation of the translation between period and cohort measures of fertility. Ryder himself was strongly committed to giving priority to the cohort perspective on fertility (Ryder 1965, cf. Ní Bhrolcháin

J.R. Goldstein
University of California Berkeley, 2232 Piedmont Avenue, Berkeley, CA 94720, USA
e-mail: josh@demog.berkeley.edu

T. Cassidy (✉)
Bucknell University, 1 Dent Drive, Lewisburg, PA 17837, USA
e-mail: tcassidy@bucknell.edu

R. Schoen (ed.), *Dynamic Demographic Analysis*,
The Springer Series on Demographic Methods and Population Analysis 39,
DOI 10.1007/978-3-319-26603-9_2

1992). The formal nature of his model is bidirectional, allowing translation from periods to cohorts or vice versa.

The second model we present is Ronald Lee's moving-target formulation. This model is distinctly more behavioral, translating between targeted cohort fertility goals and the flow of births seen in period fertility. Lee's model can be seen as translation model, but it is not translating between total cohort fertility and total period fertility, but rather between the "desired fertility" or "target" of a cohort and observed period fertility.

The modern work on fertility dynamics was pioneered by Bongaarts and Feeney (1998), whose approach has been extended by many, including Kohler and Philipov (2001), and Bongaarts and Sobotka (2012). Here the implicit behavioral model is not about changing targets of completed family size but about the timing of planned children. Bongaarts and Feeney build on Ryder's general insight about fertility timing to show that in the particular case where births are shifted from one period to another, the period fertility rate will change even if there is no change in completed fertility.

Finally, we present our own work (Goldstein and Cassidy 2014), which can be seen as the re-application of Bongaarts and Feeney's thinking to the timing of cohort fertility. Instead of relatively high-frequency variation in period birth timing, we focus on gradual changes of the timing of cohort fertility. The result is that period fertility is influenced by cohort timing in a manner similar to what Ryder first envisioned, but in the more modern context of shifting births.

All of these models have in common a simplified view of how fertility change occurs. A remarkable lesson told by all of these formulations is that small changes in timing or targets can produce large fluctuations in period fertility. This is what we see as the take-home lesson of the last half-century of fertility models. It is not that one model is right and the others are wrong, but rather that all of the models tell us period fertility is remarkably sensitive to changes in other underlying aspects of the fertility process.

The sensitivity of period fertility is seen by some (e.g. Sobotka and Lutz 2010) as reason to abandon the total fertility rate as a meaningful demographic measure. However, population processes like the size of birth cohorts and the shape of the population age pyramid are driven by period fertility. So even though period fertility may be fickle and fleeting, it is important enough to merit the attention that demographers have given it.

2.2 Fertility Models

When describing the dynamics of fertility over time, demographers distinguish between cohort and period measures. Cohort measures apply to the life-times of people born at the same time, whereas period measures are cross-sectional, and apply across the ages of people alive at the same moment in time. Cohort measures are more “real” in the sense that they summarize the lifetime experience of a group of individuals. But period measures are also important, in that the rate of birth at a given moment in time determines the unrolling of demographic history that is seen in a population’s age structure. The period total fertility rate is the most commonly used measure of fertility.

Beginning with Ryder, demographers have also distinguished between demographic change brought about by changes in “tempo” (the timing of births) and changes in “quantum” (the level or intensity of births). Quantum can be operationalized as a change in the level of fertility across all ages. In contrast, a tempo change can be thought of as an adjustment in the timing of childbirth that shifts the mean age at birth up or down without raising or lowering the fertility schedule. One can distinguish between a *period* quantum change, which is an increase or decrease between periods, and a *cohort* quantum change, where the adjustment varies by cohort. Similarly, we can separate *period* tempo changes from *cohort* tempo changes, where the former refers to a rigid shift in the period fertility schedule, while the later assumes the shift operates on the cohort fertility schedule.

Tempo and quantum are idealized concepts, and the dichotomy between the two cannot be maintained when switching between cohort and period perspectives. For example, it is possible that a sequence of changes in cohort quantum, when viewed from the period perspective, could be indistinguishable from a period tempo change. Likewise, a sequence of period quantum changes could appear from the cohort perspective to be a cohort tempo change. The aim of some models, for example the period-shift model of Bongaarts and Feeney or our own cohort-shift model, is to make tempo and quantum changes separable. In these cases, the goal is to obtain a “pure” measure of quantum that is not influenced by timing changes. While actual changes in fertility may be a blend of tempo, quantum, and other transformations, the distinction between tempo and quantum remains a valuable interpretative tool for demographers. By classifying changes into these two basic types, dynamic demographic models distinguish decisions about when to have children from decisions about how many children to have.

Each of the models we present will have some of its own notation, however some common notation can be used across all the models. Denote the fertility rate at age a and time t by $f(a,t)$. Births are the product of the fertility rate and person-years of exposure, so that a woman exposed for 1 year to a fertility rate of 0.2 would be expected to average 0.2 births. We use $c = t - a$ to denote the birth year of a cohort, so that the fertility rate at age a of the cohort born at time c is given by $g(a,c) = f(a, c+a)$. The (period) total fertility rate in year t is defined as the sum of age-specific rates over the whole age range,

$$TFR(t) := \int_{\alpha}^{\beta} f(a,t)\, da \tag{2.1}$$

where α and β are the lower and upper biological limits of childbearing. The completed total fertility rate for the cohort born in year c is given by

$$CTFR(c) := \int_{\alpha}^{\beta} f(a, c + a)\, da = \int_{\alpha}^{\beta} g(a, c)\, da. \tag{2.2}$$

We let $f_0(a)$ denote a normalized, standard baseline fertility schedule that sums to one, giving the baseline probability density of births by age. The models discussed in Sects. 2.5 and 2.6 assume the existence of period quantum effect, i.e. a level change in period fertility as a consequence of events that are independent of timing. We use $q(t)$ to denote such an effect.

2.3 Ryder's Approximation

Norman Ryder's classic result (1964) pioneered the formal analysis of the relationship between period and cohort total fertility in the context of changing age-specific rates.

Without changes in fertility timing, it is easy to imagine that cohort total fertility is a moving average of period total fertility. If each year the schedule of fertility has the same shape, with an unchanging share $f_0(a)$ having children at age a every year, and period total fertility equal to $q(t)$, then the surface of fertility over age and time will have the form

$$f(a, t) = q(t) f_0(a) \tag{2.3}$$

and cohort total fertility of the cohort born in year c will be

$$CTFR(c) = \int_{\alpha}^{\beta} f(a, c + a)\, da = \int_{\alpha}^{\beta} q(c + a) f_0(a)\, da. \tag{2.4}$$

Thus, cohort total fertility is a moving average of period total fertility with weights equal to the share of childbearing occurring at each age.

An analogous relationship could be written expressing period total fertility as a moving average of cohort totals, with the assumption of constant cohort age-distribution.[1] In either case, however, there should be no consistent divergence of cohort and period fertility – the long run averages of cohort and period fertility should be roughly equal, particularly if the appropriate correspondence between dates and ages is made.

The innovation of Ryder was to introduce the possibility of a changing age schedule. Any surface of age specific rates can be written as a Taylor series

[1] If we posit a world with fluctuating cohort quantum $q(c)$ and fertility rates given by $f(a, t) = q(t - a) f_0(a)$, then period total fertility rate is a moving average of cohort total fertility. In most countries we see greater variability in period measures than in cohort measures, suggesting that the model with fluctuating period quantum is a better approximation of reality.

$$f(a,t) = f(a,0) + f'(a)t + \dots, \tag{2.5}$$

where the $f'(a)$ term is the rate of change with respect to time in fertility at age a. Taking only these first two terms, some algebraic manipulation[2] shows that for the cohort born in year t,

$$CTFR(t) \approx TFR(t + \mu_c)/(1 - \mu_c'), \tag{2.6}$$

where μ_c is the cohort mean age of childbearing of the cohort born in year t and μ_c' is its time-derivative. Note that in this correspondence, the dating of periods and cohorts are offset. In (2.6), the fertility of the cohort born in year t is compared with period fertility in year $t + \mu_c$, several decades later. An analogous expression can be written in terms of the period age of childbearing.[3] The same translations can also be written for parity-specific fertility.

The importance of Ryder's result is that it allows us to understand how prolonged changes in fertility timing can produce a situation in which the period and cohort total fertility rates are consistently different from each other. During the Baby Boom, the mean age of birth was falling at an annual pace of roughly a tenth of a year, raising observed period fertility by about 10 % over the completed fertility of corresponding cohorts. More recently, the postponement of fertility has been at a similar pace of about a tenth per year, making period fertility some 10 % smaller than cohort fertility.

Ryder's formulation is completely general, in the sense that any continuous and differential time series in age-specific rates can be expressed as a Taylor series. His specific result relating the total fertility of periods and cohorts arises in the simple case of linear change but is completely general as to the age pattern of this linear change, allowing for changes in the level ("quantum") of completed cohort fertility as well as in period fertility. One can view this result either as an exact expression of linear change or as a first order (linear) approximation of more complicated changes. The tempo-models introduced more recently by Bongaarts and Feeney, Kohler and Philipov, and Goldstein and Cassidy (among others) add more structure as to the specific nature of age-specific change. However, as we will

[2]Set μ_c to be $\frac{\int af(a, t+a)\,da}{CTFR(t)}$ and use the first to two terms of (2.5) to show that $\mu_c' = \frac{\int (a-\mu_c)f'(a)\,da}{CTFR(t)}$. Note that $CTFR(t) = \int f(a,0)\,da + t\int f'(a)\,da + \int af'(a)\,da$. From Eq. (2.1), we get $TFR(t+\mu_c) = \int f(a,0)\,da + t\int f'(a)\,da + \int \mu_c f'(a)\,da = CTFR(t) - \int af'(a)\,da + \int \mu_c f'(a)\,da = CTFR(t)(1-\mu_c')$. It follows that $CTFR(t) = \frac{TFR(t+\mu_c)}{1-\mu_c'}$.

[3]Letting $\mu_p'(t)$ be the derivative of the period mean age of childbearing at time t, we obtain

$$CTFR(t - \mu_p) \approx TFR(t) \times (1 + \mu_p'(t)).$$

With this translation, one can estimate cohort fertility from easily obtainable period measures.

see, they produce similar, even if not identical, estimates of the relationship between period and cohort rates (or tempo-adjusted and observed period rates) to that found in Ryder's classic expression. The formal interpretation of this similarity is that Ryder's first order approximation can be applied to any model of fertility change. A more substantive interpretation is that the importance of changes in timing revealed by Ryder are of such importance that they manifest themselves in a wide family of models for fertility change.

2.4 Moving Targets

We now turn to Ronald Lee's more behavioral formulation of fertility dynamics, modeling the process by which desired family size translates into period quantum. His (1980) "moving-target" model has women updating their family size targets from one period to the next and shows the consequences of changing intentions on the flow of births as measured by fertility rates. Lee shows how this process can be modeled as a differential equation and uses his model, among other things, to show how relatively large changes in period total fertility can result from small changes in fertility goals.

Lee's moving-target model allows desired completed family size, $D(t)$, to change from year to year in response to broad socioeconomic forces, which are assumed to influence all ages equally. This framework is in contrast to what could be called a "fixed-target" model, in which women's target family size is determined early in life and does not change. The moving-target model relies on the simplifying assumption that all cohorts alive in year t share the same desired completed family size. Based on their predictions about the future socioeconomic climate and other factors, couples make decisions about an ideal family size, and changes in these decisions lead to revisions in the target. Since predictions of the future are constructed out of current experiences, this model allows women to revise their target each year. Thus D is function of the year t. Survey data indicates that for the period beginning in 1946 through the mid 1960s, the value of D for women in the U.S. rose, but then declined rapidly in the 1970s.

Lee's model for fertility can be thought of as analogous to the industrial process of inventory and manufacture, in which we can imagine large swings in production based on the gap between actual and desired inventories. When the demand for children outstrips the current supply, women can respond by increasing their fertility. There are, however, some obvious biological limitations to viewing fertility changes as a sequence of stock adjustments. In particular, if supply were to exceed demand, there is no way to reduce inventory.[4] Consequently, the analogy works

[4] Changes in additional desired fertility create an important asymmetry for the moving-target model. When desired family size is rising, women can respond with adjustments to their fertility intentions that take into account the new ideal family size. In fact, as desired family size rises,

best if we imagine that the inventory is always less than the demand, and it is the changing size of this gap that leads to variation in production.

In the moving-target model, a cohort's fertility in a given year is a function of the gap between each cohort's average achieved fertility and the value of D for that year. This gap, called additional desired fertility, varies both by year and by cohort. It is defined as the difference between the current desired family size and a cohort's average accumulated fertility to date. For example, if 2.6 children represented the average desired family size in the year 1980, and if women from the cohort of 1955 had an average of 1.9 children by 1980, then for this cohort the additional desired fertility is 0.7 children. The assumption of a single target for women of all ages means that different cohorts would experience the same goal of 2.6 children, but because each cohort can have a different accumulated fertility, additional desired fertility varies by age within a given year.

In order to make the model identifiable in a simple way, Lee's formulation makes the additional assumption that, for women aged 25 and over, the annual birth rate is a fixed fraction of the additional desired fertility. Based on a comparison between survey evidence of additional desired fertility and birth rates in the subsequent year, Lee approximates this fixed fraction, denoted λ, as 0.18. Extensions of the model could account for variation in λ with age.

We now turn to a formal development of the moving-target model. Let $C(a, c)$ denote cumulated fertility of the cohort born in year c at age a, i.e. $C(a, c) := \int_0^a g(x, c)dx$. (Inversely, one can obtain age-specific fertility rates from the cumulative cohort fertility by differentiating, with the partial derivative $\frac{\partial}{\partial a}C(a, c)$ equal to $g(a, c)$). We use $A(a, c)$ to denote the additional desired fertility for women in cohort c, i.e.,

$$A(a, c) = D(a + c) - C(a, c). \tag{2.7}$$

The fixed fraction assumption for the rate at which desired fertility is realized means that, for $a \geq 25$, we have $g(a, c) = \lambda A(a, c)$. Multiplying Eq. (2.7) by λ yields

$$g(a, c) = \lambda D(a + c) - \lambda C(a, c). \tag{2.8}$$

For a fixed cohort c, we can differentiate this equation with respect to age a to produce

$$\frac{\partial}{\partial a}g(a, c) = \lambda D'(a + c) - \lambda g(a, c). \tag{2.9}$$

women who had perhaps thought that their family was complete, might change their minds and plan for additional children. In contrast, if desired family size is declining, some women who have completed childbearing may find themselves with children that were wanted when conceived, but are now part of a larger family size than the current ideal. Lee calls this phenomenon "the irreversibility of fertility," and formulates an alternate version of the moving-target model to address this issue.

This first order linear differential equation can be solved for $g(a, c)$ to obtain

$$g(a, c) = \lambda e^{-\lambda a} \left[\int_0^a e^{\lambda u} D'(u + c) du + D(c) \right]. \tag{2.10}$$

This formula presents fertility rates as an outcome of previous change in desired family size. For our purposes, the importance of this result is not that it allows us to calculate fertility rates from known targets. Rather, this model allows us to understand, in a formal way, the complex interplay of changing desires, past behaviors, and period measures. It presents observed fertility rates as the outcome of an evolving history of target family sizes. The model implies that women can change their fertility intentions each year and that their fertility choices are determined by the gap between the current ideal size and the fertility already achieved as a consequence of their pursuit of previous targets. This means that observed fertility rates are not simply a lagged version of desired family size, with period total fertility replicating the desired family size of a few years earlier. Instead, the model presents period total fertility as a function of the rate of change in the desired family size. We explore implications of this relationship next.

Implications of the moving-target model The moving-target model has several interesting consequences, provided that desired fertility D changes smoothly over time. These consequences are illustrated in detail in Lee (1980), where the implications of both a linear and a sinusoidal shape for D are investigated. We discuss just a few of these consequences here, including one that is, at least superficially, counterintuitive.

First, the underlying assumptions in the moving-target model have implications for the analysis of fertility change over time. If we accept the central tenet that desired family size is a function of period rather than of cohort, then a cohort's completed fertility should not be interpreted as a measure of that cohort's fixed intentions throughout its reproductive years. Rather, the completed fertility of a cohort is the outcome of changing desires, and may tell us nothing about the aspirations of the cohort when it began its reproductive years.

Second, when desired family size D is rising, TFR is greater than D, and when desired family size is falling, TFR is less than D. Although this phenomena can be demonstrated by substituting various specific functions of desired family size into Eq. (2.10), it can also be understood as a consequence of the fact that additional desired fertility is the difference between current desires and previously accumulated fertility. When desired family size is rising, cumulative fertility C will lag behind current desires because C is the consequence of behavior in previous years when desired family size was lower. It follows that women at all ages, in order to catch up with the latest desired target, will need to increase their fertility by more than would have been necessary had the latest value of D been in place in previous years. Similarly, if family size is falling, then cumulative fertility will represent births that occurred in years when the goal was higher, and women will need to reduce births to avoid overshooting the new target.

Third, when desired family size fluctuates, *TFR* will fluctuate with greater amplitude, so that in effect changes in *TFR* are exaggerated versions of changes in *D*. This fact follows from the previous result, since *TFR* will be larger than desired family size when desired family size is increasing but become smaller than desired family size when desired family size is decreasing.

Fourth, when desired family size fluctuates, turning points in *TFR* may precede those in desired family size by several years. Intuitively, we might have thought of *TFR* as responding to changes in the desired family size, and so we might expect that turning points in *TFR* would follow turning points in *D*. However, *TFR* is a sum of fertility rates that are determined not by the value of *D*, but rather by the difference between *D* and the *C*. From Eq. (2.7), we see that fertility rates, and consequently *TFR*, will be highest when the difference between desired family size and cumulative fertility is greatest. If desired family size has been increasing but begins to slow down, this gap will shrink, and *TFR* will begin to drop. By the time that desired family size peaks, *TFR* will already be dropping.

2.5 Period Shifts

In this section and Sect. 2.6 we present two models in which births are shifted, either postponed or advanced. One approach is period-based and the other is cohort-based. Much of the material in these sections is drawn directly from Goldstein and Cassidy (2014), where a more detailed comparison of these models can be found.

The period-shift model of fertility change appears in the appendix to Bongaarts and Feeney (1998). An alternate derivation is provided in Rodriguez (2006). Bongaarts and Feeney were interested in eliminating distortion in *TFR* caused by changes in the timing of births. Their influential paper introduced a method for correcting these distortions, and inaugurated a debate that continues to this day about how to adjust period measures of fertility and whether such adjustments provide a meaningful picture of fertility behavior. Some objections to the Bongaarts-Feeney adjustment technique have been raised (Kim and Schoen 2000), and modifications to the adjustment procedure have been proposed (Kohler and Philipov 2001; Bongaarts and Feeney 2006). In this chapter, our focus is on the assumptions underlying this adjustment procedure and the consequences of these assumptions, not on the arguments for or against tempo adjustment.

The period-shift model aims to get at a “true” period quantum $q(t)$ that is exogenous to the model. Bongaarts and Feeney's formulation assumes that the timing of fertility is a function only of the year t, so that women of all ages in a given year will postpone or advance the timing of childbirth by the same amount. This shift in timing acts on a base-line fertility schedule f_0, and thus the shape of the fertility schedule is assumed to be constant, although both its position and level are allowed to change. Period-based postponement is represented by $R(t)$, which denotes the total number of years by which women in period t have shifted their fertility, i.e., if $R(2010) = 3$ then women in the year 2010 have postponed childbirth by 3 years relative to the baseline schedule. Period fertility rates are determined both by the timing changes, via R, and by changes in the period quantum $q(t)$. R' denotes the derivative $\frac{dR}{dt}$.

The period-shift model of fertility Although originally presented as an adjustment procedure, it is possible to formalize Bongaarts and Feeney's approach in terms of a model of fertility change over time, with the age-period surface of fertility described as

$$f(a,t) = f_0(a - R(t))(1 - R'(t))q(t). \tag{2.11}$$

Notice that the period-shift model includes the term $(1 - R'(t))$ that lowers or raises the fertility level depending on whether women have delayed or advanced fertility. This term appears because timing changes that occur within cohorts spread or consolidate period births.

A derivation of the model, due to Rodriguez (2006), is as follows. Let $F_0(a) = \int_0^a f_0(x)dx$ be the cumulative fertility for the baseline schedule. We assume $F_0(\beta) = 1$, where β is sufficiently large so as to be well beyond the last age of fertility, leaving room for shifts. Under the period-shift model, the shifted cumulative fertility schedule for women born in year c and currently of age a is $F_0(a - R(c + a))$. Differentiating cumulative fertility with respect to age a gives us $f_0(a-R(c+a))(1-R'(c + a))$. To obtain the observed fertility rate $f(a,t)$, we include period quantum effects $q(t)$ and rewrite cohort in terms of period and age.

If $f(a,t)$ is given by Eq. (2.11), then the total fertility rate will simplify to a product of independent quantum and tempo terms.

$$TFR(t) = \int_\alpha^\beta f(a,t)da = q(t)(1 - R'(t)). \tag{2.12}$$

If we wish to determine the value of q that would have been observed in period t had there been no postponement, we need to estimate R'. This turns out to be relatively easy, since Eq. (2.11) implies that $R' = \mu_p'(t)$, where $\mu_p(t)$ denotes mean age at birth in period t. This derivative can be approximated by comparing the mean age at birth in year t with that in adjacent periods. Thus we have the shift-adjusted total fertility rate

$$TFR^*(t) := TFR(t)/(1 - \mu_p'(t)). \tag{2.13}$$

If the hypotheses of the period-shift model hold, then $TFR^*(t)$ will be equal to $q(t)$, and this adjustment technique will reveal the true period quantum. In practice, the period mean age at birth is sometimes influenced by changes in parity composition, since first births tend to occur to women at younger ages while higher parity births occur to older women. For example, if women in year t choose to have smaller family sizes, then they will forgo higher order births, and this will lower the mean age at birth even though there may be no change in the timing of births of any fixed order. For this reason, Bongaarts and Feeney recommend that (2.13) be used to calculate TFR_i^* separately for each birth order i, with adjusted total fertility of all orders being the sum of the adjusted parity-specific total fertility rates.

Bongaarts and Feeney's tempo adjustment Eq. (2.13) is strikingly reminiscent of Ryder's period-cohort translation relationship given by Eq. (2.6). Indeed, when Ryder's relationship is re-expressed in terms of changes in the period mean age μ_p, one obtains

$$CTFR(t - \mu_p) \approx TFR(t) \times (1 + \mu_p'(t)).$$

These expressions are nearly the same, but with a minor and a major difference. The minor difference is that one expression is *divided* by $1 - \mu_p'$, whereas the other is *multiplied* by $1 + \mu_p'$, but for small values of μ_p' this numerical difference is very small. The major difference is that the period-shifts approach is giving us a formula for tempo-adjusted period fertility, whereas the Ryder approach gives us a formula for cohort fertility. The lesson we draw from this similarity is not that tempo-adjusted period fertility is a measure of cohort fertility. Indeed, Bongaarts and Feeney repeatedly caution against this interpretation. Rather, we take the approach that the first-order approximation of age-specific fertility change is a good representation of a wider class of models, including tempo-shifts. The other commonality, as we have emphasized, is that both approaches show the influence of timing changes, in which small changes in the pace of timing change are consistent with large changes in the summary measure of interest.

The story implicit in the period-shift model has several appealing features. Fluctuations in fertility rates are represented as the result of two forces, both period driven. The first is quantum, which is understood to be the outcome of period specific factors outside of this model. The second force is the timing of births, which can be postponed or advanced in response to changing socioeconomic conditions. Since both of the forces are functions of period only, this model does not require one to know the complete fertility history for any of the cohorts active in year t. Rather, one can examine fertility rates in adjacent years to measure the mean age at birth, and from this demographers can uncover the value of the quantum q and then measure the size of the timing influence as the difference between TFR and q.

The hypothesis that changes in timing are a purely period-driven phenomenon means that a recent history of postponement has no long-term impact on the total fertility rate. For example, if we assume that q is constant, then the end of a period of shifting means that $R'(t)$ will be zero after 1 year, and so all fluctuations in the total fertility rate will cease. In this way, the period-shift model asserts that timing decisions made in the past do not impose on the future. In contrast, the moving target model in Sect. 2.4 and the cohort-shift model in Sect. 2.6 each imply that decisions made in the past can continue to influence fertility rates for years.

Any useful formal model will require some simplifying assumptions. The strongest, and therefore most questionable, assumption in the period-shift model is the idea that women of all ages will shift their timing by the same amount in a given year. However, it is plausible that women would respond to a period shock in different ways, with, for example, women at younger ages choosing to postpone more in response to an economic slowdown, while women at older ages would postpone less or not at all. A further challenge to this assumption is that it allows for the complete independence of behavior at different ages across a given cohort. So, for example, postponement at younger ages has no direct implication for the fertility of the same cohort as it ages. For example, a decision to postpone childbirth until after completing college or after establishing a career has implications for years beyond the time when the decision was made. It is with these criticisms in mind that we turn to the cohort-shift model.

2.6 Cohort Shifts

The cohort-shift model (Goldstein and Cassidy 2014) decomposes observed fertility into an interaction of cohort-based decisions about fertility timing and period-based decisions about the intensity of fertility. Fluctuations in period fertility are thus understood as the result of both cohort tempo, i.e. changes in period fertility that results from the timing of births in cohorts, and period quantum, i.e. level changes in period fertility as a consequence of events that are independent of age and cohort. Like the period-shift model, the cohort-shift model treats quantum q as an exogenous function of period t. However, timing changes are now seen as a cohort, rather than a period, phenomenon.

Cohort-based postponement is introduced through a shift function $S(c)$, the cohort analog of the period shift R. The shift $S(c)$ indicates the amount by which women from cohort c have shifted their fertility relative to the baseline schedule $f_0(a)$.[5] For example, if "33 is the new 30" for the cohort of 1960, then $S(1960) = 3$.

The cohort-shift model of fertility The model of the fertility age-period surface consistent with the cohort-shift model is

$$f(a,t) = f_0(a - S(t-a))q(t). \tag{2.14}$$

This model is derived as follows. The shifted cumulative fertility schedule for the cohort-shift model is $F_0(a - S(c))$. Differentiating cumulative fertility with respect to age a gives us $f_0(a - S(c))$, the fertility rate for cohort c at age a in the absence of any period quantum effects. To obtain the observed fertility rate $f(a,t)$, we include period quantum effects and rewrite in terms of period and age. Note that unlike the case in the period-shift model, there is no need for an R' term or its cohort equivalent in the cohort-shift description of the fertility surface. This is because, unlike the case with period shifts, cohort shifts do not spread or consolidate births within cohorts.

[5] Simultaneous period and cohort postponement can be modeled as a sum of R and S. This combined model is discussed in Goldstein and Cassidy (2014).

The cohort-shift model implies its own tempo-adjusted measure of period fertility, denoted $TFR^{\dagger}(t)$. This is a period measure, and should not be used to approximate levels of cohort fertility. Rather, $TFR^{\dagger}(t)$ represents the total fertility rate that would have been observed if the cohorts active during year t had neither postponed nor accelerated their fertility schedules.

Under the assumptions of the cohort-shift model (2.14), $q(t)$ can be recovered from observed rates by defining the shift-adjusted period total fertility rate as

$$TFR^{\dagger}(t) := \int_0^{\omega} f(a,t)(1+S'(t-a))da, \tag{2.15}$$

where $S'(t-a)$ is the derivative of $S(c)$ evaluated at $t-a$, the incremental shift relevant for the cohort aged a at time t. To show that $TFR^{\dagger}(t) = q(t)$, use (2.14) to replace $f(a,t)$ with $f_0(a-S(t-a))q(t)$. Substituting $w = a - S(t-a)$ and $dw = (1+S'(t-a))da$ gives the desired result. Thus, the shift-adjusted $TFR^{\dagger}(t)$ recaptures the period TFR that would have been observed in the absence of cohort shifts.

An interpretation of the effect of cohort shifts is that they compress (or dilate) the period cross-section of the age-period fertility surface. The adjustment factor $(1+S'(t-a))$ allows the passage of time in the period to return to its original pace, as if there had been no changes in timing.

The tempo-adjusted measure $TFR^{\dagger}$ bears some resemblance to the Average Completed Fertility (ACF) measure developed by Butz and Ward (1979) and Ryder (1980), and subsequently analyzed in Schoen (2004). ACF, a weighted average of completed cohort fertility rates, is equal to the TFR divided by a quantity called the timing index. The timing index $TI(t)$ is the sum of the ratios of fertility rates at each age in period t to the completed fertility rate for that particular cohort, i.e. $TI(t) = \int_{\alpha}^{\beta} f(a,t)/CTFR(t-a)\, da$. In the special case of constant period quantum q, the hypotheses of the cohort-shift model imply that $ACF(t)$ and $TFR^{\dagger}(t)$ will both be equal to q. However these two measures are not in general equivalent.

A drawback of the cohort approach is that estimation of the adjustment factor $1+S'$ is not as easy as is estimation of the corresponding factor R' for the period-shift model. Here we mention just one of two estimation methods developed in Goldstein and Cassidy (2014). If completed cohort fertility data is available, S' can be calculated as the change in the cohort mean age at birth. If the available cohort fertility data is truncated, it is still possible to estimate S', as detailed in the appendix to this chapter.

The cohort-shift model shares with the period-shift model a strong assumption about the age-distribution of fertility. Both use an unchanging baseline fertility age-schedule $f_0(a)$, although in the case of the cohort-shift model, the realized cohort age-schedule is also modifiable via period quantum, $q(t)$. Neither the period-shift assumptions nor the cohort-shift assumptions hold perfectly. Goldstein and Cassidy (2014) use goodness-of-fit comparisons to argue that at least in the case of the Netherlands, the cohort shift description appears to fit the observed fertility surface better. On the other hand, Bongaarts and Sobotka (2012) argue for the superiority of the period-shift model.

The cohort-shift model represents the timing of births as the outcome of shifts that describe how much each cohort may have advanced or delayed its fertility schedule. Different generations are allowed to have different plans for the timing of childbearing, and these plans play out over the course of their lives. These underlying schedules of intended fertility then encounter period driven events or shocks that may ultimately reduce or increase the cohort's total fertility. This model, a mixture of cohort and period influences, captures both the lifetime implications of cohort fertility intentions and the immediate responses to unanticipated period events. The interplay of cohort plans and period events then produces the variety in the observed fertility surface.

The cohort-shift model allows postponement choices made in the past to contribute to a 'fertility momentum' that plays out in terms of evolving fertility rates over the life of cohorts. Fertility momentum characterizes how fertility rates might change even after fertility quantum and shifts are fixed at current levels. We can define fertility momentum as the ratio of a future stationary total fertility rate to the total fertility rate in year t_0. In the numerator we have the total fertility rate that would eventually occur if quantum and shifts (period or cohort) are fixed at the most recent levels. Cohort shifts are fixed by setting $S(c) = S(t_0 - 15)$ for all $c \geq t_0 - 15$. In either case, the numerator is evaluated when the timing of fertility is constant. Thus in a period shift world we have fertility momentum

$$\frac{q(t_0)}{TFR(t_0)} = \frac{TFR^*(t_0)}{TFR(t_0)},$$

while the cohort-shift model yields momentum

$$\frac{q(t_0)}{TFR(t_0)} = \frac{TFR^\dagger(t_0)}{TFR(t_0)}.$$

The two shift models of postponement differ significantly in the lag time until fertility rates have stabilized. If shifts in the timing of births are a function purely of period, then fixing current quantum and shifts implies a stabilization of fertility rates within 1 year. In contrast, if shifts are represented as a cohort driven process, then fertility rates will continue to evolve for 30 or more years before stabilizing. Just as with population momentum, this evolution under fixed conditions is a result of the age structure within the population. The cohort-shift model represents postponement as a process played out over a lifetime, and this implies a continuing transformation in future fertility rates until all active cohorts follow the same fertility schedule.

The cohort formulation of fertility postponement has inherent implications for future fertility rates, and so can be used to forecast *TFR* under the assumption of no changes in quantum. For all currently active cohorts, we look at the timing of births to deduce the intended fertility schedule for these cohorts. We then assume that the newer cohorts who have not yet begun to give birth will follow the timing of the most recent cohorts. By projecting fertility schedules for each cohort individually, we can predict *TFR* in subsequent years. If the currently active cohorts have experienced different schedules, then we will see fertility rates evolve for several years even if quantum and shifts are now fixed.

2.7 An Illustrative Application to the United States

In this section, we apply the moving-target model, the period-shift model, and the cohort-shift model to the last several decades of period fertility change in the United States. We base our calculations on data from the Human Fertility Database (HFD, www.humanfertility.org), an excellent source of accurate and detailed fertility information for a number of low and moderate fertility countries.

Before presenting the results, we first provide a few more details about fitting the models to observed fertility rates. As discussed in Lee (1977), the moving-target model allows us to use period fertility rates to track changes in desired family size over time. One can calculate implied period level of desired fertility (D) by solving Eq. (2.8) to obtain

$$D(t) = \frac{g(a, t-a)}{\lambda} + C(a, t-a). \tag{2.16}$$

This provides a distinct estimate of desired fertility for each cohort in a given year. We reduce these estimates to a single period value by averaging across ages 25–35.

For the Bongaarts-Feeney tempo adjustment, we use their recommended approach of calculating tempo-adjusted total fertility for each parity separately and then let the estimate of tempo-adjusted quantum TFR^* be the sum of these parity-specific quantities. For the cohort shift estimator, we do use all-parity fertility rates in combination with the truncated mean estimator derived in Goldstein and Cassidy (2014) and described in the appendix to this chapter. We do not show any estimates from the Ryder model. However, if one considered the Ryder-like relationship $TFR = CTFR \times (1 - \mu_p')$, then the Bongaarts-Feeney results could also be interpreted as first-order estimates of the Cohort *TFR*.

Figure 2.1 shows the period total fertility rate observed in the United States from 1970 to 2010 along with the period measures from the models under consideration. We see in panel (a) that period fertility fell rapidly at the onset of the 1970s, a continuation of the baby-bust that began in the mid-1960s. This rapid fertility decline occurred at a time of new contraceptive technologies (the pill), liberalization of abortion law (Roe v. Wade), rapid change in female education and career ambitions, and economic change (the Oil Shock), a perfect storm of forces that understandably produced changes in birth rates. However, the continuation of low fertility throughout the 1970s and 1980s is less understandable in terms of forces influencing birth rates, particularly since the upswing in fertility at the end of the 1980s did not coincide with much in the way of remarkable social or economic

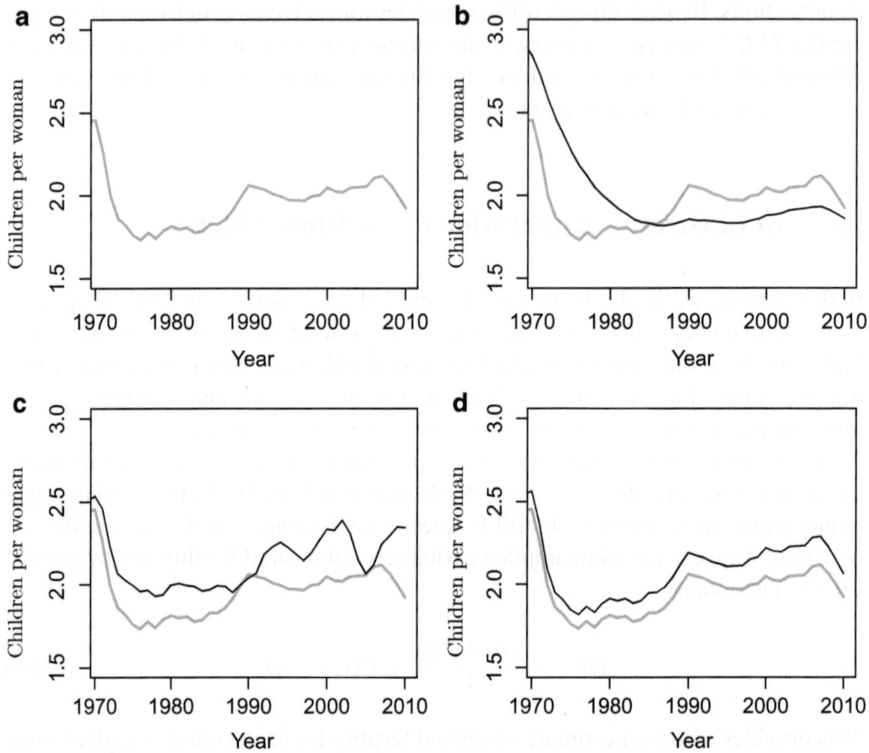

Fig. 2.1 *TFR*, desired family size, *TFR** and *TFR*† for the United States from 1970 to 2010. (**a**) Observed TFR. (**b**) Period target fertility D. (**c**) Period tempo-adjusted TFR*. (**d**) Cohort tempo-adjusted TFR†

changes. Since 1990 the slight dip of the Bush recession and the slight increase during the Clinton years and into the bubble economy of the second Bush's second term are all consistent with broad economic trends, as is the sudden decline of fertility since the onset of the Great Recession.

Panel (b) shows the estimated values of *D*, period desired fertility, obtained by applying Lee's moving-target model. The historical unfolding of desired fertility is considerably simpler than the *TFR*, with a steady decline from 1970 to 1980, and nearly three decades of stability thereafter, until the onset of the recession. According to Lee's model, the level of period fertility will be related to the rate of change in *D*. Thus, the rise in period fertility in around 1990 corresponds to an end to the decline in *D*. Likewise, the low level of births during the 1970s and 1980s is attributable to steady decline in desired fertility. We note also that the bottoming out of period fertility occurs before the minimum of *D*, a feature of Lee's model that we discussed above.

The level of desired fertility since about 1980 is quite low, well below the observed *TFR*. This results because we estimate *D* using Lee's value, from the

1960s, for the fraction λ of unachieved, desired fertility that occurs in each year. More plausible *levels* of D could be obtained by using an updated estimate, and, perhaps, by including age-specific variation in λ. In the context of fertility postponement, it seems likely to us that λ might well vary over time, in general taking on smaller values when fertility shifts to older ages.

Panel (c) shows the estimated values of TFR^*, period-shift-adjusted period total fertility, obtained from the Bongaarts-Feeney formula. These estimates show a markedly higher period fertility rate during the baby-bust years of the 1970s, on the order of 2.0 children per woman. The story the period-shift model tells is that the below-replacement period fertility of the 1970s and 1980s was to a large extent the result of fertility postponement, which, once taken into account, suggests that period fertility remained at near replacement levels. The period since 1990 suggest that there was a substantial increase in the quantum of fertility in the 1990s, up to as high as 2.3 children per woman, before fluctuating somewhat eratically since 2000. These "seemingly random fluctuations" (Bongaarts and Feeney 2006) are a feature of the TFR^* adjustment method, addressable by smoothing or by alternative calculation procedures (Bongaarts and Sobotka 2012) which have the effect of smoothing.

Panel (d) shows the estimated values of $TFR^\dagger$, cohort-shift-adjusted period total fertility, obtained from the Goldstein-Cassidy formula. The cohort adjustment, because it combines tempo effects from multiple cohorts in a single period, is considerably smoother than the period adjustment. We see during the Baby Bust years that its estimated value lies between the observed *TFR* and the period-tempo adjusted value. The cohort-shift adjusted measure tracks the observed period variation but at an offset equal to the average cohort shift effect. For the case of the United States this suggests that there is still a detectable tempo effect depressing fertility rates. Increases of 0.1 children would be achieved simply from a slowdown in postponement. A disadvantage of the cohort formulation is that by construction it cannot attribute much of the decline in period fertility during the recession to a sudden increase in postponement. In contrast to the Bongaarts-Feeney approach, which is perhaps oversensitive to period changes in timing, the Goldstein-Cassidy approach will tend to dampen period-to-period changes.

Taken together, we see that each of these models tells a somewhat different story about the last half century of fertility change in the United States. While the tempo-shift models attribute the low fertility of the 1970s and 1980s to changes in fertility timing, the moving-target model shows that low levels of period fertility can also result from changing fertility goals even in the absence of timing changes. For more recent decades, the period postponement approach suggest that without postponement we would have seen quite a high demand for children in the 1990s, whereas the cohort approach suggests a shrinking effect of postponement after 1990 such that observed and tempo-adjusted period fertility come close to convergence.

In terms of the recession, both the cohort model and the moving-target approach suggest a marked decline in underlying period quantum, whereas the period-shifts approach offers at least the possibility that much of the decline was "just" the postponement of births.

2.8 Discussion

Period fertility can vary substantially even when there are relatively small changes in actual completed family sizes. The Baby Boom and Bust were both larger in period terms than cohort changes would imply. "Lowest low" fertility (Kohler et al. 2002) and its apparent end in some countries (Goldstein et al. 2009), both represent larger swings in period fertility than changes in cohort fertility (Myrskylä et al. 2013) would imply.

The models of fertility dynamics that we have presented here offer different explanations for variation in period fertility. They all share in common the feature that the flow of period fertility is sensitive to the rate of change (the derivative) of other quantities like fertility timing or fertility targets. Table 2.1 summarizes the four models, showing for each model the mathematical formulation of period fertility rates and a related property discussed in this chapter.

Each model has something new to teach us. The generality of Ryder's formulation is useful in showing us the fundamental dependency of period fertility on changes in fertility timing. The formal nature of period-cohort translation lacks any behavioral basis, but for this reason Ryder's logic is applicable to a wide range of different reasons for fertility change. Lee's model of moving targets, which we have included here despite its general absence in the modern tempo-shifts literature, is more behavioral in its conception. The particular version that Lee develops, and the application we make of it, is based on fairly stylized assumptions, which could be relaxed in future applications. Nonetheless, even in its stylized form, the moving-target model gives us a valuable perspective on how the timing of period fertility changes will relate to changes in fertility desires, with amplification, lags, and other features.

The models of tempo-shifts in births give us another way of thinking about the relationship between individual behavior and aggregate fertility. The sum of individual decisions to time births differently, postponing or advancing otherwise planned births, changes the stream of births arriving in a time period, producing changes in period fertility simply as a result of timing changes. Both the period and cohort formulations of these shift-models include the simplifying assumption that all ages shift together. These can be relaxed (e.g. Kohler and Philipov 2001) but there is a tradeoff between model flexibility and estimation. More flexible models can be more sensitive to small violations of assumptions, producing erratic estimates

Table 2.1 Models of dynamic fertility change

Model	$f(a,t)$ equation	Property
Ryder	$f(a,0) + f'(a)t + \cdots$	$CTFR(t) \approx TFR(t+\mu_c)/(1-\mu_c')$
Moving-target	$\lambda e^{-\lambda a}\left[\int_0^a e^{\lambda u} D'(u+t-a)du + D(t-a)\right]$	$D(t) = f(a,t)/\lambda + C(a,t-a)$
Period-shift	$q(t)f_0(a-R(t))(1-R'(t))$	$TFR^*(t) = TFR(t)/(1-R')$
Cohort-shift	$q(t)f_0(a-S(t-a))$	$TFR^\dagger(t) = \int f(a,t)(1+S')\,da$

of tempo-adjusted fertility. This is the fertility modelers version of the statistician's bias-variance tradeoff.

There are also attempts to improve estimation of shift models by looking at different measures of fertility rates, including parity-specific hazards and new kinds of rates taking different sub-populations as the exposure to risk (Bongaarts and Sobotka 2012). These approaches may result in better estimators, but they can also run into difficulties in interpretation. Developments are still occurring, and so we should not rule out the possibility that there are better approaches. But we are skeptical that there is a "magic bullet" that will allow us to get substantially more precise or accurate estimates of the effect of tempo changes. The general principle found by Ryder that period fertility would be proportional to changes in mean ages will hold in all of these models and beyond this there is, we believe, only limited room to obtain a more detailed description.

On the other hand we do believe, as of this writing in 2015, that there are still many fruitful directions to explore in the dynamic modeling of fertility. The model introduced by Alkema (2011) of fertility transitions, currently being used as the basis for the United Nations probabilistic population forecasts, could benefit from more behavioral foundations. The possibility of changing timing needs to be incorporated into Lee's moving-target model in order to have it apply to the modern context. Improved methods of cohort forecasting, building on Myrskylä et al. (2013) and Schmertmann et al. (2014), should make it possible to include the analysis of more recent cohorts. The possibilities for gaining further insight into individual aspects of fertility change have not been exhausted. The recent literature on fertility intentions has focused on prospective plans (Morgan et al. 2011). One avenue worth consideration is to ask people about their plans retrospectively and about how the unfolding of their lives changed their behavior, in terms of timing, in terms of desired total fertility, and in other dimensions. Finally, the recent- and in some countries ongoing- recession offers a chance to see the inadequacies in current formulations and to create new theories to help us make sense of changing fertility.

Appendix: Tempo Adjustment for the Cohort-Shift Model

Tempo adjustment for the cohort-shift model of fertility requires estimation of the adjustment factor $1+S'$. The first step in this estimation is to control for the influence of changes in the period quantum q on the cohort mean age at birth. Our 'quantum-normalized fertility rates' $\tilde{f}(a,t)$ are calculated as follows. In each period t there is an age (or ages) with the peak fertility rate, and we denote this rate by $m(t)$. We can calculate $m(t)$ from observed data as $\text{Maximum}_a\{f(a,t)\}$. We then define the quantum-normalized rate as

$$\tilde{f}(a,t) := \frac{f(a,t)}{m(t)}.$$

Let m_0 denote the peak value of the baseline schedule f_0. Under the assumptions of purely cohort shifts, $m(t) = m_0 q(t)$ and the quantum-normalized rates are

$$\tilde{f}(a,t) = \frac{1}{m_0} f_0(a - S(t-a)). \tag{2.17}$$

If the cohort born in year c has completed its years of fertility, we can use the change in the cohort quantum-normalized mean age at birth to estimate $S'(c)$. For cohorts that have not yet completed the fertile years, we can estimate $S'(c)$ using the change in mean age at birth truncated to the latest year of data.

Our formula depends on several quantities. Let $\mu(c)$ be the quantum-normalized mean age at birth for cohort c as of latest available age, i.e.

$$\mu(c) = \frac{\int_l^h x\tilde{f}(x, c+x)\, dx}{\int_l^h \tilde{f}(x, c+x)\, dx} \tag{2.18}$$

where l is the lowest age of available fertility data for cohort c, and h is the highest age of available fertility data for cohort c. Let $v(a, c)$ be the proportion of cohort c's truncated fertility that occurs at age a calculated from the quantum-normalized schedules, i.e.

$$v(a,c) = \frac{\tilde{f}(a, c+a)}{\int_l^h \tilde{f}(x, c+x)\, dx}, \tag{2.19}$$

Let $\mu'(c)$ be the derivative of $\mu(c)$. In practice this can be estimated using

$$\frac{\mu(c+1) - \mu(c-1)}{2}$$

provided the same l and h are used for cohorts $c + 1$ and $c - 1$.

We then use this estimate of $S'(c)$:

$$S'(c) = \frac{\mu'(c)}{1 + v(h,c)(\mu(c) - h) + v(l,c)(l - \mu(c))}. \tag{2.20}$$

A detailed derivation of this formula can be found in Goldstein and Cassidy (2014).

References

Alkema, L. (2011). Probabilistic projections of the total fertility rate for all countries. *Demography, 48*(3), 815–839.

Bongaarts, J., & Feeney, G. (1998). On the quantum and tempo of fertility. *Population and Development Review, 24*, 271–291.

Bongaarts, J., & Feeney, G. (2006). The quantum and tempo of life-cycle events. In *Vienna yearbook of population research*, vol. 4 (pp. 115–151). Austrian Academy of Sciences Press.
Bongaarts, J., & Sobotka, T. (2012). A demographic explanation for the recent rise in European fertility. *Population Development Review, 38*, 83–120.
Butz, W., & Ward, M. (1979). The emergence of countercyclical U.S. fertility. *American Economic Review, 69*, 318–328.
Goldstein, J., & Cassidy, T. (2014). A cohort model of fertility postponement. *Demography, 51*(5), 1797–1819.
Goldstein, J., Sobotka, T., & Jasilioniene, A. (2009). The end of "Lowest-low" fertility? *Population and Development Review, 35*(4), 663–699.
Kim, Y., & Schoen, R. (2000). On the quantum and tempo of fertility: Limits to the Bongaarts-Feeney adjustment. *Population and Development Review, 26*, 554–559.
Kohler, H., & Philipov, D. (2001). Variance effects in the Bongaarts-Feeney formula. *Demography, 38*(1), 1–16.
Kohler, H., Billari, F., & Ortega, J. (2002). The emergence of lowest-low fertility in Europe during the 1990s. *Population and Development Review, 28*(4), 641–680.
Lee, R. (1977). Target fertility, contraception, and aggregate rates: Toward a formal synthesis. *Demography, 14*(4), 455–479.
Lee, R. (1980). Aiming at a moving target: Period fertility and changing reproductive goals. *Population Studies, 34*, 205–226.
Morgan, P., Sobatka, T., & Testa, M. (Eds.). (2011). *Vienna yearbook of population research* (Special issue on reproductive decision-making). Vienna: Austrian Academy of Sciences.
Myrskylä, M., Goldstein, J., & Cheng, Y. (2013). New cohort fertility forecasts for the developed world: Rises, falls, and reversals. *Population and Development Review, 39*(1), 31–56.
Ní Bhrolcháin, M. (1992). Period paramount? A critique of the cohort approach to fertility. *Population and Development Review, 18*(4), 599–629.
Rodriguez, G. (2006). Demographic translation and tempo effects. *Demographic Research, 14*(article 6), 85–110.
Ryder, N. (1964). The process of demographic translation. *Demography, 1*(1), 74–82.
Ryder, N. (1965). The cohort as a concept in the study of social change. *American Sociological Review, 30*(6), 843–861.
Ryder, N. (1980). Components of temporal variations in American fertility. In R. Hiorns (Ed.), *Demographic patterns in developed societies* (pp. 15–54). London: Taylor and Francis.
Schmertmann, C., Zagheni, E., Goldstein, J., & Myrskylä, M. (2014). Bayesian forecasting of cohort fertility. *Journal of the American Statistical Association, 109*(506), 500–513.
Schoen, R. (2004). Timing effects and the interpretation of period fertility. *Demography, 41*(4), 801–819.
Sobotka, T., & Lutz, W. (2010). Misleading policy messages from the period TFR: Should we stop using it? *Comparative Population Studies, 35*(3), 637–664.

Part II
Dynamic Mortality and Morbidity

Chapter 3
Am I Halfway? Life Lived = Expected Life

Vladimir Canudas-Romo and Virginia Zarulli

"Nel mezzo del cammin di nostra vita. Mi ritrovai per una selva oscura. Ché la diritta via era smarrita. [In the middle of the journey of our life, I came to myself in a dark wood, for the straight way was lost.]"

(Dante 1308–1320, translated by Durling (1996)).

This is how the most famous of Dante's poems, the Inferno of the Divine Comedy (Dante 1308–1320), begins. Dante is 35 years old, half of the biblical life span of 70 and writes about his imaginary trip. Midlife was then placed in the mid-thirties.

The desire to classify the stages of life has always accompanied human beings. Beginning with the ancient Greek physician Hippocrates who described seven stages of life, each of them a multiple of 7: a young boy until age 7, a boy from 8 to 14, a lad from 15 to 21, a young man from 22 to 28, a man from 29 to 49, an elderly man from 50 to 56 and an old men after 56. The 5th stage (man) represented the "prime of life" and the stage of leadership (Overstreet 2009). Later on, this classification inspired another great poet, William Shakespeare, who used the metaphor of a stage where all men and women are merely players playing seven acts, to indicate the parabola of life as a succession of seven stages, from birth to death (Shakespeare 1989).

Stages of life are still of interest to individuals, institutions and societies. For individuals, life stages are meaningful for those wanting to plan their future, while for institutions and societies classification helps in the planning of important population events such as education policies, retirement, and social aid. Throughout childhood, adolescence and the enthusiasm of young adulthood, life progresses to its summit of maturity, before continuing onto old and oldest-old age, finally ending at death. Typically, the peak of this path is considered to be the middle stage, the stage between youth and old age, the prime of life, the age of maturity in a mental rather than a biological way. As The New York Times declared in 1881, midlife is

V. Canudas-Romo (✉) • V. Zarulli
Max-Planck Odense Center on the Biodemography of Aging, University of Southern Denmark, J.B. WinsloewsVej 9, 5000 Odense C, Denmark
e-mail: vcanudas@health.sdu.dk; vzarulli@health.sdu.dk

R. Schoen (ed.), *Dynamic Demographic Analysis*,
The Springer Series on Demographic Methods and Population Analysis 39,
DOI 10.1007/978-3-319-26603-9_3

the prime of life, the age when "our powers are at the highest point of development and our power of disciplining these powers should be at their best."

If, on one hand, these classifications share the idea of middle age as the "prime of life," on the other hand there is no agreement about which is the age or age-range that represents the middle phase. Thus, what is the midlife age? In the ancient Greek and Roman worlds, following Hippocrates classification, this stage would be placed approximately between ages 30 and 50. Today, the literature in psychology, sociology and psychiatrics identifies the mid-life period in the range 40–60 (Golembiewski 1978; Dannefer 1984; Brown 1995; Willis and Reid 1998). Books are published (Cohen 2012), and surveys are conducted (Midlife in the United States 2013) about this stage of life that draw a lot of attention from society and around which there seems to be a convergence of opinions that 50 is the starting age.

Demographically speaking, midlife means something very specific: it is the age that equals our remaining life expectancy at that age. Literally, it is the age when we are halfway, when we have lived half of our life and we can expect to live as much as that. Did ancient societies place this stage at the right point? Is the social perception of midlife today right?

The concept of midlife is not new to demography. Lotka has already addressed his attention to it (Lotka 1939). Lotka (1998, pp. 75–82) discussed halfway-age in the context of the first cumulant, or ratio of the moments of the survival function, and as an approximation to the mean age in the stationary population. Lotka presented the halfway-age as a "property" of the first cumulant, and he used it to obtain a better estimate of the relation between growth rate and birth rate in a stationary population. More recently, the concept was mentioned again when it was shown that the average age of a stationary population equals the average remaining life expectancy in that population (Goldstein 2009) and, later, that the age at which remaining life expectancy equals itself is approximately the mean age of the stationary population (Goldstein 2012). A more general expression was shown by Vaupel (2009) on the symmetry between life lived and life left, proving Carey's inequality, which states that the age composition and distribution of remaining life spans are identical. Although, the concern since Lotka has been on examining the relations in a stationary population between the averages, lived and left, Lotka (1998, pp. 75–81) also researched the relation at the age that equals its remaining life expectancy at that age. Here we will further the latter aim to describe and explain the trends over time of the halfway-age.

In the rest of the text we refer to this age as the halfway-age in life. It must be pointed out that we do not intend to confute the categorizations used by other disciplines, as they look at this phenomenon from different perspectives that require other definitions. Our aim is to bring into the midlife debate the demographic perspective.

This study provides a demographic contribution to the halfway-age debate with respect to three dimensions: (1) by assessing the halfway-age under human mortality models; (2) by showing its empirical trends over time, by sex, for periods and cohorts and in comparison with life expectancy at birth; (3) by showing a forecast value for cohorts born today.

3.1 Methods and Data

3.1.1 *Halfway-Age*

We refer to halfway in life when our age equals the remaining life expectancy at that age. Formally, this can be represented by the equation

$$x = e_x, \tag{3.1}$$

where life expectancy at age x, denoted as e_x, is defined in terms of the force of mortality at age y, denoted as μ_y, as $e_x = \int_x^\omega e^{-\int_x^a \mu_y dy} da$. Thus, halfway-age only reflects mortality at ages beyond the halfway-age.

To compare halfway-age with life expectancy at birth, we partition the latter into a component of mortality before and after halfway-age x, as

$$e_0 = {}_xe_0 + e_x \ell_x, \tag{3.2}$$

where ${}_xe_0$ is the temporary life expectancy between ages 0 and x, and $e_x \ell_x$ is the product of the probability of surviving to age x and the remaining life expectancy beyond this age. At the halfway-age, both right-hand terms of Eq. (3.2) are below the halfway-age value x, therefore $2x$ is an upper bound for life expectancy at birth, or $e_0 \leq 2x$. The perfect equality of life expectancy at birth and $2x$ is reached only when mortality below age x does not exist. On the other hand, it is possible for life expectancy at birth to be below the halfway-age (when temporary life expectancy ${}_xe_0$ is very low), e.g., Icelandic females in 1860 had a life expectancy at birth of 21.5 but their halfway-age was 30. However, both cases are unusual and, in the results presented in the next section they do not occur.

3.1.2 *Mortality Models*

Equation (3.1) is mentioned by Lotka (1998, pg. 75), and shown for some selected times and countries (see Lotka 1998, Table 7). Also, procedures for estimating the halfway-age are discussed. Among the derivations of the halfway-age relation studied by Lotka (1998, pg. 79) is the study of De Moivre's mortality model implying a linear survival function. Here we follow that procedure by analyzing the halfway-age under two mortality models: Logistic-Makeham and Siler-Makeham. The Logistic-Makeham arises from a fusion of the models by Gompertz (1825), Makeham (1860) and Thatcher et al. (1998), while the Siler-Makeham model combines the models of Makeham (1860), Siler (1979) and Thatcher et al. (1998). These models are widely considered by demographers as a good approximation of the force of mortality (Canudas-Romo 2008; Engelman et al. 2014; Gavrilov and Gavrilova 1991; Horiuchi et al. 2013; Missov et al. 2015; Schoen et al. 2004). We also include a component of mortality change over time.

Let the force of mortality at age x and time t, denoted as $\mu_{x,t}$, be:

(i) Logistic-Makeham,

$$\mu_{x,t} = c + \frac{\alpha e^{\beta x - \rho t}}{1 + \alpha e^{\beta x - \rho t}}, \tag{3.3}$$

(ii) Siler-Makeham,

$$\mu_{x,t} = \alpha_2 e^{-\beta_2 x - \rho_2 t} + c + \frac{\alpha e^{\beta x - \rho t}}{1 + \alpha e^{\beta x - \rho t}}, \tag{3.4}$$

where the constants α, β, ρ, and c represent the mortality level at the initial age, the rate of mortality increase by age, the rate of mortality decline over time, and the level of external mortality, respectively. Equation (3.4) models separately the decline of infant mortality in the early ages, therefore it includes also the parameters α_2, β_2 and ρ_2. The parameters α and α_2 combined represent the initial mortality level, while β_2 and ρ_2 model the rate of infant mortality decrease by age and its rate of decline over time.

By assuming reasonable values for the parameters of the equations, observed in human populations in historical times, we were able to calculate the halfway-age and life expectancy at birth: $\alpha_2 = 0.135$, $\beta_2 = 1$, $\rho_2 = 0.02$, $\alpha = 0.00005$, $\beta = 0.1$, $\rho = 0.01$, and $c = 0.0005$ (see Canudas-Romo 2008 and Engelman et al. 2014). We then assessed the behavior of the halfway-age and compared it with the life expectancy at birth within the frameworks of the two mortality models considered.

3.1.3 *Empirical Time Trends*

To show the empirical trends we used data from the Human Mortality Database (2015), HMD. Using selected 1 × 1 life tables (single year and single age) available in the HMD, we calculated for each life table the age when remaining life expectancy is equal to its age. Then, for each given year, we calculated the mean of all the obtained halfway-ages for the selected populations and their standard deviation. This is calculated independently for females and males as well as for period and cohorts life tables. The time span is from 1816 to 1920 for cohorts (1816 being the first year for which cohort information is available for at least two populations) and from 1850 to the year 2010 for periods (The country selection-criteria was based on countries with long enough information to compare pseudo-cohorts, with the shortest series starting in 1956. Table 3.1 in the Appendix includes the detail information of the period and cohort data used for each country).

We estimated halfway-age with decimal point precision by comparing the difference between age x and life expectancy at this age, or e_x. Given the linearity

Table 3.1 Countries/regions included in the analysis of halfway ages

Country	Period data	Cohort data
Australia	1921–2010	
Austria	1947–2010	
Bulgaria	1947–2010	
Canada	1921–2010	
Czech Republic	1950–2010	
Denmark	1850–2010	1835–1920
Finland	1878–2010	1878–1920
France	1850–2010	1816–1920
Germany East	1956–2010	
Germany West	1956–2010	
Iceland	1850–2010	1838–1919
Ireland	1950–2009	
Italy	1872–2009	1872–1918
Japan	1947–2010	
Netherlands	1850–2009	1850–1918
New Zealand	1948–2008	
Norway	1850–2009	1846–1918
Portugal	1940–2010	
Spain	1908–2010	
Sweden	1850–2010	1816–1920
Switzerland	1876–2010	1876–1920
UK: England & Wales	1850–2010	1841–1920
UK: Northern Ireland	1922–2010	
UK: Scotland	1855–2010	1855–1920
USA	1933–2010	

Source: Human Mortality Database (2015)

of the function $\Delta x = e_x - x$ at the halfway-age, we calculated the line between two points $(x_0, \Delta x_0)$ and $(x_1, \Delta x_1)$. These two points are the values of the minimum difference Δx in absolute value, say at the age and difference $(x_0, \Delta x_0)$, and the adjacent age either $x_1 = x_0 - 1$ or $x_1 = x_0 + 1$. Depending on whether Δx_0 is positive or negative, the adjacent age x_1 is selected as the one with Δx_1 having the opposite sign. Finally, halfway-age is calculated with decimal precision based on those two points as $x = x_0 - \frac{\Delta x_0 (x_1 - x_0)}{(\Delta x_1 - \Delta x_0)}$.

3.1.4 Forecasting Cohort Halfway-Age

We performed a forecast of the halfway-age for cohorts born today by applying the standard forecasting procedure of the Lee-Carter (1992) model. We forecasted independently the available period mortality information between years 1950 and

2010 for each of the countries in the HMD. Then we reconstructed the complete cohort mortality experience in each country. Finally, we calculated the projected mean halfway-age for cohorts born later than 1920 combining as before the results for all countries.

As shown in Eq. (3.1), halfway-age only depends on mortality beyond this age. Therefore we tried two versions of the Lee-Carter model: one starting at age 0 and the other at age 30 (approximately the age of the first halfway-ages in our reported results). Both procedures returned practically the same results for cohort halfway-ages, so here only those starting at age zero are presented. In the Appendix we include an alternative simplistic forecasting procedure based on extrapolation of the gap between cohort and period halfway-ages as a contrasting alternative forecast.

3.2 Results

Figure 3.1 shows the decreasing function of remaining life expectancy by age and the increasing identity function for age, for American females in the year 2010. As defined in Eq. (3.1), the crossing of the two lines occurs at age 40.8 when

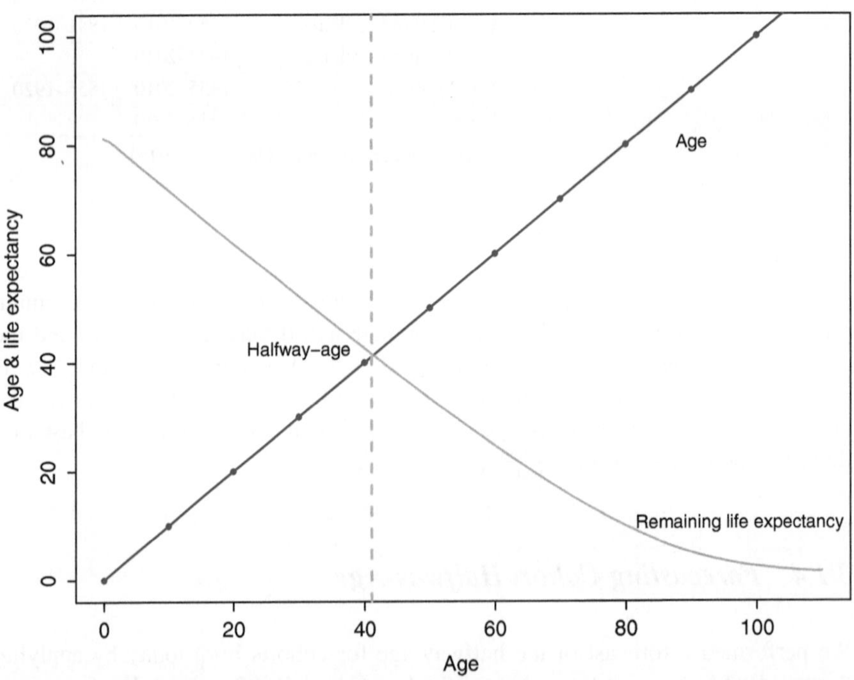

Fig. 3.1 Remaining life expectancy by age and the identity function for age, American females 2010 (Source: HMD)

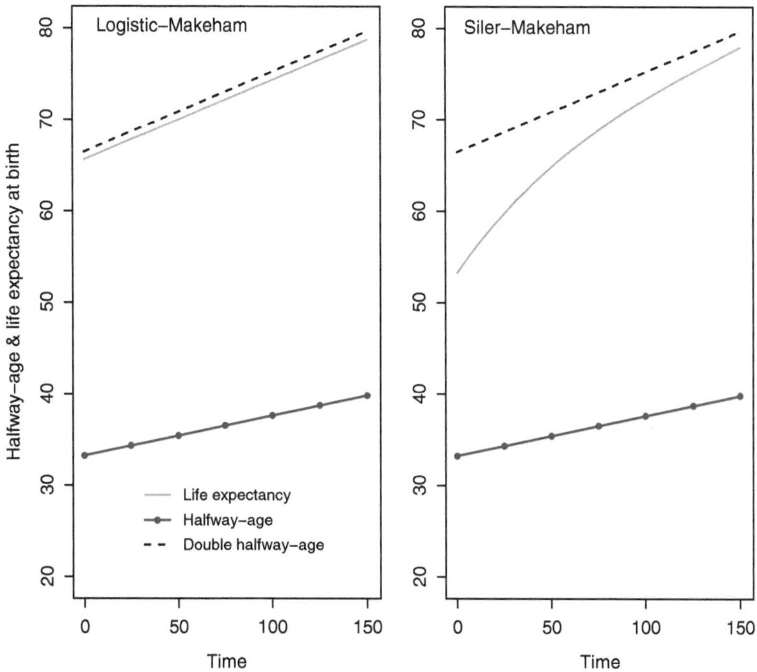

Fig. 3.2 Halfway-age and life expectancy at birth estimated under two mortality models: Logistic-Makeham (*left panel*) and Siler-Makeham (*right panel*) over a period of 150 years (Source: The values for the parameters of Eqs. (3.3) and (3.4) were: $\alpha_2 = 0.135$, $\beta_2 = 1$, $\rho_2 = 0.02$, $\alpha = 0.00005$, $\beta = 0.1$, $\rho = 0.01$, and $c = 0.0005$)

the population has reached its halfway-age. To assess changes in the halfway-age when mortality is declining, we study simulated time trends based on the mortality models.

Figure 3.2 compares the behavior of halfway-age and life expectancy at birth under two mortality models, the Logistic-Makeham (left panel) and the Siler-Makeham (right panel), over a time span of 150 years. Under the first model, life expectancy and halfway-age show a similar pace of increase. These common trends in increase are due to the fact that most of the mortality improvement happens at ages older than halfway-age and both measures are influenced by these changes. Thus, life expectancy at birth is practically the same as double of the halfway-age, although not exactly the same. Such a scenario is currently present in low mortality countries.

Under the Siler-Makeham model on the right panel, the two indicators show diverging patterns over time, even though the difference between them widens at a decreasing pace. In the initial years, life expectancy grows rapidly and later on tends to increase more slowly, due to the gradually fading impact of the mortality reductions at younger ages. Thus, life expectancy has asymptotically approached

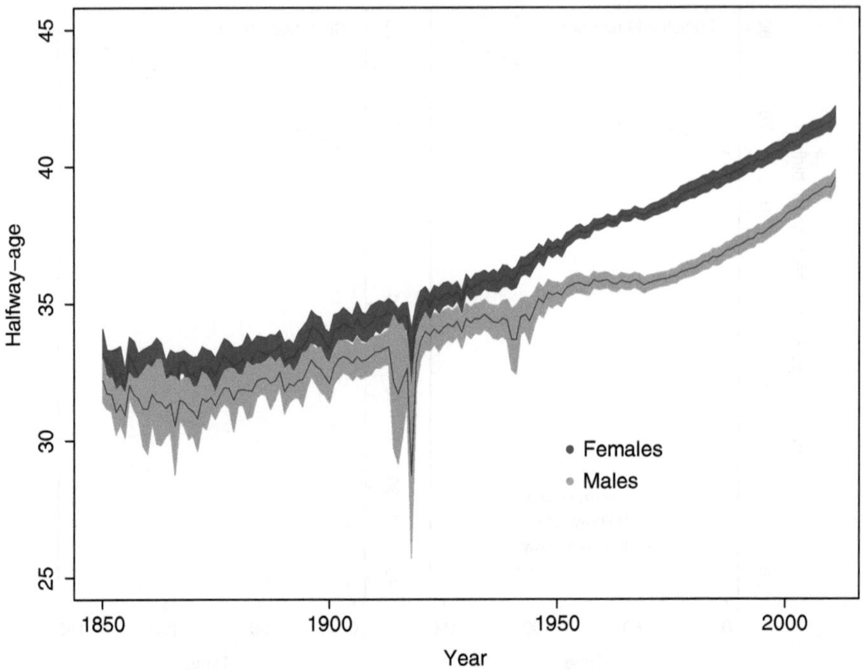

Fig. 3.3 Female and male period halfway-age, mean of HMD countries 1850–2010 (Source: Author's calculations, based on HMD. Notes: Together with the mean halfway-age also bands of 95 % CI are shown)

its upper limit, namely the double of halfway-age. This pattern follows closely the trends seen in countries with historical data as shown in our results of empirical trends.

Similar to the halfway-age for American females in 2010 in Fig. 3.1, the halfway-ages are calculated for all the available years and all the countries included in the HMD. Figure 3.3 includes the mean halfway-age trends over time for all the HMD countries and their 95 % confidence intervals (95 % CI = mean $\pm$ 1.96 standard error). In this Figure, the increase in halfway-age for both females and males and the reduction in the CI-band around the mean across time are observable. As such, all countries have transitioned to relatively similar values. Moreover, it is possible to see how, since 1950, the difference between male and female halfway-age widened. In 1850 the halfway-age mean values for females and males were found at ages 33.3 and 32.2 respectively, but by year 2010 females have halfway-ages of 41.8 and males lag behind by 2 years at 39.5 years. While the female pace of increase seems invariant over the 160 years of observation, the male measure stagnated in the 1950s, and only increases after the 1970s when it starts to move at a similar pace as that of its female counterpart.

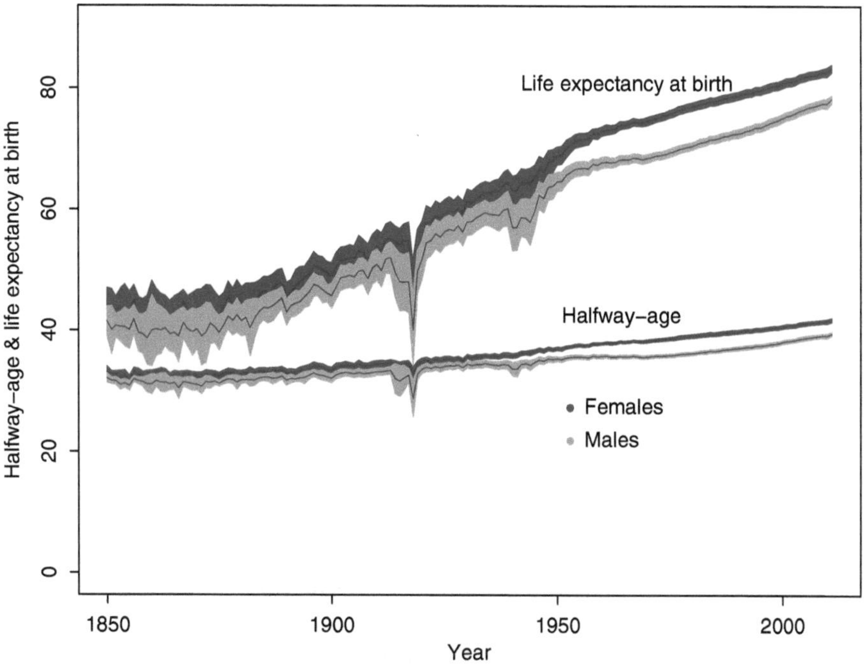

Fig. 3.4 Female and male period halfway-age and life expectancy at birth, means of HMD countries 1850–2010 (Source: Authors' calculations, based on HMD)

As halfway-age is a measure that focuses only on mortality in the second half of life, its trend over time differs from that of a measure that includes the entire age range, like, life expectancy at birth. Figure 3.4 shows the trend of the mean and 95 % CI of halfway-age and life expectancy at birth for period life tables from 1850 to 2010. Life expectancy at birth in 1850 starts in the early forties, at 44.3 years for females and 41.4 years for males, and not far below are the first halfway-ages found at ages 33.3 and 32.2 respectively. At the end of the studied period, in 2010, life expectancy at birth practically doubled at levels of 83.3 and 78.9 years respectively for females and males (it should be noted that there are more countries included in the mean values in 2010 than in 1850). As mentioned in the methods section, the values of life expectancy at birth are upper bounded by the double of the halfway-age which by 2010 has values of 83.7 years for females and 79.1 years for males. Thus, mortality in the first half of life, i.e., before the halfway-age, is practically at its minimum, bringing life expectancy to almost equal its upper bound (double halfway-age).

Figure 3.5 presents halfway-ages for cohort life tables (note that this refers to completed cohorts). The cohort information spans from 1816 to 1920 while the period data used in Figs. 3.3 and 3.4 goes from 1850 to 2010. Both series show

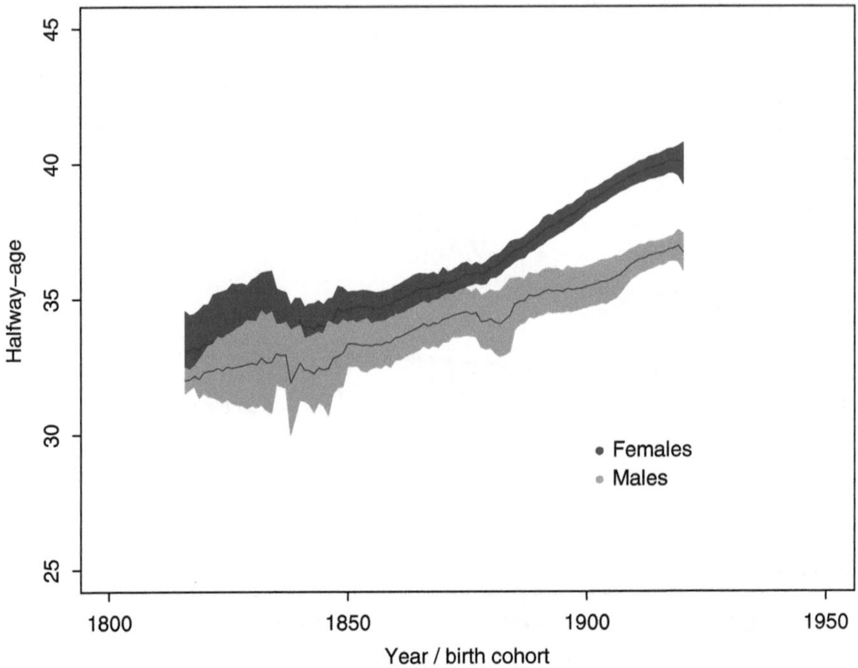

Fig. 3.5 Female and male cohort halfway-age, means of HMD countries 1816–1920 (Source: Authors' calculations, based on HMD)

increasing halfway-ages over time, but the increase is observed much earlier in cohorts than periods. Considering only the last 20 years of each series, the annual rate of increase has values of around 20 % for females in the period perspective and 22 % for the cohort, and for males 28 % in the period and 20 % in the cohort (see Figs. 3.3 and 3.5). Males are catching up with their females counterparts in the period perspective, but in the cohort perspective are still behind in trends.

The cohort halfway-series presented in Fig. 3.5 is extended to recent years using the Lee-Carter model and included in Fig. 3.6a, b for females and males respectively. The Lee-Carter model estimates levels of mean female halfway-age of 44.9 years for the cohort born in 2010, which corresponds to a period-cohort gap of 3 years, slightly less than the gap observed at the beginning of the twentieth century of 4 years. For males the halfway-age value is at 42.3 years for the cohort born in 2010, with a period-cohort gap of 2.7 years. This modest increase in the cohort halfway-ages contrasts with the previous fast upward trend seen in the cohort measures in Fig. 3.5. In the Appendix we show a simplistic model relating the current values of the period perspective with the observed gaps between cohort and period halfway-ages from the past. The latter forecasting procedure suggested that cohort halfway-age for females born in 2010 might be almost 7 years more than the Lee-Carter perspective at 52.2 years.

a

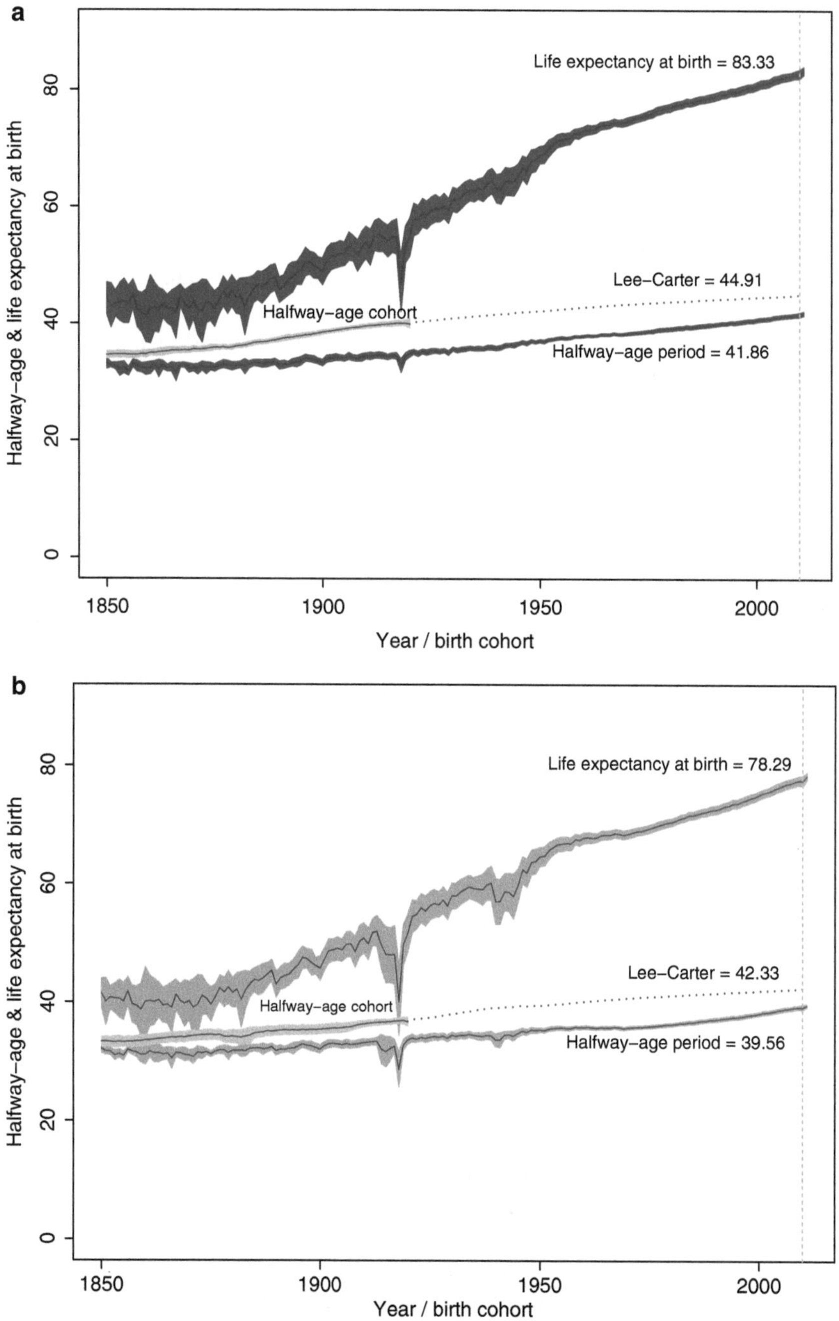

b

Fig. 3.6 (**a**) Female period life expectancy at birth and halfway-age, cohort halfway-age and Lee-Carter predicted cohort halfway-age, means of HMD countries 1850–2010 (Source: Authors' calculations, based on HMD). (**b**) Male period life expectancy at birth and halfway-age, cohort halfway-age and Lee-Carter predicted cohort halfway-age, means of HMD countries 1850–2010 (Source: Authors' calculations, based on HMD)

3.3 Discussion

This paper focused on the demographic concept of halfway-age. This is the age that equals our remaining life expectancy at that age. The analysis first assessed the halfway-age under mortality models, then investigated the empirical trends over time, from both the period and the cohort perspective. Finally, it provided some likely future values of this indicator for cohorts born today.

Partitioning life into two equal parts might seem, and actually is, an arbitrary division. Why would one be interested in halfway-age rather than in a third of life or a quarter of life? In these cases, the analysis would focus on the age at which remaining life expectancy is twice as much as that age (a third of life) or the age when remaining life expectancy is three times that age (a quarter). However, the stage of midlife has always been considered the "prime" in the life of human beings in ancient societies as well as in modern times.

In this study we observed that life expectancy and halfway-age have increased over time. However, from a theoretical point of view, the way they do so can differ depending on the mortality model used to describe the age-schedule of mortality. By definition, halfway-age focuses only on the second half of life and ignores younger ages, while life expectancy at birth takes into account the whole age range. Therefore, by assuming a Logistic-Makeham model, halfway-age and life expectancy at birth show a very similar pace of increase, because the trend of life expectancy reflects mostly changes at old ages. On the contrary, by assuming a Log-Siler-Makeham model, which explicitly models the component of infant and childhood mortality, thus fully reflecting the changes happening in the survival at those ages, halfway-age and life expectancy show remarkably different trends. In particular, their difference tends to increase with time, as life expectancy is boosted by the mortality reduction at younger ages, but at a decreasing pace, as the progress reaches very low levels of mortality at these ages. In populations under a low mortality regime, such as most of the developed countries today the double of halfway-age is practically the same as life expectancy at birth. But in the past the situation was completely different, life expectancy being at similar values as halfway-age. This means that it was very difficult to survive the hazards of young age, but those who succeeded had about the same life left as today (see Fig. 3.4). At the same time halfway-age, has been increasing and keeps moving upward.

In a declining mortality scenario, as the one observed in the last centuries in developed countries, the old-age modal age at death, or most frequent age at death, is only influenced by changes in mortality at ages older than the mode (Canudas-Romo 2010; Horiuchi et al. 2013). Similarly, halfway-age is a measure that is only influenced by mortality beyond this age. The modal age at death is higher than life expectancy at birth, and since the latter is practically the same as the double of halfway-age, then the modal age at death is also higher than the double of halfway-age. For example, for American females in 2010 the modal age at death is 89.5 years, life expectancy at birth 81.21 years, and the double of halfway-age is 81.6 years, i.e.

halfway-age depends on ages 40.8 and above, and the mode only from ages 89.5 and up. Thus, the modal age at death changes depend on a narrower age-group than the halfway-age.

In this study we also observed that halfway-age has been increasing over time, for both the period and cohort perspectives. Period halfway-age in 2010 was shown to be slightly above 40 for women and slightly below 40 for men. This value of 40 corresponds to the low bound of what is today considered as "middle life" by most of the socio-psychological and psychiatric literature (Golembiewski 1978; Dannefer 1984; Brown 1995; Willis and Reid 1998), which places this stage of life somewhere between 40 and 60. In general, the public debate seems to agree on the age 50 as the "division line," showing a tendency to overestimating this age compared to the period demographic observation obtained in our calculations.

Looking at the phenomenon from a cohort perspective (which represents the "true" life experience of individuals) instead, we can see that social perceptions might be right. As is the case with life expectancy, cohort is higher than period halfway-age. The Lee-Carter method, a standard forecasting procedure, predicts a halfway-age value of 44.9 years for women born today. However, one must be aware of the fact that such a method is known to have the tendency to underestimate survival at older ages, given its constant improvement in mortality rate (Li and Lee, 2005). Since halfway-age is a measure completely dependent on the second part of life, it is reasonable to think that the value produced by the Lee-Carter model is somewhat underestimated. In fact, female cohorts born today might actually reach the middle point of the journey of life exactly around the age 52, as estimated by the period-cohort gap extrapolation model (see Appendix). It must be pointed out that the latter prediction is based on rather simplistic calculation, but the different estimation results clearly deserve further analyses and careful investigation. Thus, our results for the first time shed a demographic light on the important and strongly socially perceived concept of middle life.

Thanks to the improvements in survival that have taken place in the last centuries the frontier of the middle point of human life has continuously moved upward. Perhaps it is not surprising that also other important events in the life cycle, that typically happen in the first half of life, have also shifted to older ages. As shown by the wide literature on the "Tempo" of life cycle events, in fact, age at first marriage and age at first birth have increased. For instance, from 1960 to 2000, the age at first marriage for French females moved from ages 22 to 28, while age at first birth for American women increased from age 22 to 25 (Bongaarts and Feeney 2008).

The debate about halfway-age can also be embedded within the framework of the "Characteristics Approach" to population aging of Sanderson and Scherbov (2005, 2007, 2010, 2013). They advocate the need to change paradigms in measuring population aging, shifting from measures based in terms of elapsed years since birth, to measures that look at a forward age in terms of remaining life expectancy. They propose a set of indicators that capture the changing age-based characteristics. Among these, there is the rising age at which populations have certain constant characteristics, such as a fixed value of remaining life expectancy or a certain mortality rate, which can help in redefining the threshold of old age. Halfway-age adds to the picture another very important dimension of the life course: the time when we have lived half of our life and we still have as much to live.

However, even though halfway-age partitions the life into two parts of the same duration, these are likely to be two different halves. Not only, as we have seen, different life cycle events take place during each of them, but they are also characterized by different health and economic profiles. It is reasonable to assume that health in the second half is worse than health in the first half, as it is reasonable that the income produced in the second half is higher than the income produced in the first half. The latter is because, given the current halfway-age of around 40, the first half of life is a period spent half in the dependent population – from age 0 to 20 – and half in the active population – from 20 to 40 – while the second half of life, instead, is mostly spent in the active population – with 25 years as active population (from age 40 to 65, or older with moving retirement age). In fact, the data from the National Transfer Accounts project seems to confirm these dynamics. For example, in the USA for the year 2003, per capita labor income from 0 to 40 was $721,357 US while from 41 onward it was $1,241,490; total health consumption, both public and private, from 0 to 40 was $100,775 while from 41 onward it was $568,612 (Lee et al. 2011).

To conclude, halfway-age is not only a measure that is interesting from the formal demography point of view, but it also has several socio-economic and political implications that make it an age worthy of consideration even by policy makers. An age that can be helpful in public policy planning, with potential applications to pensions, health care, labor market regulations and other social dimensions.

Acknowledgments We would like to thank Robert Schoen, Michel Guillot, Carlo Giovanni Camarda and the Max Planck Odense Center group for their comments and suggestion on how to improve our study.

Appendix: The Forecasting "Gap Method"

Here we present a simplistic forecasting procedure of the cohort halfway-age which consists in a linear extrapolation of the observed gap between cohort and period halfway-ages (cohort minus period) for the time in which the two data sets overlap (1850–1920). Similar relations between cohort and period life expectancies have been previously studied (Goldstein and Wachter 2006; Canudas-Romo and Schoen 2005). Given the linearity of the cohort minus period halfway-ages trend, a basic linear extrapolation was the best model fitting the data and which returned much higher halfway-ages than the standard forecasting of the Lee-Carter model. The model links the value of halfway-age for period t, $x_{p,t}$ with the value of halfway-age for cohorts, $x_{c,t}$, as:

$$x_{c,t} = x_{p,t} + \kappa_t, \tag{3.A1}$$

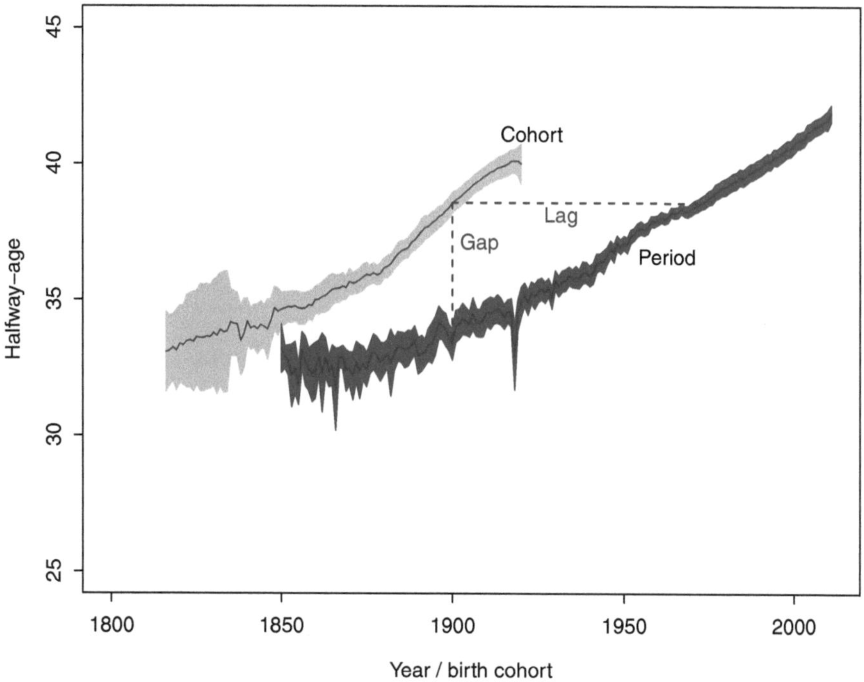

Fig. 3.7 Female period and cohort halfway-ages and lines depicting the gaps and lags between these measures, means of HMD countries 1816–2010 (Source: Authors' calculations, based on HMD)

where κ_t is the line fitted to the cohort minus period halfway-age gap (1850–1920), and used for calculating this gap between (1921–2010) as:

$$\kappa_t = 5.64 + 0.05\ (t - 1920)\,. \tag{3.A2}$$

Figure 3.7 presents the lags and gaps between period and cohort perspectives of halfway-age for females. As shown in Fig. 3.7, the female cohort halfway-age is higher than the period one for any of the available years. However, we can observe a linearly increasing trend in the cohort-period gap, which we extrapolate, as described in Eqs. (3.A1) and (3.A2), and use to assert the simplistic prediction on the halfway-age for cohorts born in 2010.

Figure 3.8 presents the mean female life expectancy at birth and halfway-age for cohorts and periods. Also included in this Figure are the time trends for the two forecasting results, namely from the Lee-Carter model and from the "gap-method." As shown in Fig. 3.8, the female cohort halfway-age is higher than the period one for all available years, and the linear prediction of this gap is higher than those values obtained by using the Lee-Carter model. Under the gap-model, the mean cohort female halfway-age will be the result of adding to the period halfway-age in 2010

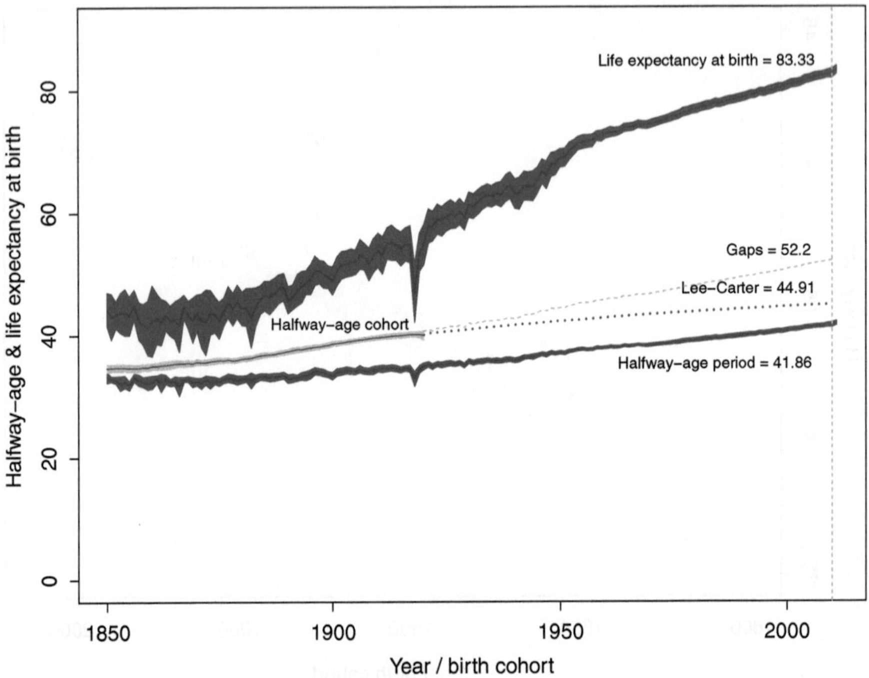

Fig. 3.8 Female period life expectancy at birth and halfway-age, cohort halfway-age and predicted cohort halfway-age under two scenarios, means of HMD countries 1850–2010 (Source: Authors' calculations, based on HMD. Notes: The two predicting scenarios are based on the Lee-Carter model and the gap method explained in the Appendix)

the predicted cohort-period gap (see Eqs. (3.A1) and (3.A2) above): by 2010, this gap amounts to more than 10 years and, as shown in Fig. 3.8, the predicted halfway-age for the 2010 cohort is 52.2 years (41.8 years for the period 2010). Alternatively, the Lee-Carter model estimates levels of mean halfway-age of 44.9 years in 2010, which corresponds to a period-cohort gap of 3 years, similar to the gaps observed at the beginning of the twentieth century.

References

Bongaarts, J., & Feeney, G. (2008). How long do we live? In E. Barbi, J. W. Vaupel, & J. Bongaarts (Eds.), *The quantum and tempo of life-cycle events* (pp. 29–65). Berlin/Heidelberg: Springer.

Brown, S. (1995). Life begins at 40? Further thoughts on marketing's "mid-life crisis". *Marketing Intelligence & Planning, 13*(1), 4–17.

Canudas-Romo, V. (2008). The modal age at death and the shifting mortality hypothesis. *Demographic Research, 19*(30), 1179–1204.

Canudas-Romo, V. (2010). Three measures of longevity: Time trends and record values. *Demography, 47*(2), 299–312.

Canudas-Romo, V., & Schoen, R. (2005). Age-specific contributions to changes in the period and cohort life expectancy. *Demographic Research, 13*(3), 63–82.

Cohen, P. (2012). *In our prime: The invention of middle age*. New York: Simon and Schuster.

Dannefer, D. (1984). Adult development and social theory: A paradigmatic reappraisal. *American Sociological Review, 49*(1), 100–116.

Durling, R. M. (1996). *The divine comedy of Dante Alighieri*. Oxford: Oxford University Press.

Engelman, M., Caswell, H., & Agree, E. (2014). Why do lifespan variability trends for the young and old diverge? A perturbation analysis. *Demographic Research, 30*(48), 1367–1396.

Gavrilov, L. A., & Gavrilova, N. S. (1991). *The biology of life span: A quantitative approach*. Chur: Harwood Academic Publications.

Goldstein, J. R. (2009). Life lived equals life left in stationary populations. *Demographic Research, 20*(2), 3–6.

Goldstein, J. R. (2012). Historical addendum to life lived equals life left in stationary populations. *Demographic Research, 26*(7), 167–172.

Goldstein, J. R., & Wachter, K. W. (2006). Relationships between period and cohort life expectancy: Gaps and lags. *Population Studies, 60*(3), 257–269.

Golembiewski, R. T. (1978). Mid-life transition and mid-career crisis: A special case for individual development. *Public Administration Review, 38*(3), 215–222.

Gompertz, B. (1825). On the nature of the function expressive of the law of human mortality, and on a new mode of determining the value of life contingencies. *Philosophical Transactions of the Royal Society of London, 115*, 513–583.

Horiuchi, S., Ouellette, N., Cheung, S. L. K., & Robine, J.-M. (2013). Modal age at death: lifespan indicator in the era of longevity extension. *Vienna Yearbook of Population Research, 11*, 37–69.

Human Mortality Database. (2015). University of California, Berkeley (USA), and Max Planck Institute for Demographic Research Germany. Available at http://www.mortality.org.

Lee, R., & Carter, L. R. (1992). Modeling and forecasting US mortality. *Journal of the American Statistical Association, 87*(419), 659–671.

Lee, R. et al. (2011). NTA country report, US, 2003.N.T.A. (http://www.ntaccounts.org)

Li, N., & Lee, R. (2005). Coherent mortality forecasts for a group of populations: An extension of the Lee-Carter method. *Demography, 42*(3), 575–594.

Lotka, A. J. (1939). *ThéorieAnalytique des Associations Biologiques*, II. Paris, Hermann et Cie.

Lotka, A. J. (1998). *Analytical theory of biological populations*. London: Plenum Press.

Makeham, W. M. (1860). On the law of mortality and the construction of annuity tables. *The Assurance Magazine, and Journal of the Institute of Actuaries, 8*(6), 301–310.

Midlife in the United States. (2013). MIDUS. University of Wisconsin – Madison, Institute on Aging.

Missov, T. I., Lenart, A., Laszlo, N., Canudas-Romo, V., & Vaupel, J. W. (2015). The Gompertz force of mortality as a function of the mode. *Demographic Research, 32*(36), 1031–1048.

Overstreet, L. R. (2009). The Greek concept of the "seven stages of life" and its new testament significance. *Bulletin for Biblical Research, 19*(4), 537–563.

Sanderson, W. C., & Scherbov, S. (2005). Average remaining lifetimes can increase as human populations age. *Nature, 435*(7043), 811–813.

Sanderson, W. C., & Scherbov, S. (2007). A new perspective on population aging. *Demographic Research, 16*(2), 27–58.

Sanderson, W. C., & Scherbov, S. (2010). Remeasuring aging. *Science, 329*(5997), 1287–1288.

Sanderson, W. C., & Scherbov, S. (2013). The characteristics approach to the measurement of population aging. *Population and Development Review, 39*(4), 673–685.

Schoen, R., Jonsson, S. H., & Tufis, P. (2004). *A population with continually declining mortality* (Working Paper 04–07), Population Research Institute, Pennsylvania State University, PA.

Shakespeare, W. (1989). *William Shakespeare: the complete works*. New York: Barnes & Noble Publishing.

Siler, W. (1979). A competing-risk model for animal mortality. *Ecology, 60*(4), 750–757.
Thatcher, A. R., Kannisto, V., & Vaupel, J. W. (1998). *The force of mortality at ages 80 to 120.* (Odense monographs on population aging, Vol. 5, 104p) Odense: Odense University Press.
Vaupel, J. W. (2009). Life lived and left: Carey's equality. *Demographic Research, 20*(3), 7–10.
Willis, S. L., & Reid, J. B. (1998). *Life in the middle: Psychological and social development in middle age*. San Diego: Academic.

Chapter 4
Revisiting Life Expectancy Rankings in Countries that Have Experienced Fast Mortality Decline

Michel Guillot and Vladimir Canudas-Romo

4.1 Introduction

It is well known that deaths occurring today in a population are not only the product of today's mortality conditions, but also the product of past exposures and behaviors that have accumulated over the entire life span of individuals (Barker 2007; Elo and Preston 1992; Forsdahl 1977; Guillot 2011; Myrskylä 2010; Preston et al. 1998). Examples of exposures and behaviors that have lagged effects on mortality include smoking (Doll et al. 2004; Wang and Preston 2009) as well as exposure to infections and malnutrition in infancy (Almond 2006; Crimmins and Finch 2006; Doblhammer and Vaupel 2001; Frost 1995). Mortality selection is another mechanism through which past conditions and behaviors may affect today's mortality rates (Vaupel et al. 1979).

Because of the lagged effect of past conditions on current mortality rates, current levels of period life expectancy are difficult to interpret in terms of current conditions (Vaupel 2002; Guillot 2011). This is particularly problematic in the presence of rapid mortality decline. When simulating a synthetic cohort for calculating current levels of life expectancy, the current mortality experience of individuals who were born, say, 30 years ago, is combined with the current experience of individuals born, say, 80 years ago. If mortality levels have changed rapidly, these

M. Guillot (✉)
Department of Sociology and Population Studies Center, University of Pennsylvania, 3718 Locust Walk, Philadelphia, PA 19104, USA
e-mail: miguillo@sas.upenn.edu

V. Canudas-Romo
Max-Planck Odense Center on the Biodemography of Aging, University of Southern Denmark, J.B. WinsloewsVej 9, 5000 Odense C, Denmark
e-mail: vcanudas@health.sdu.dk

R. Schoen (ed.), *Dynamic Demographic Analysis*,
The Springer Series on Demographic Methods and Population Analysis 39,
DOI 10.1007/978-3-319-26603-9_4

two groups of individuals are likely to have been exposed, at comparable ages, to a drastically different set of historical conditions and behaviors. This makes the assumption of homogeneity – a key assumption underlying the use of synthetic cohorts – particularly difficult to defend. Yet period life expectancy at birth remains the primary indicator for international mortality comparisons.

Cohort life expectancy, by linking age-specific mortality rates in a more theoretically-coherent fashion, better reflects this accumulation of exposures in determining levels of longevity (Guillot 2011; Richards et al. 2006; Willets 2004). However, this indicator can be observed only for cohorts that are now extinct, i.e., those born at least about 90–100 years ago. This seriously limits the usefulness of cohort life expectancy for making international comparisons, as it ignores large amounts of more recent mortality information, i.e., mortality information pertaining to cohorts born more recently but not yet extinct.

In this chapter, we propose a simple procedure for making international comparisons of life expectancy. This procedure builds on the theoretical advantages of actual cohorts (as opposed to synthetic cohorts) for building life tables, but uses all the available mortality information up to the present. Specifically, for each non-extinct cohort present in the population at time t, we calculate the cohort's truncated life expectancy at birth, with the truncation age being the age reached by the cohort at time t. In symbolic terms, for a cohort aged x at time t, we calculate ${}_{x}e_{0}^{C}$ (t – x), i.e., the life expectancy at birth truncated at age x for the cohort born at time $t - x$.

The advantages of using ${}_{x}e_{0}^{C}$ are the following. First, by relying on real cohorts rather than synthetic cohorts, ${}_{x}e_{0}^{C}$ links together mortality rates pertaining to individuals who have shared the same set of average conditions and exposures. As a result, this measure does not need to invoke the assumption of homogeneity to remain unbiased. Whatever heterogeneity occurs, it is intrinsic to the cohort's conditions and exposures and therefore does not generate bias in life expectancy values. This is particularly relevant for populations that have experienced fast mortality decline and for whom the mortality experience of different cohorts will be particularly heterogeneous. (Note that in cohorts with large flows of international migration, the homogeneity assumption still needs to be invoked to produce unbiased summary measures, because international migration makes the frailty distribution of a cohort not completely intrinsic to local exposures and behaviors.) Second, while ${}_{x}e_{0}^{C}$ does not reflect current conditions, it reflects the actual set of conditions to which a particular cohort has been exposed. As such, it is a summary measure of the cohort's accumulated exposures and behaviors, which is important to document given the impact of past exposures and behaviors on later mortality. Third, provided that enough historical mortality information is available, this measure can be observed for all cohorts present in the population, included those that are not yet extinct. This allows the comparison of cohort mortality across populations without having to resort to mortality projections. In light of the advantages of ${}_{x}e_{0}^{C}$, we argue that international comparisons based on this indicator are informative, alongside comparisons of period mortality.

This chapter builds on an existing body of literature that recognizes the importance of cohort mortality patterns and seeks to examine them in a way that applies to the current population. This line of research has led to the development of a number of summary mortality indicators, including the Cross-Sectional Average Length of Life (CAL) (Brouard 1986; Guillot 2003), Truncated CAL or TCAL (Canudas-Romo and Guillot 2015), Average Cohort Life Expectancy (ACLE) (Schoen and Canudas-Romo 2005), and Lagged Cohort Life Expectancy (LCLE) (Bongaarts 2005; Guillot and Kim 2011). These summary mortality indicators are calculated using different cohort mortality quantities: survival probabilities for multiple cohorts in the case of CAL or TCAL; life expectancies for multiple cohorts in the case of ACLE; life expectancy for one cohort in the case of LCLE. The approach proposed in this chapter has many parallels with these other approaches, but it also differs in several ways. First, the main mortality quantity examined in this chapter, ${}_xe_0^C$, includes more information than the cohort survival probabilities used in CAL or TCAL. Second, the focus of this chapter is not on the development of a summary indicator. Instead, by focusing on ${}_xe_0^C$, we seek to convey important age-specific information. Third, unlike ACLE and LCLE, the use of truncated cohort life expectancies does not require mortality projections. We thus believe that the approach proposed here provides a new angle for the examination of cohort vs. period mortality patterns.

This chapter is organized as follows. We first discuss the data sources that we use and the methodological approach we rely on. We then present results for French and Swedish females as an illustration of the procedure we use, followed by a discussion of results for 17 countries. Finally, we discuss results more broadly in a discussion section in which we also introduce the concept of "momentum of cohort mortality disadvantage."

4.2 Data and Methods

We use mortality information from the Human Mortality Database (HMD 2015). We focus here on countries with sufficient historical information such that ${}_xe_0^C\,(t-x)$ can be calculated up to an age of at least 75 years, allowing a more complete examination of country discrepancies in this indicator. Seventeen countries of the HMD meet this selection criteria: Australia (1921–2009); Belgium (1919–2009); Canada (1921–2009); Denmark (1898–2009); England & Wales (1898–2009); Finland (1898–2009); France (1898–2009); Iceland (1898–2009); Italy (1898–2009); Netherlands (1898–2009); New Zealand Non-Maori (1901–2008); Norway (1898–2009); Scotland (1898–2009); Spain (1908–2009); Sweden (1898–2009); Switzerland (1898–2009); and the US (1933–2009). For the US, additional mortality information prior to 1933 is taken from official life tables for Death Registration Area (DRA) states, available for cohorts born starting in 1900 (Bell and Miller 2005). Since DRA states are likely to have more favorable mortality outcomes than non-DRA states, this mortality information is probably biased downwards as an estimate of mortality for the entire US. However, we will use this information only for cohorts born between 1900 and 1930, and only for their time spend prior to the year 1933. Thus any bias in ${}_xe_0^C\,(t-x)$ values for the current period is likely to be small.

Truncated life expectancies (also called temporary life expectancies in the demographic literature) are defined with the following equations. Equation (4.1) shows the equation for ${}_xe_0^P(t)$, the life expectancy at birth for period t, truncated at age x:

$$ {}_xe_0^P(t) = \int_0^x p^P\,(a,t)\;da = \int_0^x e^{-\int_0^a \mu(y,\,t)dy}\,da \qquad (4.1) $$

Where $p^P\,(a,t)$ is the probability of surviving from birth to age a in the period life table for time t, and where $\mu\,(y,\;t)$ is the mortality rate at age y and time t.

In terms of classic life table notation, ${}_xe_0^P(t)$ can be calculated as follows:

$$ {}_xe_0^P(t) = \frac{\sum_{0,n}^{x-n}\,{}_nL_a^P(t)}{l_0^P(t)} \qquad (4.2) $$

where ${}_nL_a^P(t)$ is the number of person-years lived between age a and $a{+}n$ in a life table for period t with radix $l_0^P(t)$.

Equation (4.3) shows the equation for ${}_xe_0^C\,(t-x)$, the life expectancy at birth for the cohort born at time $t-x$, truncated at age x (or, equivalently, truncated at time t):

$$ {}_xe_0^C\,(t-x) = \int_0^x p^C\,(a,t-x)\;da = \int_0^x e^{-\int_0^a \mu(y,\,t-x+y)dy}\,da \qquad (4.3) $$

where $p^C\,(a,t-x)$ is the probability of surviving from birth to age a in the life table for the cohort born at time $t-x$, and where $\mu\,(y,\;t)$ is the mortality rate at age y and time t.

${}_xe_0^C\,(t-x)$ can be expressed in terms of quantities of a cohort life table:

$$ {}_xe_0^C\,(t-x) = \frac{\sum_{0,n}^{x-n}\,{}_nL_a^C\,(t-x)}{l_0^C\,(t-x)} \qquad (4.4) $$

where ${}_nL_a^C\,(t-x)$ is the number of person-years lived between age a and $a{+}n$ in the life table for a cohort born at time $t-x$ with radix $l_0^C\,(t-x)$.

It should be noted that ${}_xe_0^P(t)$ monotonically increases with age x until reaching a constant value corresponding to $e_0^P(t)$, the life expectancy at birth for period t. Indeed, $\lim_{x\to\infty}{}_xe_0^P(t) = e_0^P(t)$. However, ${}_xe_0^C\,(t-x)$ may not monotonically increase with age at a given time t. Typically, in populations that have experienced steady mortality decline, ${}_xe_0^C\,(t-x)$ will initially increase until reaching a maximum value, followed by a steady decrease. This is because when examining age trajectories of ${}_xe_0^C\,(t-x)$ at time t in such populations, two contradicting forces are operating: (1) an increase in x means a higher truncation age in the life table, which affects ${}_xe_0^C$ positively; (2) an increase in x means that ${}_xe_0^C\,(t-x)$ will refer to cohorts born earlier and exposed to higher mortality, which affects ${}_xe_0^C$ negatively. The maximum value of ${}_xe_0^C\,(t-x)$ will be reached for the cohort with the most favorable combination of truncation age and mortality history.

In this chapter, we use the maximum value at time t of ${}_xe_0^P(t)$ and ${}_xe_0^C\,(t-x)$ as a way to summarize the trajectories of these two functions. In the case of ${}_xe_0^P(t)$, the maximum value is simply $e_0^P(t)$, the period life expectancy at birth at time t. In the case of cohorts, the maximum value of ${}_xe_0^C\,(t-x)$ is the truncated life expectancy for the cohort with the most favorable combination of truncation age and cohort mortality rates. As we will see in the results section, there is little cross-over in ${}_xe_0^C\,(t-x)$ across countries. This means that a higher value of $max\left\{{}_xe_0^C\,(t-x)\right\}$ will be indicative of a more favorable overall trajectory of cohort mortality, even if this maximum occurs at different ages across countries. As an alternative, we summarize ${}_xe_0^C\,(t-x)$ trajectories by calculating the unweighted mean of ${}_xe_0^C\,(t-x)$ between ages 75 and 88, i.e., an age interval within which $max\left\{{}_xe_0^C\,(t-x)\right\}$ occurs in the countries we study. We also present ${}_{76}e_0^C\,(t-76)$, 76 being the highest age at which ${}_xe_0^C\,(t-x)$ can be calculated in the US using the fully-representative HMD data. Results presented later show that these three ways of summarizing ${}_xe_0^C\,(t-x)$ trajectories produce virtually identical country rankings, reflecting the fact that there is indeed little cross-over in ${}_xe_0^C\,(t-x)$ across countries.

We calculated values of ${}_xe_0^P(t)$, and ${}_xe_0^C\,(t-x)$ in the 17 countries (by sex), using observed age-specific deaths rates by single calendar year and single completed age. Period and cohort life table survivors and corresponding person-years lived below age x were calculated with the assumption that deaths occurred on average half-way within the interval (${}_1a_x = x+.5$), except for the first age group for which ${}_1a_0$ was estimated using the Coale and Demeny equations (Preston et al. 2001).

4.3 Results

4.3.1 France vs. Sweden

As an illustration of the procedure we use in this chapter, we first discuss the example of two countries with contrasting period vs. cohort experience. Figure 4.1 shows trajectories of ${}_xe_0^P(t)$ vs. ${}_xe_0^C\,(t-x)$ among females in France vs. Sweden in 2011, the most recent year for which mortality information is available in both countries. Period life expectancy at birth among females was slightly higher in France (85.03 years) compared to Sweden (83.68 years) in 2011. The trajectories of ${}_xe_0^P$ in these two countries show how such levels of period life expectancy are reached. Initially ${}_xe_0^P$ increases quickly, almost at a slope of 1.00, reflecting very low mortality at younger ages. (A slope of 1.00 would indicate that $x - {}_xe_0^P = 0$, i.e., that no amount of life is lost due to mortality below age x, or equivalently, that there is zero mortality below age x.) Eventually, both ${}_xe_0^P$ trajectories stabilize at a constant level, but France's ${}_xe_0^P$ stabilizes at a level that is slightly higher, reflecting

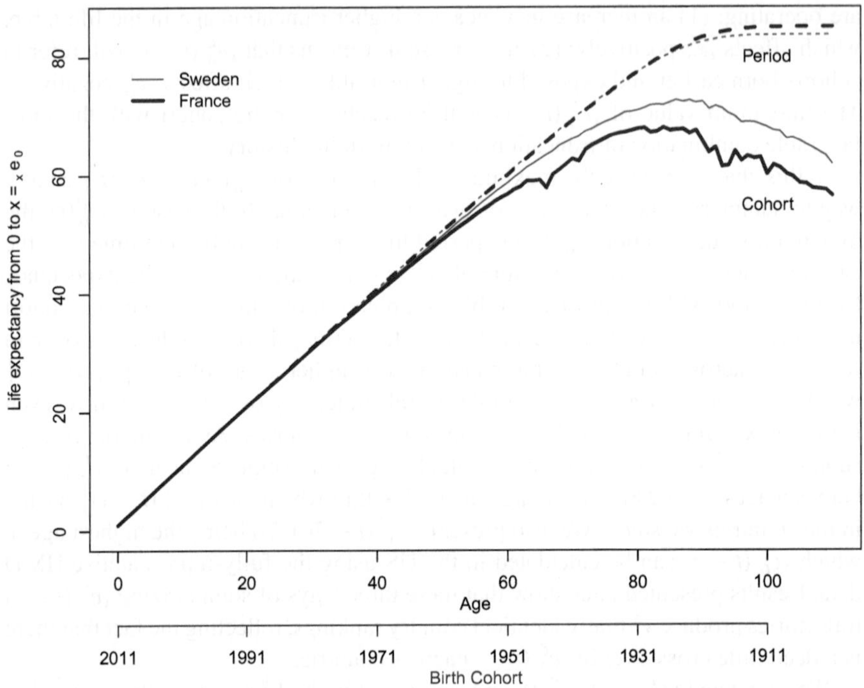

Fig. 4.1 Truncated period ($_xe_0^P$) and cohort ($_xe_0^C$) life expectancies, France and Sweden, Females, 2011 (Source: HMD)

France's higher level of period life expectancy at birth. The $_xe_0^P$ trajectories in France vs. Sweden show that period mortality differences between the two countries are relatively minor, and that accumulated person-years start to diverge between the two countries relatively late in the life of the synthetic cohorts.

Trajectories of $_xe_0^C$, also shown on Fig. 4.1, tell a different story. When looking at past cohort mortality for each country, we find that the two countries start diverging at an earlier age, around 50. (Corresponding birth cohorts are indicated next to the age axis in this and subsequent figures.) More significantly, we observe that truncated cohort life expectancies happen to be systematically higher in Sweden, by contrast with what period mortality tells us. In other words, the slight advantage that French females have today in terms of period life expectancy (in comparison with Sweden) masks much larger and systematic disadvantages in terms of cohort mortality. The existence and scale of this disadvantage can be summarized by looking at the maximum value of $_xe_0^C$, which is 67.94 in France (reached at age 82) vs. 72.66 years in Sweden (reached at age 86).

Another way of comparing the period vs. cohort experience of these two countries is to examine *ratios* of $_xe_0^P$ in France vs. Sweden, and to compare them with corresponding $_xe_0^C$ ratios. This is shown in Fig. 4.2. In this figure, we can see more clearly how late in life (above age 80) the French advantage in period

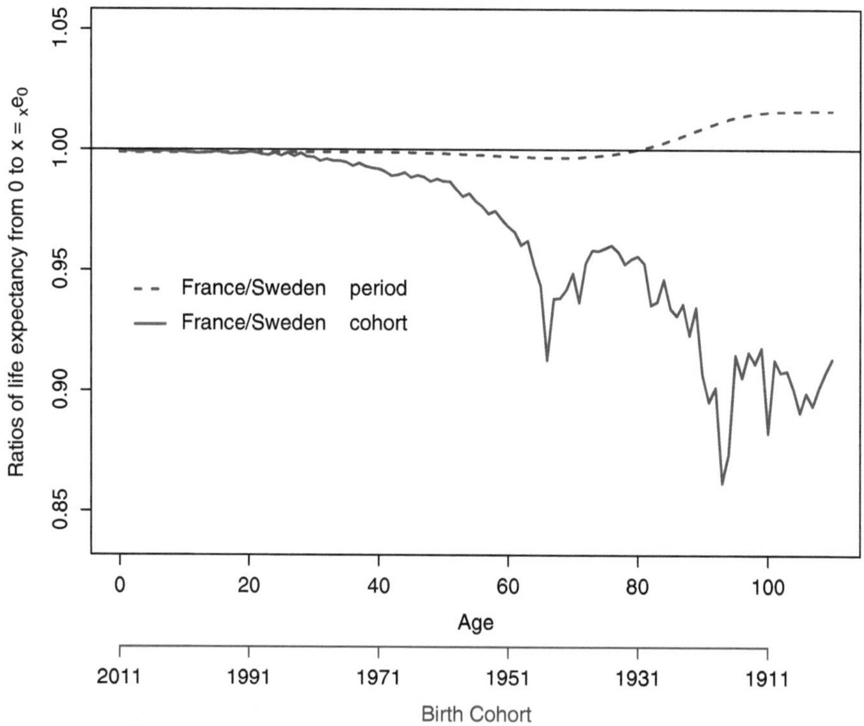

Fig. 4.2 France/Sweden ratios of truncated period ($_xe_0^P$) and cohort ($_xe_0^C$) life expectancies, Females, 2011 (Source: HMD)

life expectancy emerges, and how it stabilizes at a level that is only 2 % above Sweden's e_0^P. By contrast, every single cohort in France vs. Sweden, even those born recently, has a lower level of $_xe_0^C$. (Indeed, the ratio of $_xe_0^C$ in France vs. Sweden is always below 1, even if by a tiny amount initially.) The relative level of $_xe_0^C$ in France vs. Sweden starts decreasing more significantly at age 30, with low values at ages 67 and 94 years reflecting particularly high mortality in France vs. Sweden for the cohorts born in 1918 and 1945. War mortality and other fluctuations aside, the France/Sweden $_xe_0^C$ ratio stabilizes at a level of about 0.90.

The reason for France's lower levels of $_xe_0^C$ is that even though period mortality is now lower in France, cohorts in France have for most of their lives passed through years during which mortality was actually higher than in Sweden. Indeed, period life expectancy became higher in France only starting in 1987. In other words, France's advantage in period mortality is *too recent* to translate into any meaningful cumulative cohort mortality advantage from birth onwards. It is in fact interesting to note that although period life expectancy in France has been higher than in Sweden for the past 25 years or so, no single cohort in France has accumulated a superior level of life expectancy. Even cohorts born during the last 25 years display a disadvantage, albeit tiny. This is due to the fact that France's higher level of period life expectancy since 1987 arises primarily from an advantage in old-age mortality. As for French cohorts born earlier, the mortality advantages that they have experienced in recent years (i.e., at older ages) has not been sufficient to compensate for the mortality disadvantages they experienced earlier in life.

Examining trajectories of ${}_{x}e_{0}^{C}$ thus allows us to expand the scope of mortality comparisons by studying whether period advantages have lasted long enough and have affected enough age groups to actually translate into real cohort advantages. ${}_{x}e_{0}^{C}$ comparisons summarize complex dynamics of time- and age-specific mortality advantages vs. disadvantages in a way that is meaningful for real cohorts of individuals.

4.3.2 Countries of the HMD

In this section, we expand country comparisons to all 17 countries discussed in the data section. Trajectories of ${}_{x}e_{0}^{P}$ vs. ${}_{x}e_{0}^{C}$ are shown in Fig. 4.3 for females and in Fig. 4.4 for males. Results are also summarized in Tables 4.1 and 4.2, which show for each country (by sex) levels of e_{0}^{P} as well as the three summary measures of

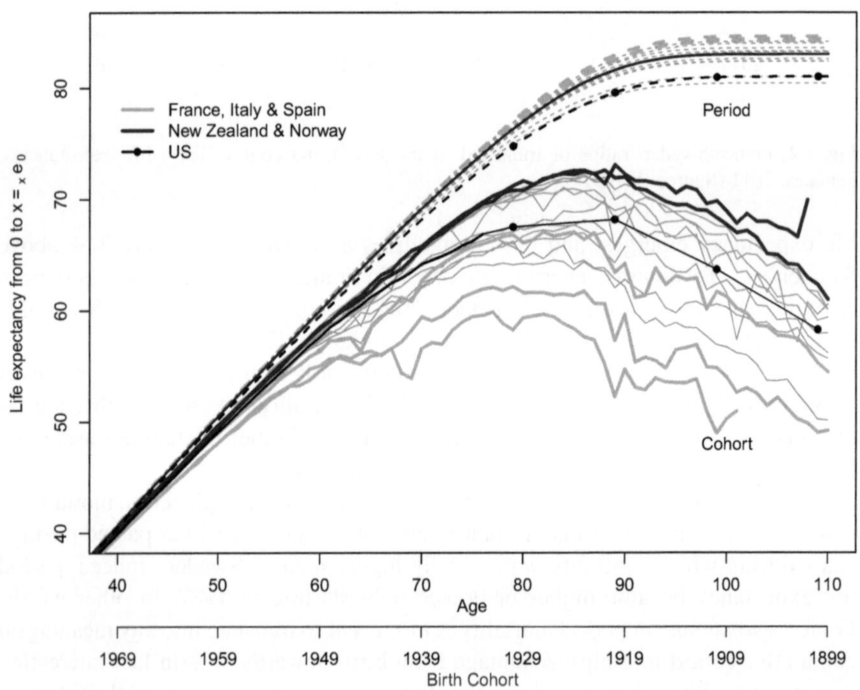

Fig. 4.3 Truncated period (${}_{x}e_{0}^{P}$) and cohort (${}_{x}e_{0}^{C}$) life expectancies, 17 countries of the HMD, Females, 2009 (Source: HMD & USA cohort life tables from Death Registration Area states for birth cohorts of 1900, 1910, 1920 and 1930)

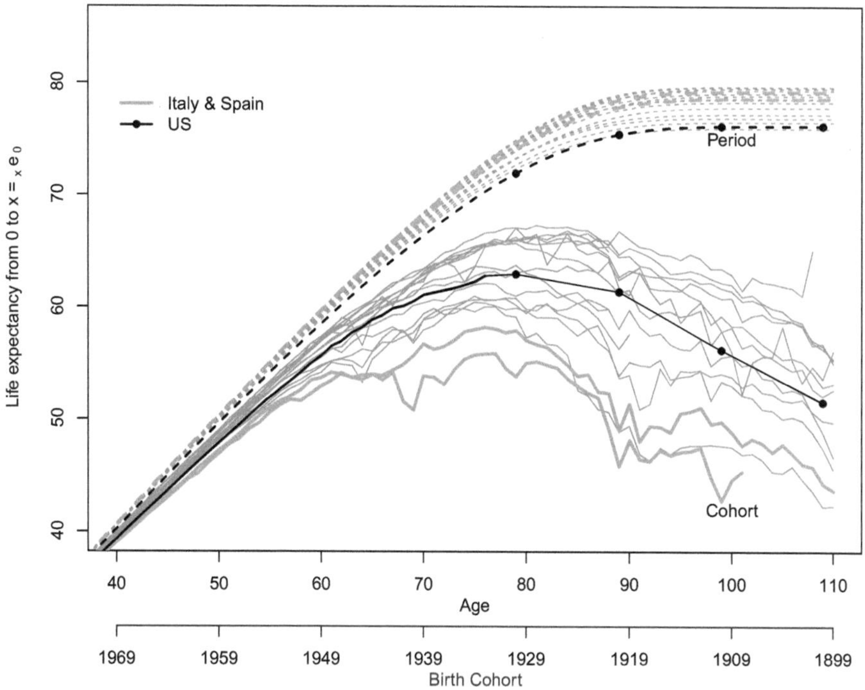

Fig. 4.4 Truncated period ($_xe_0^P$) and cohort ($_xe_0^C$) life expectancies, 17 countries of the HMD, Males, 2009 (Source: HMD & USA cohort life tables from Death Registration Area states for birth cohorts of 1900, 1910, 1920 and 1930)

country-specific $_xe_0^C\,(t-x)$ trajectories discussed earlier. These results refer to the year 2009, the most recent year for which mortality information is available in the selected countries. Tables 4.1 and 4.2 also show country rankings in terms of $_xe_0^P$ vs. $_xe_0^C$.

We find that cohort mortality experiences represented with $_xe_0^C$ exhibit much greater variability than period mortality experiences represented with $_xe_0^P$. As clear from Figs. 4.3 and 4.4, the range of variation in $_xe_0^C$ across countries is much greater than the range of variation in $_xe_0^P$.

It is also clear that every single country fares worse when examining $_xe_0^C$ rather than $_xe_0^P$, reflecting mortality decline over time. However, the amount of loss varies greatly by country. As a result, country rankings vary depending on whether the focus is on $_xe_0^P$ vs. $_xe_0^C$, as shown in Tables 4.1 and 4.2.

Some countries have particularly divergent period vs. cohort rankings. For females, the following groups of countries, highlighted in Fig. 4.3, emerge:

1. France, Italy and Spain standout as countries that have above average levels of period life expectancy. In terms of $_xe_0^P$, these three countries remain consistently

Table 4.1 Female life expectancies, under period and cohort $_{x}e_{0}$ perspectives and their rankings in 2009

	A = max {period $_{x}e_{0}$} = Period e_0	Rank A	B = max {cohort $_{x}e_{0}$}	Rank B	C = mean {cohort $_{75}e_{0}$ to $_{88}e_{0}$}	Rank C	D = cohort $_{76}e_{0}$[b]	Rank D
Spain	84.65	1	59.85	17	58.92	17	59.46	17
France	84.46	2	67.03	12	66.51	12	65.69	12
Italy	84.24	3	62.13	16	61.60	16	61.54	16
Switzerland	84.20	4	72.22	4	70.82	3	68.66	3
Australia	84.05	5	72.49	3	70.67	4	68.43	6
Iceland	83.61	6	71.05	6	69.76	7	66.78	7
Sweden	83.33	7	72.03	5	70.67	5	68.64	4
Canada	83.28	8	67.69	11	66.77	11	65.97	11
Finland	83.11	9	66.19	14	65.05	14	64.91	14
Norway	83.08	10	72.53	2	71.15	2	68.94	2
New Zealand[a]	83.06	11	73.10	1	71.31	1	69.11	1
Netherlands	82.64	12	70.93	7	70.07	6	68.54	5
Belgium	82.44	13	66.93	13	66.02	13	65.19	13
England & Wales	82.43	14	68.09	10	67.44	10	66.66	8
USA	81.04	15	68.15	9	67.50	9	66.02	10
Denmark	81.03	16	69.88	8	67.76	8	66.28	9
Scotland	80.45	17	64.50	15	64.15	15	64.12	15

Source: HMD & USA cohort life tables from Death Registration Area states for birth cohorts of 1900, 1910, 1920 and 1930

Note: Only countries from HMD that had historical period data at least starting in 1933 were included; Columns refer to A = period life expectancy at birth; B = maximum cohort $_{x}e_{0}$ over ages; C = mean of cohort $_{x}e_{0}$ from x = 75 to x = 88; D = cohort $_{76}e_{0}$

Note[a]: New Zealand data is for 2008;

Note[b]: Age 76 is the last value for the USA with fully-representative mortality information

Table 4.2 Male life expectancies, under period and cohort $_{x}e_0$ perspectives and their rankings in 2009

	A = max {period $_{x}e_0$} = Period e_0	Rank A	B = max {cohort $_{x}e_0$}	Rank B	C = mean {cohort $_{75}e_0$ to $_{88}e_0$}	Rank C	D = cohort $_{76}e_0$[b]	Rank D
Iceland	79.67	1	67.07	2	65.43	6	63.56	8
Switzerland	79.63	2	66.20	5	65.75	5	64.84	6
Australia	79.51	3	66.14	6	65.76	4	64.98	5
Sweden	79.33	4	66.82	3	66.09	2	65.24	4
Italy	79.22	5	58.20	16	56.22	16	58.01	16
New Zealand[a]	79.05	6	67.32	1	66.87	1	66.14	1
Canada	78.85	7	62.10	11	61.25	11	62.00	10
Norway	78.62	8	66.53	4	65.94	3	65.33	3
Spain	78.59	9	55.87	17	54.32	17	55.75	17
Netherlands	78.53	10	65.70	7	64.83	7	65.42	2
England & Wales	78.29	11	63.73	9	62.24	10	63.63	7
France	77.80	12	60.83	13	59.79	13	60.65	13
Belgium	77.16	13	60.95	12	60.14	12	60.95	12
Denmark	76.83	14	64.05	8	63.18	8	62.76	9
Finland	76.48	15	60.03	15	56.75	15	59.58	15
USA	76.13	16	62.93	10	62.31	9	61.99	11
Scotland	75.87	17	60.43	14	59.09	14	60.43	14

Source: HMD & USA cohort life tables from Death Registration Area states for birth cohorts of 1900, 1910, 1920 and 1930

Note: Only countries from HMD that had historical period data at least starting in 1933 were included; Columns refer to A = period life expectancy at birth; B = maximum cohort $_{x}e_0$ over ages; C = mean of cohort $_{x}e_0$ from x = 75 to x = 88; D = cohort $_{76}e_0$

Note[a]: New Zealand data is for 2008;

Note[b]: Age 76 is the last value for the USA with fully-representative mortality information

above the other countries. When examining cohorts, however, these countries exhibit a clear disadvantage. Spain is a particularly striking case, as it moves from the top rank in terms of ${}_xe_0^P$ to last rank in terms of ${}_xe_0^C$ (whatever summary measure of ${}_xe_0^C$ is used) as indicated in Table 4.1.
2. New Zealand and Norway are two countries that have average levels of ${}_xe_0^P$, but are ranking at the top in terms of ${}_xe_0^C$.
3. The US is a country that is significantly disadvantaged in terms of ${}_xe_0^P$, but has average levels of ${}_xe_0^C$.

For males (Fig. 4.4 and Table 4.2), two sets of countries stand out:

1. Here also, Spain and Italy rank high in terms of ${}_xe_0^P$, but rank at the bottom in terms of ${}_xe_0^C$.
2. Like in the case of females, US males are significantly disadvantaged in terms of period life expectancy, but exhibit average levels of ${}_xe_0^C$.

4.4 Discussion

The results presented above show that period mortality comparisons mask vastly heterogeneous experiences in terms of the mortality conditions to which cohorts present in a population have been exposed. To the extent that past mortality conditions matter for a cohort's future health outcomes, this heterogeneity deserves to be factored in when making international mortality comparisons.

The results also show that in a number of cases, country rankings vary greatly when examining ${}_xe_0^C$ rather than ${}_xe_0^P$. In countries that lose rankings in terms of cohorts (e.g., Spain, Italy, France), period mortality advantages have been too recent to translate into actual cohort mortality advantages. In countries that gain rankings in terms of cohorts (e.g., the US), there is a history of cohort mortality advantage that has not been much jeopardized by relatively recent mortality disadvantages.

Why does it matter to document, for example, that cohorts in the US have so far experienced relatively advantageous cumulative mortality, embodied in higher levels of ${}_xe_0^C$, if today's survivors are experiencing higher mortality rates? Isn't what is happening now more important than what has happened in the past? It matters because without full knowledge of a cohort's entire mortality trajectory, these higher period mortality rates are difficult to interpret. As said in the introduction, current mortality rates represent a mix of past and current influences that are difficult to tease out. Some arguments for explaining relative country rankings in current mortality are made with explicit reference to the past. For example, it is often argued that the US current mortality disadvantage is in part due to the lagged effect of past smoking patterns, which in the US peaked earlier and at a higher level than in other Western countries (Wang and Preston 2009). It could also be argued that due to relatively lower mortality in the US in the past, mortality selection forces may not have been operating as strongly in that country, bringing current mortality rates up. These types of arguments make it difficult to interpret today's US mortality disadvantage.

With ${}_{x}e_0^C$, comparisons are in many ways more straightforward. When we see in Fig. 4.4 that every single cohort in the US has accumulated more person-years than in Italy, in spite of Italy's current mortality advantage, it means that there is an entire history of mortality conditions in the US which may or may not explain today's US mortality disadvantage, but which has definitely favored US cohorts. Whatever excess mortality has been experienced among US adults as a result of smoking or other behaviors and exposures, or whatever the effect of mortality selection on current mortality, these detrimental effects have not been large enough to reverse the advantages that were experienced at various points in the life course of cohorts. While seeing a US disadvantage in period life expectancy is a concern (Ho and Preston 2010; Murray and Frenk 2010), examining ${}_{x}e_0^C$ trajectories puts this disadvantage in perspective by expanding the time frame of the mortality comparison and by linking current disadvantages with earlier experiences in a way that makes sense substantively and methodologically. A period mortality disadvantage that has not lasted long enough to generate cohort mortality disadvantages is obviously less significant than a period mortality disadvantage that has persisted long enough to generate clear cohort mortality disadvantages. Mortality comparisons based on ${}_{x}e_0^C$ place current mortality advantages and disadvantages in a perhaps more proper time frame.

4.4.1 The Momentum of Cohort Mortality Disadvantage

It could be argued that in some countries, while period mortality advantages (disadvantages) have not lasted long enough to generate cohort mortality advantages (disadvantages), they may very well produce just that if period advantages (disadvantages) remain in the future. While this can be examined empirically with mortality projections, we propose here to study the impact of keeping future mortality constant at the levels currently observed. This scenario introduces the concept of "momentum of cohort mortality disadvantage," which posits that some cohorts have already accumulated such mortality disadvantage in the past that they may not be able to reverse this tendency, even if they spend the rest of their life course in the advantageous situation reflected by their current, lower mortality rates.

Momentum values of cohort life expectancy at birth, e_0^{C*}, are defined as follows:

$$e_0^{C*}\,(t-x) = {}_{x}e_0^C\,(t-x) + p^C\,(x,\; t-x)\,\bullet\,e_x^P(t) \tag{4.5}$$

where $e_0^{C*}\,(t-x)$ is the momentum value of life expectancy at birth for the cohort born at time t-x (i.e., aged x at time t), and where $e_x^P(t)$ is the period life expectancy at age x at time t.

$e_0^{C*}\,(t-x)$ is the cohort life expectancy at birth (untruncated) for the cohort aged x at time t, calculated with the assumption that period mortality at time t remains constant after time t. Thus $e_0^{C*}\,(t-x)$ combines real cohort mortality experience prior to time t together with period mortality at time t (embodied in $e_x^P(t)$ in Eq. (4.5)) utilized for completing the life course of cohorts that are truncated at time t.

(Like the classic population momentum, which combines past demographic regimes embodied in the current age distribution together with current demographic regimes, momentum values of cohort life expectancies combine past mortality regimes with current ones.) The purpose of calculating momentum cohort life expectancies is not to propose a realistic forecast of levels of life expectancy for these cohorts, but rather to frame current period mortality advantages in a broader perspective by showing what they would imply for cohorts were they there to stay in the future.

The concept of momentum of cohort mortality disadvantage can be illustrated by comparing again French vs. Swedish females. As discussed earlier, while French females are currently (i.e., in 2011) in an advantageous situation in terms of period mortality, their levels of ${}_xe_0^C$ are systematically below those of Swedish females. This is illustrated in Fig. 4.5, which reproduces ${}_xe_0^C$ trajectories for Swedish vs. French females shown earlier in Fig. 4.1.

In addition, Fig. 4.5 shows momentum values of cohort life expectancy at birth, $e_0^{C*}\ (t-x)$, as defined in Eq. (4.5), for these two populations. As seen on this figure,

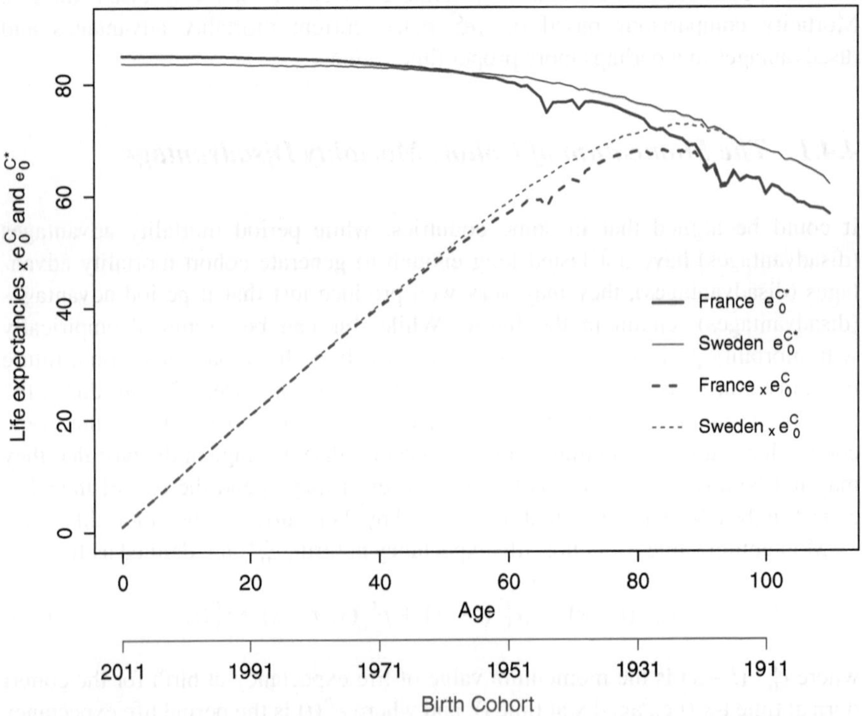

Fig. 4.5 Truncated cohort life expectancies (${}_xe_0{}^C(t-x)$), and untruncated "momentum" cohort life expectancies ($e_0{}^{C*}(t-x)$), France and Sweden, Females, 2011 (Source: HMD. Note: $e_0{}^{C*}(t-x)$ is the cohort life expectancy at birth that would eventually be observed for the cohort born at time *t-x* if mortality in France and Sweden remained constant in the future at their respective levels observed at time $t=2011$)

the gap between ${}_{x}e_0^C\ (t-x)$ and $e_0^{C*}\ (t-x)$ for a given population decreases with age. This occurs because as age x increases, the portion of a cohort's life course that is truncated at time t becomes smaller. Therefore, as x increases, $e_0^{C*}\ (t-x)$ tends towards ${}_{x}e_0^C\ (t-x)$. Above age 85, there is virtually no difference between ${}_{x}e_0^C\ (t-x)$ and $e_0^{C*}\ (t-x)$ in each country.

Figure 4.5 also shows that values of $e_0^{C*}\ (t-x)$ are slightly higher in France for cohorts born currently, reflecting France's higher level of period life expectancy. Indeed, if mortality regimes in France and Sweden stayed constant in the future at the levels currently observed, female cohorts born currently would experience a life expectancy of 85.03 years in France vs. 83.68 years in Sweden. (In fact, the interpretation of the momentum value of cohort life expectancy for a cohort born today corresponds to a classic interpretation of today's value of period life expectancy, i.e., the number of years that a baby born today can expect to live if age-specific mortality rates remained constant at today's level in the future. In Eq. (4.5), if $x=0$ then $e_0^{C*}\ (t-x) = e_0^P(t)$.) As age x increases, however, the gap between France vs. Sweden's $e_0^{C*}\ (t-x)$ decreases, and a cross-over occurs around age 50. For all cohorts aged 50 and above, values of $e_0^{C*}\ (t-x)$ remain lower in France vs. Sweden.

These results indicate that for French cohorts born in the last 50 years, who have so far experienced relatively minor mortality disadvantages relative to Sweden and who still have many person-years of exposure left to be lived in the future, current mortality advantages would translate into superior levels of life expectancy. (Of course, this scenario does not raise the possibility that France's mortality advantage may be only temporary, in which case French cohorts may at the end remain disadvantaged. But at least, Fig. 4.5 indicates that given France's current advantage in period mortality, the possibility is real that these cohorts may indeed experience superior levels of life expectancy.) For French cohorts currently aged above 50, however, momentum values of life expectancy at birth remain below those for Sweden. This pattern indicates that for these cohorts, current advantages in period mortality (relative to Sweden) are unlikely to translate into superior levels of life expectancy at birth. These older cohorts have already accumulated such mortality disadvantage (as shown with the lower ${}_{x}e_0^C$ values), that even if France's current advantages remained in the future, there is too little time remaining in the life course of these cohorts to reverse this tendency. In other words, cohorts above age 50 in France are experiencing a significant momentum of mortality disadvantage.

4.5 Conclusion

In this chapter, we argue that the examination of ${}_{x}e_0^C$ patterns offers a simple way of summarizing complex time- and age-specific mortality trajectories and enriches international mortality comparisons. This approach is particularly relevant for countries that have experienced mortality change and for whom it is likely that cohort vs. period mortality will offer a significantly different picture. We also introduce the concept of "momentum of mortality disadvantage," which shows that cohort mortality patterns that have already occurred and can be already observed will play an important role in determining how cohorts will eventually rank in terms of their life expectancy at birth once they eventually become extinct. With these concepts in hand, we believe that analysts are better equipped for situating period mortality comparisons in a broader context, and for understanding the implications of current mortality levels for actual cohorts of individuals.

References

Almond, D. (2006). Is the 1918 Influenza pandemic over? Long-term effects of in utero Influenza exposure in the post-1940 US population. *Journal of Political Economy, 114*(4), 672–712.

Barker, D. J. (2007). The origins of the developmental origins theory. *Journal of Internal Medicine, 261*(5), 412–417. doi:10.1111/j.1365-2796.2007.01809.x.

Bell, F., & Miller, M. (2005). Life tables for the United States social security area 1900–2100. *Actuarial Study No. 120*. Office of the Chief Actuary, Social Security Administration.

Bongaarts, J. (2005). Five period measures of longevity. *Demographic Research, 13*, 547–558.

Brouard, N. (1986). Structure et dynamique des populations. La pyramide des années à vivre, aspects nationaux et exemples régionaux. *Espace, Populations, Sociétés, 4*(2), 157–168.

Canudas-Romo, V., & Guillot, M. (2015). Truncated cross-sectional average length of life: A measure for comparing the mortality history of cohorts. *Population Studies*, 1–13. doi:10.1080/00324728.2015.1019955

Crimmins, E. M., & Finch, C. E. (2006). Infection, inflammation, height, and longevity. *Proceedings of the National Academy of Sciences of the United States of America, 103*(2), 498–503.

Doblhammer, G., & Vaupel, J. W. (2001). Lifespan depends on month of birth. *Proceedings of the National Academy of Sciences, 98*(5), 2934–2939.

Doll, R., Peto, R., Boreham, J., & Sutherland, I. (2004). Mortality in relation to smoking: 50 years' observations on male British doctors. *BMJ, 328*(7455), 1519.

Elo, I. T., & Preston, S. H. (1992). Effects of early-life conditions on adult mortality: A review. *Population Index, 186–212*.

Forsdahl, A. (1977). Are poor living conditions in childhood and adolescence an important risk factor for arteriosclerotic heart disease? *British Journal of Preventive & Social Medicine, 31*(2), 91–95.

Frost, W. H. (1995). The age selection of mortality from tuberculosis in successive decades. *American Journal of Epidemiology, 141*(1), 4–9.

Guillot, M. (2003). The cross-sectional average length of life (CAL): A cross-sectional mortality measure that reflects the experience of cohorts. *Population Studies-a Journal of Demography, 57*(1), 41–54. doi: 10.1080/0032472032000061712.

Guillot, M. (2011). Period versus cohort life expectancy. In R.G. Rogers & E.M. Crimmins (Eds.), *International handbook of adult mortality*. New York: Springer.

Guillot, M., & Kim, H. S. (2011). On the correspondence between CAL and lagged cohort life expectancy. *Demographic Research, 24*. doi:10.4054/Demres.2011.24.25.

Ho, J. Y., & Preston, S. H. (2010). US mortality in an international context: Age variations. *Population and Development Review, 36*(4), 749–773.

Human Mortality Database. (2015). University of California, Berkeley (USA), and Max Planck Institute for Demographic Research (Germany). Available at: www.mortality.org

Murray, C. J., & Frenk, J. (2010). Ranking 37th—Measuring the performance of the US health care system. *New England Journal of Medicine, 362*(2), 98–99.

Myrskylä, M. (2010). The effects of shocks in early life mortality on later life expectancy and mortality compression: A cohort analysis. *Demographic Research, 22*(12), 289–320.
Preston, S. H., Hill, M. E., & Drevenstedt, G. L. (1998). Childhood conditions that predict survival to advanced ages among African–Americans. *Social Science & Medicine, 47*(9), 1231–1246.
Preston, S., Heuveline, P., & Guillot, M. (2001). *Demography: Measuring and modeling population processes*. Malden: Blackwell Publishers.
Richards, S. J., Kirkby, J., & Currie, I. D. (2006). The importance of year of birth in two-dimensional mortality data. *British Actuarial Journal, 12*(01), 5–38.
Schoen, R., & Canudas-Romo, V. (2005). Changing mortality and average cohort life expectancy. *Demographic Research, 13*, 117–142.
Vaupel, J. W. (2002). Life expectancy at current rates vs. current conditions: A reflexion stimulated by Bongaarts and Feeney's "How long do we live?". *Demographic Research, 7*(8), 366–376.
Vaupel, J. W., Manton, K. G., & Stallard, E. (1979). The impact of heterogeneity in individual frailty on the dynamics of mortality. *Demography, 16*(3), 439–454.
Wang, H., & Preston, S. H. (2009). Forecasting United States mortality using cohort smoking histories. *Proceedings of the National Academy of Sciences, 106*(2), 393–398.
Willets, R. (2004). The cohort effect: Insights and explanations. *British Actuarial Journal, 10*(04), 833–877.

Chapter 5
Changing Mortality Patterns and Their Predictability: The Case of the United States

Christina Bohk and Roland Rau

5.1 Introduction: Challenges in Mortality Forecasting

The probable length of life spans affects not only each individual, but also the socio-economic and political dimensions of societies. While previous gains in life expectancy have primarily led to increases in the number of people of working ages, recent improvements have enabled more people to enjoy additional years in retirement. Despite the ongoing progress in survival in many developed countries, forecasting how long people are likely to live is challenging, and is prone to errors. These challenges are mainly due to dynamic mortality reduction, which tracks the progress of mortality improvements from younger to older ages over time. Two trends are particularly apparent in the methodological development of recent efforts to capture these dynamic mortality developments. In order to capture and quantify the uncertainty of future developments, researchers appear to be increasingly using probabilistic forecasts instead of deterministic forecasts. However, many researchers are also supplementing extrapolative methods with additional information and expert judgment. At the one end of this extended spectrum, detailed information, such as obesity and smoking rates (Wang and Preston 2009; Soneji and King 2010; Janssen et al. 2013; Stewart et al. 2009), are used in approaches designed to forecast mortality and the forces driving its increase/decline. At the other end of this spectrum are approaches that are mainly data-driven, but which include some extra information. For example, Torri and Vaupel (2012) modeled country-specific life expectancy at birth as deviations from best-practice life expectancy. Our model (Bohk and Rau 2014a, b), along with a number of coherent approaches (Li and Lee

C. Bohk • R. Rau (✉)
Department of Sociology and Demography, University of Rostock, Ulmenstrasse 69, 18057 Rostock, Germany
e-mail: christina.bohk@uni-rostock.de; Roland.Rau@uni-rostock.de

R. Schoen (ed.), *Dynamic Demographic Analysis*,
The Springer Series on Demographic Methods and Population Analysis 39,
DOI 10.1007/978-3-319-26603-9_5

2005; Cairns et al. 2011; Hyndman et al. 2013), appear to be between the two ends of this spectrum, as they allow researchers to complement data-driven extrapolation with mortality trends of other (sub)populations. In this article, we investigate how forecasting approaches from different sides of the spectrum deal with regular and irregular mortality developments in the United States. Moreover, we discuss what additional information could be used to substantially increase the plausibility of U.S. mortality forecasts, and how this knowledge could be incorporated adequately from a methodological perspective.

5.2 Mortality Development in the United States

5.2.1 Life Expectancy at Birth

Life expectancy at birth has increased steadily in many highly developed countries for more than 150 years (Oeppen and Vaupel 2002; White 2002), and record life expectancy has been rising by about 2.5 years every decade. Figure 5.1 illustrates

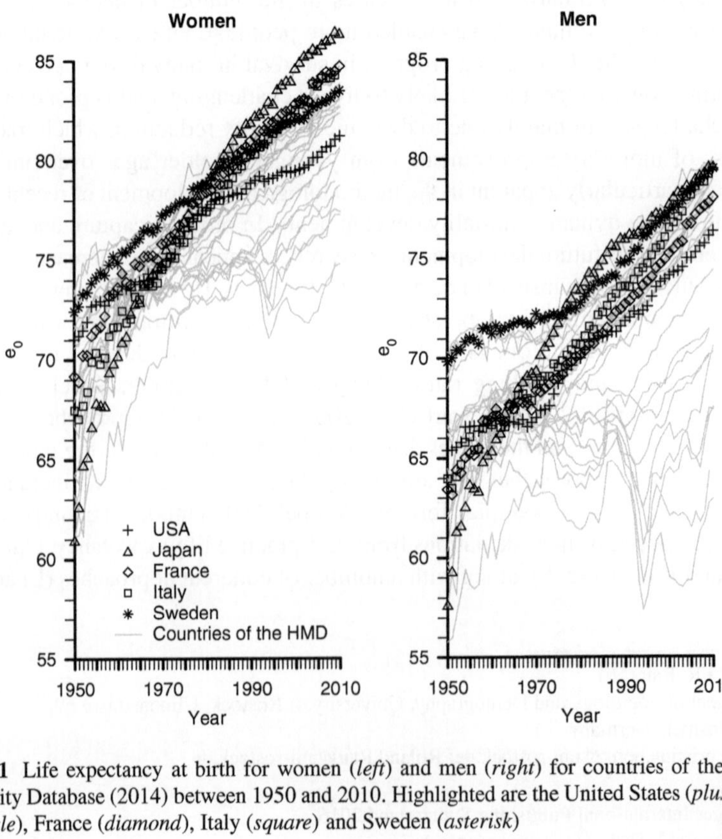

Fig. 5.1 Life expectancy at birth for women (*left*) and men (*right*) for countries of the Human Mortality Database (2014) between 1950 and 2010. Highlighted are the United States (*plus*), Japan (*triangle*), France (*diamond*), Italy (*square*) and Sweden (*asterisk*)

that Japanese women and men currently have the longest life spans worldwide: in 2010, a newborn girl could expect to live 86.30 years, while a boy can expect to live 79.56 years (Human Mortality Database (2014)), assuming death rates remain constant over the next 100–120 years. People in Italy, France and Sweden are likely to have long lives as well, though their life spans are expected to be up to three years shorter for women and up to 1.5 years shorter for men compared to the life spans of their Japanese peers. In the U.S., average life expectancy lags behind these values: In 2010, a newborn girl could expect to live 81.21 years, whereas a boy could expect to live 76.37 years (Human Mortality Database (2014)). Moreover, the course of U.S. life expectancy appears to be irregular, as periods of stronger increases are often followed by periods of weaker increases (and almost stagnation). Since the 1990s in the U.S., men have experienced greater increases in life expectancy than women.

5.2.2 Death Rates and Change of Mortality

The developments in U.S. mortality are shown in greater detail in Fig. 5.2, on so-called Lexis surfaces (Caselli et al. 1985; Gambill and Vaupel 1985; Vaupel et al. 1985). In the upper two panels, the death rates of men and women are depicted for the single ages zero to 100 on the y-axis, and for the calendar years 1950–2010 on the x-axis. In the lower two panels, the rates of mortality improvement–i.e., the rate of change in mortality at a given age over time, expressed in percent per year–are shown. In our figures, the level of mortality itself and of its annual change are depicted with a color gradient, which ranges from blue for lower levels over green and yellow to red for higher levels. Moreover, mortality increases are depicted in gray and black for the rates of mortality improvement.

Smooth color shifts over time – e.g., from red to yellow or from green to blue for a given age–suggest that death rates decreased between 1950 and 2010. This finding of a broad mortality decline is also supported by slightly increasing contour lines, which indicate that certain mortality levels gradually proceed to older ages over time. Apart from mortality changes, the surfaces depicting the rates of mortality improvement allow us to easily observe cohort and period effects. Since their aspect ratio of 1:1 denotes that a calendar year has the same length as an age year, a cohort effect is visible along a 45° line, whereas a period effect can be identified along a vertical line. For instance, a cohort effect of decreasing mortality improvement is visible as a gray area for adult women between 1980 and 2005. This development is clearly responsible for the slow increase in female life expectancy at birth in this period. Meanwhile, a period effect of increasing survival improvement is visible as a green and partly red area for women at almost all ages between 1970 and 1980. This effect is probably responsible for the relatively large gains in female life expectancy in this period.

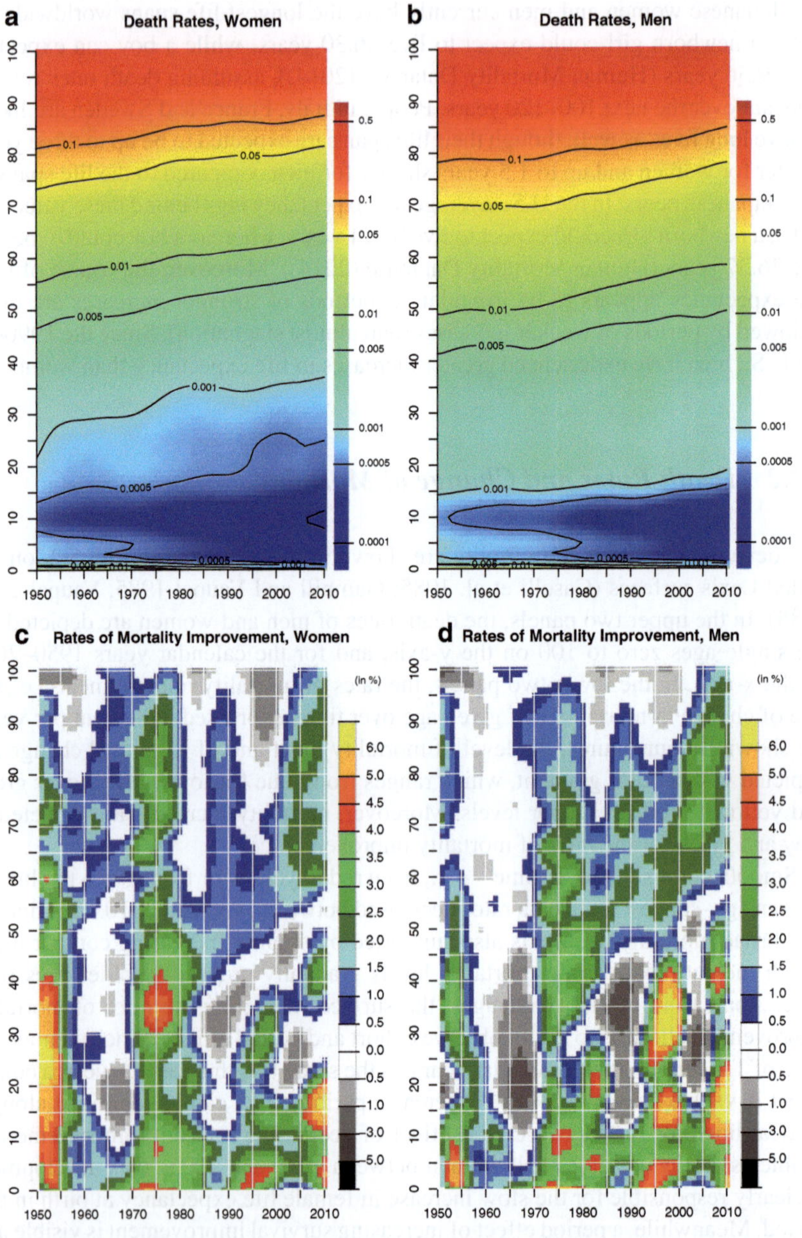

Fig. 5.2 *Upper Panel*: U.S. death rates for women (**a**) and men (**b**). *Lower Panel*: Rates of mortality improvement in the United States for women (**c**) and men (**d**) (Source: Own estimation based on data from the Human Mortality Database (2014))

5.2.3 *Driving Factors*

As the gap in mortality between the U.S. and other highly developed countries has grown during the last five decades of the twentieth century, researchers (Crimmins et al. 2011) have been seeking explanations for why U.S. mortality is higher at almost all ages for both women and men. Many preventable deaths have been attributed to modifiable lifestyle, dietary, and metabolic risk factors (Danaei et al. 2009): namely, cigarette smoking, high blood pressure, obesity, physical inactivity, and high sodium intake (*Healthy People 2020* framework).

Although we think that such behavioral factors play a dominant role, other factors are also highly relevant. For example, recent initiatives like the Healthy People 2020 framework and the Affordable Care Act (ACA)–also known as Obama Care–are meant to increase the awareness of determinants of health, as well as the availability, access, and affordability of health care in the U.S. population. Progress in medical treatments can only reduce mortality at the population level if it is available and affordable for a large share of people. Comparative studies have shown that such a universal access (and insurance coverage) is one of the main drawbacks in the U.S. health care system compared to other industrialized countries (Davis et al. 2014). If the U.S. population could reduce modifiable risk factors and (socio-economic) disparities in the access to health care via, for example, expanding prevention programmes and health insurance coverage, the disadvantage in U.S. life expectancy to other highly developed countries might decline.

5.2.3.1 Cigarette Smoking

Cigarette smoking is often cited as a lifestyle risk factor in studies that seek to explain differences in mortality between sexes and countries (Janssen et al. 2013; Wang and Preston 2009; Crimmins et al. 2011). It is the behavioral risk factor that is responsible for the greatest number of U.S. deaths (Danaei et al. 2009), most notably from lung cancer (Preston et al. 2010). The top row of Fig. 5.3 depicts rates of mortality improvement for lung cancer among U.S. women and men between 1960 and 2010. These rates reveal a cohort pattern of decreasing survival improvements (gray and black colors) which, however, seems to have disappeared in recent years. Among subsequent cohorts smoking-related mortality reductions have been higher among males than among females. This trend could explain why the sex differential in U.S. life expectancy has narrowed in recent years. As smoking-related mortality strongly depends on smoking duration, mortality reductions are expected to reach higher ages in the future. In terms of the future course of U.S. mortality, we expect to see greater mortality reductions for males than for females in the *short run*, because smoking prevalence has been decreasing for a longer period of time among men (since the birth cohorts 1915–1919) than among women (since the birth cohorts 1945–1949) (Preston and Wang 2006). But we also anticipate that smoking prevalence will further decrease among women as well, as they are likely to catch up to international trends in the *medium and long run*.

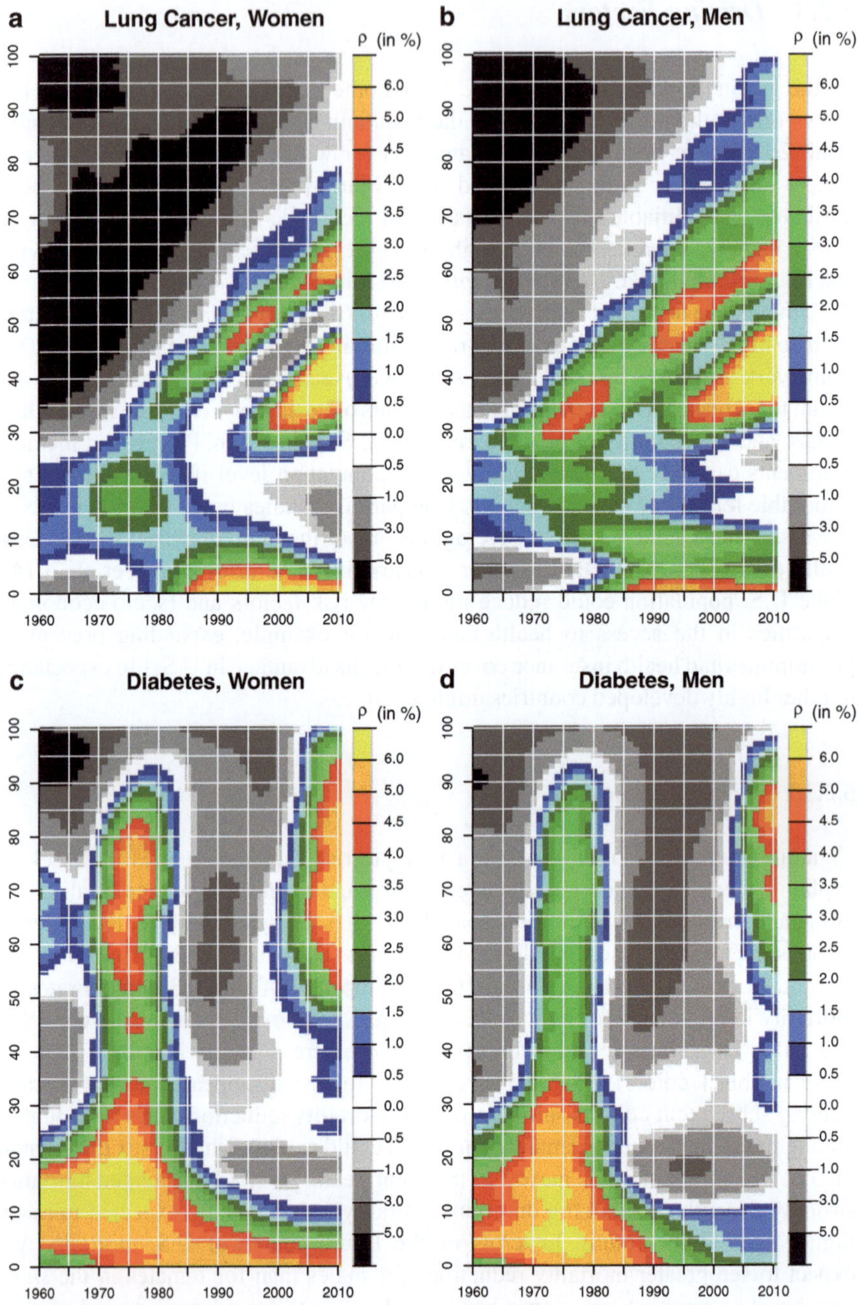

Fig. 5.3 Rates of improvement for lung cancer-related mortality (*top row*) and for diabetes-related mortality (*bottom row*) for U.S. women (*left*) and men (*right*) (Source: Own estimation based on data from the Human Mortality Database (2014) and NCHS's Multiple Cause of Death Data (2013))

5.2.3.2 Obesity

High blood pressure and overweight are among the metabolic risk factors that cause many deaths in the U.S. due to, for example, circulatory diseases or diabetes (Danaei et al. 2009). According to the World Health Organization (2000), people are defined as obese if their body mass index (BMI) is equal to or above 30, which implies that they have excessive fat accumulation. Although the usefulness of this definition/measure is controversial, the BMI is applied in many studies due to its simple data requirements. Using BMI, Ezzati et al. (2006) found an almost linear increase in the prevalence of obesity, which reached 28.7 % for men and 34.5 % for women in the United States in 2002. Between 2009 and 2012, 35.3 % of the U.S. population aged 20 or older were obese according to the *Healthy People 2020* framework. Crimmins et al. (2011) attributed this development in part to a lack of preventive intervention by the U.S. health care system. Since diabetes appears to be closely linked to obesity, the bottom row of Fig. 5.3 depicts the rates of mortality improvement for this cause of death among women and men from 1960 to 2010. These rates indicate that there was a strong period effect of decreasing survival improvements (gray and black colors) for all people over age 10 between 1985 and the early 2000s, and of survival improvements for both men and women in recent years, though the effect was slightly more pronounced for women than for men. If obesity continued to be an important mortality risk factor, it may have offset the expected survival improvements as smoking-related mortality declined (Crimmins et al. 2011). However, since the influence of obesity appears to have decreased recently, and the link between obesity and life expectancy has been called into question (Crimmins et al. 2011; King and Soneji 2011), we expect that obesity will have a much weaker influence on survival improvements than cigarette smoking.

5.2.3.3 High Sodium Intake

Although high sodium intake is a primary dietary risk factor, it plays a much smaller role in U.S. mortality than cigarette smoking or obesity. Nevertheless, it can also cause high blood pressure, which is the factor responsible for the second-largest number of U.S. deaths (Danaei et al. 2009). Given the available information, it remains unclear through which pathways this dietary risk factor might affect future U.S. mortality.

5.3 Mortality Forecasting Approaches

Based on these findings, plausible U.S. mortality forecasts must fulfill three requirements. First, while we may expect to see further gains in life expectancy for both sexes, we should avoid predicting that men will outlive women. Second, considering the mortality risk factors discussed above, we may assume that in a short run mortality will decline faster among men than among women, but we should also assume that in the long run mortality will decline (faster) among women as well, particularly in light of the expected trends in tobacco smoking. Third, given the recent changes in mortality patterns, a plausible forecast should also predict a dynamic shift of relatively large mortality improvements from younger to older ages.

5.3.1 *Common Approaches*

Predicting dynamic age shifts in mortality is challenging for many forecasting approaches. For instance, the widely accepted model of Lee and Carter (1992) extrapolates past mortality trends using an inflexible age schedule of mortality change, which induces systematic bias, especially in the long run. Although the original Lee-Carter model has been modified and extended in various ways (Booth et al. 2006; Shang et al. 2011; Shang 2012) so that it can, for example, account for extra cohort and period effects (Renshaw and Haberman 2003, 2006), these shortcomings still apply (Mitchell et al. 2013). Recently developed approaches use new methodological strategies to cope with this problem. For example, Li et al. (2013) let the age pattern rotate with time in the predictor structure of the Lee-Carter model, Hyndman and Ullah (2007) used multiple principal components, and Haberman and Renshaw (2012) and Mitchell et al. (2013) used rates of improvement rather than death rates to forecast variable mortality changes. Moreover, scholars like Janssen et al. (2013), King and Soneji (2011), Wang and Preston (2009), and Stewart et al. (2009) developed approaches that include etiological data, such as prevalence of smoking and/or obesity (and their attributable mortality), to account for major risk factors when forecasting (overall) mortality. Comparative studies on the forecasting performance of such extrapolative and explanatory approaches have suggested that the major differences in the outcomes of these approaches appear to depend on their explicit assumptions (like the base period and jump-off rates) (Stoeldraijer et al. 2013). The remaining minor method-related differences in the outcomes appear to be larger in the presence of non-linear mortality trends. The coherent approaches of Hyndman et al. (2013), Cairns et al. (2011), and Li and Lee (2005) are able to address co-moving as well as changing trends by jointly forecasting the mortality of multiple (sub)populations who might be defined by sex or nationality. For instance, they make it possible to increase the survival improvements for a single country if its own extrapolated trend is below the joint trend (and vice versa).

5.3.2 *Our Model*

Using our model (Bohk and Rau 2014b), our goal is to generate U.S. mortality forecasts that combine some methodological advances in a *single* framework.

To model dynamic age shifts of survival improvement from younger to older ages, our model applies rates of mortality improvement instead of death rates. In addition to predicting these dynamic age shifts, our model can also forecast accelerating and decelerating changes in mortality by optionally complementing an extrapolated mortality trajectory of a country of interest (*COI*) with the joint trend of reference countries (*RC*). *RC* should exhibit similar conditions regarding health and mortality due to, for example, cultural and political proximity with the *COI*. If these conditions are met, this optional feature might increase the plausibility of a purely extrapolated mortality trend for a *COI*. Since we expect to see an accelerating mortality decline among U.S. women in the medium and long run due to a progressive decline in tobacco smoking, we will use reference countries which have experienced lower mortality and a lower prevalence of cigarette smoking. For instance, women in France, Italy, Sweden, and Japan have had substantially lower overall and smoking-related mortality than U.S. women (Preston et al. 2010).

5.4 Forecasting Mortality

In this section we forecast how U.S. mortality is likely to develop in the future, and how many additional years of life women and men in the U.S. are likely to gain. In addition to forecasting the probable length of future life spans, we pay special attention to how forecasting approaches deal with the irregular mortality development of U.S. women (compared to the more regular trend of U.S. men), and how they meet our expectation of an accelerating increase in female life expectancy. We therefore forecast U.S. mortality for both sexes from 2011 to 2050 using our model (Bohk and Rau 2014b), the robust benchmark model of Lee and Carter (1992, *LC*), and four of its variants proposed by Lee and Miller (2001, *LM*), Booth et al. (2002, *BMS*), Hyndman and Ullah (2007, *HU*) and Hyndman et al. (2013, *H coh*). The comparison of these prospective forecasts can be regarded as credible as long as concrete assumptions like the base period, the forecast horizon, and the (raw) input data are equal (Stoeldraijer et al. 2013). Since we forecast mortality with each approach from 2011 to 2050, given U.S. death rates by single age and sex from the Human Mortality Database (2014) from 1970 to 2010, we can assume that the requirement of equal assumptions is met. As a consequence, differences in the prospective forecasts can only be due to differences in the method and in their method-related assumptions. For instance, though all of these variants rely on the predictor structure of the original Lee-Carter model, they differ depending on, for example, whether raw, logarithmized, and/or smoothed mortality data are used as inputs, which parameter is used to adjust fitted data, and whether observed or fitted jump-off data are used to yield a smooth transition between present and forecasted data (Booth et al. 2006). Moreover, the variants differ in the time series model which is applied to forecast the time index. *LC*, *LM*, and *BMS* use a random walk with drift; whereas *HU* and *H coh* automatically select optimal ARIMA models for each principal component. While these mainly data-driven approaches typically extrapolate past trends, they are often challenged to model long-term trend changes. In contrast, coherent approaches can include some extra information to overcome this drawback. The coherent approach of Hyndman et al. (2013) forecasts co-moving mortality for women and men; our model combines mortality trends of U.S.-American women with those of Japanese, French, Italian and Swedish women due to their lower overall and smoking attributable mortality. For instance, Preston et al. (2010) show that 20 % of all female deaths at age 50 and older are attributable to smoking in the U.S. in 2003. This fraction is substantially lower for women in Japan, France, Italy, and Sweden, ranging only between two and nine percent at the same time. Hence, the data-driven forecasts of our model can be flexibly strengthened with information of vanguard countries. We expect for U.S. women that they will approach the more favorable mortality conditions of the *RC*, resulting in an accelerated increase in U.S. female life expectancy in the forecast years. Moreover, our model forecasts the annual change of age-specific death rates with a two-level normal model that automatically captures dependencies among rates of mortality improvement of neighboring ages over time. Our model is implemented in the statistical software *R* (2014) and uses Markov chain Monte Carlo simulation to explore the forecasting distribution of mortality change, i.e. it lets the Gibbs sample algorithm (Geman and Geman 1984) run in *JAGS* (Plummer 2011) with five parallel chains for a total of 5200 iterations, including a burn-in period of 200 iterations. The run-length relies, for example, on trace plots and the Raftery-Lewis diagnostic (Raftery and Lewis 1992). The forecasted rates of mortality improvement can then be retransformed into death rates and additional mortality parameters such as life expectancy at birth. To conduct the forecasts with the Lee-Carter model and its four variants, we use the R-package *demography* of Hyndman (2014). Validating mortality forecasts for the U.S. as well as for 29 other countries of the Human Mortality Database (2014) are also generated to assess the robustness and the forecasting performance of the six approaches; all of them forecast mortality from 1991 to 2010 with data from 1970 to 1990. We compare the predictive ability of their median point estimates as well as the calibration of their prediction intervals (Gneiting et al. 2007; Raftery et al. 2013).

5.4.1 *Prospective Forecasts*

The prospective forecasts generated by all of the approaches in Fig. 5.4 suggest an extension of the life span for both men and women in the U.S. up to 2050. According to the median point estimates (solid lines) of our model, U.S. life expectancy at birth is likely to reach 88.8 years for women and 85 years for men in 2050. The Lee-Carter model, as well as its variants proposed by *LM*, *BMS*, and *HU*, forecast fewer additional years of life for both women and men. On average, their life expectancy forecasts are two to three and a half years lower than the forecasts of our model. It is striking that, compared to our model, these models not only predict a considerably slower increase for both sexes, but that this increase will be much

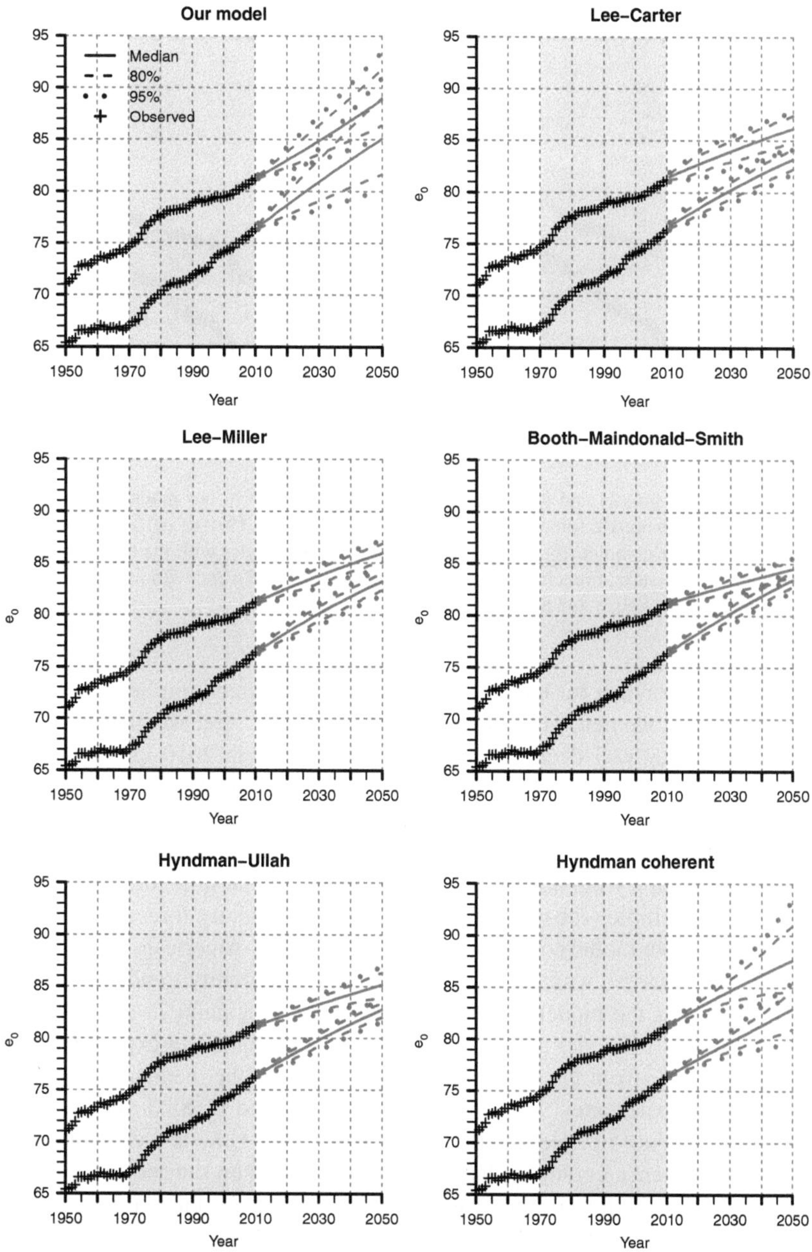

Fig. 5.4 Observed (*circle*) and forecasted life expectancy at birth of our model (*top, left*), as well as of the Lee-Carter model (*top, right*) and four of its variants (*center and bottom*) for U.S. women (*upper graph*) and men (*lower graph*). Besides the median (*solid line*), we also depict 80 % (*dashed line*) and 95 % (*dotted line*) prediction intervals. For these prospective forecasts from 2011 to 2050, we take mortality data from 1970 to 2010 (*gray rectangle*). In our model, we complement the mortality trend for U.S. females with the mortality trends of U.S.-American, Japanese, French, Italian and Swedish women

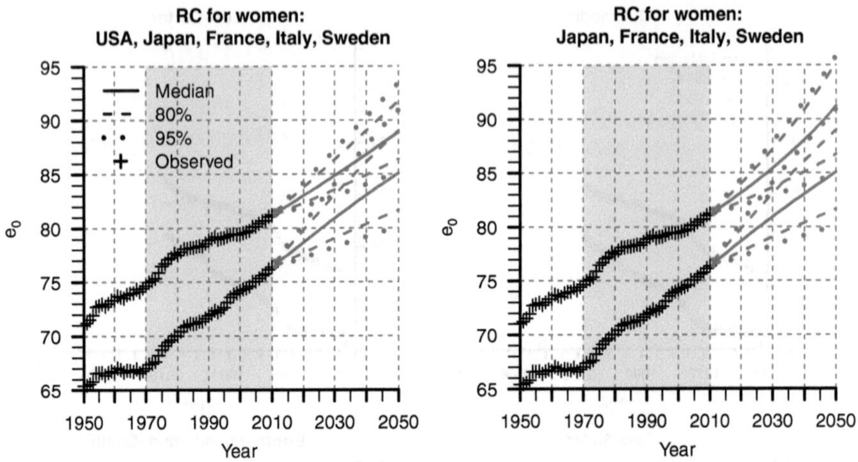

Fig. 5.5 Observed (*black*) and forecasted life expectancy at birth of our model for U.S. women and men. In this prospective forecast from 2011 to 2050, data from 1970 to 2010 (*gray box*) are the basis. Moreover, we complement the mortality trend for U.S. females with the mortality trends of U.S.-American, Japanese, French, Italian and Swedish women (*left*); and with the mortality trends of Japanese, French, Italian and Swedish women only (*right*)

slower for women than for men. A continuation of this trend implies that men will probably outlive women. For instance, the *BMS* model forecasts that the sex gap in life expectancy will decline from almost five years in 2010 to only one year in 2050. Although the effect of a crossover of female and male mortality trends is less pronounced in the forecasts of the other variants of the Lee-Carter model, they also predict that the sex gap will decline in the medium run. Our model and the coherent model of Hyndman et al. avoid making this implausible forecast by linking the trends of multiple (sub)populations. While *H coh* jointly forecasts female and male mortality, our model combines the mortality of U.S.-American, French, Italian, Swedish and Japanese women. Combining multiple mortality trends in our model not only prevents the model from forecasting decreasing gains in life expectancy, it also allows us to model accelerating increases in female life expectancy. If we excluded the U.S. from the pool of reference countries, the median point estimate of female life expectancy would be 91.2 years (instead of 88.8 years) in 2050, as depicted in Fig. 5.5. Hence, the sluggish trend of U.S. female mortality (in the pool of reference countries) has a dampening effect on our mortality forecast. Although our prediction for the growth in female life expectancy might appear to be exceptionally large, it should be noted that the models of *LC*, *LM*, *BMS*, and *HU* predict that the life span of U.S. women in 2050 will be still below the life span of Japanese women in 2010 (86.3 years).

Jointly forecasting the mortality trends of multiple populations affects not only the level and pace of forecasted life expectancy, but also its assigned uncertainty. While the 95 % prediction intervals (dotted lines) of our model and of the coherent model of Hyndman et al. become progressively broader, reaching a span of 10 years for women and men in 2050; the prediction intervals of the other models are considerably narrower, of approximately three years for women and 2.5 years for men in 2050.

5.4.2 *Validating Forecasts*

To assess the forecasting performance of the six approaches, we forecast U.S. life expectancy at birth from 1991 to 2010 using observed mortality data between 1970 and 1990 of the Human Mortality Database (2014). Figure 5.6 depicts their median point estimates (solid lines), as well as their 80 % (dashed lines) and 95 % (dotted lines) prediction intervals. To allow for a fair comparison, we abstain from using mortality trends of reference countries in our model forecasts.

The predictive ability of the U.S. validating forecasts can be assessed using forecast errors that quantify the difference between the median point estimates and the actually observed life expectancy values. Figure 5.7 depicts such forecast errors between 1991 and 2010. It is striking that the forecast errors are similar for all of the approaches, and that the deviations are relatively small for men, but relatively large for women. For instance, in 2010 they fluctuate around a value of one for women and minus half for men. Moreover, the forecast errors suggest that there is a systematic overestimation of female life expectancy. Since the increase in female life expectancy was slower in the forecast years than in the base period, the approaches tend to overestimate the actual progress.

These validating U.S. mortality forecasts suggest that our model performs similarly well as the other approaches. However, if we compare the ability of each approach to forecast mortality from 1991 to 2009 for women and men in 30 countries of the Human Mortality Database (2014), namely in Australia, Austria, Belarus, Belgium, Bulgaria, Canada, the Czech Republic, Denmark, Finland, France, East Germany, West Germany, Hungary, Italy, Japan, Latvia, Lithuania, the Netherlands, New Zealand, Norway, Poland, Portugal, Russia, Slovakia, Spain, Sweden, Switzerland, the United Kingdom, the U.S., and the Ukraine, the results indicate that our model performs on average better, particularly for men. We use the root mean square error (RMSE) as a measure of accuracy, which gives non-negative values that can be interpreted as the standard deviation of the simple forecast errors. The box plots in Fig. 5.8 depict for each approach the level and dispersion of the country-specific RMSEs (circles) for women (left) and men (right). The bottom and the top of the boxes represent the 0.25 and the 0.75 quantile, the inner line represents the median, and the ends of the whiskers represent the 0.025 and the 0.975 quantile. It is striking that the approaches have very similar RMSEs in the female forecasts; the median values range only between 0.48 and 0.67 among the models. By contrast, the RMSEs differ more among men; the median values range between 0.4 for our model and between 0.96 and 1.21 for the other approaches. It is also conspicuous that our model has a few outliers that are relatively large; they

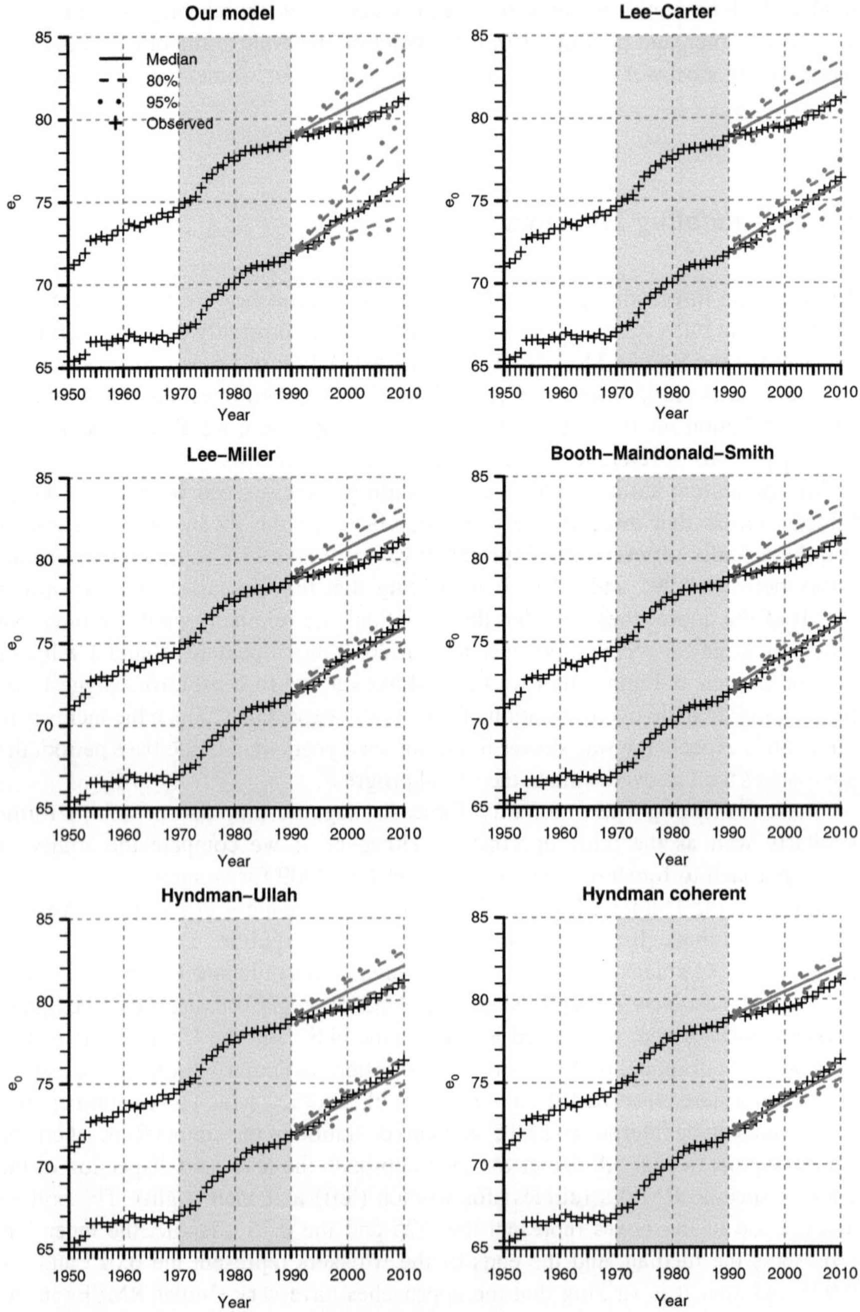

Fig. 5.6 Observed (*circle*) and forecasted life expectancy at birth of our model (*top, left*), as well as of the Lee-Carter model (*top, right*) and four of its variants (*center and bottom*) for women (*upper graph*) and men (*lower graph*). Besides the median (*solid line*), we also depict 80 % (*dashed line*) and 95 % (*dotted line*) prediction intervals. For these validating forecasts from 1991 to 2010, we take mortality data from 1970 to 1990 (*gray rectangle*)

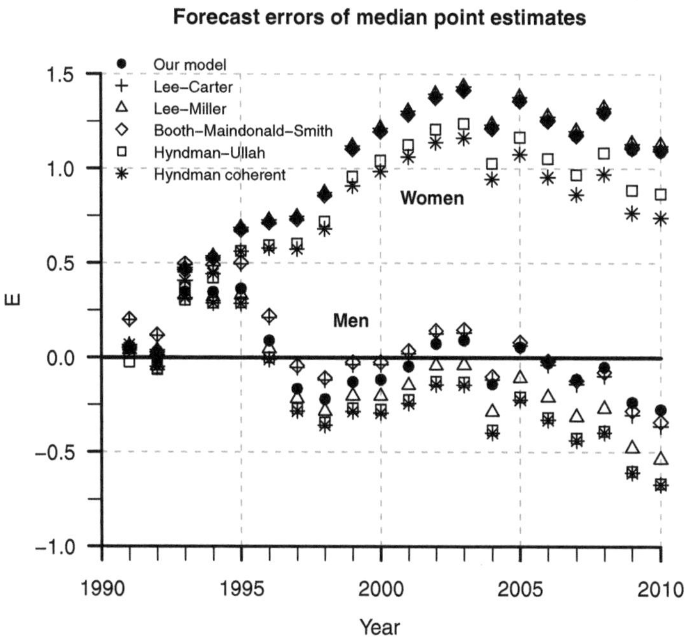

Fig. 5.7 Forecast errors (E) of the median point estimates of our model (*filled circle*), of the Lee-Carter model (*plus*) and its variants of Lee and Miller (*triangle*), of Booth, Maindonald, and Smith (*diamond*), of Hyndman and Ullah (*square*), and of Hyndman et al. (*asterisk*) for women (*upper graphs*) and men (*lower graphs*) between 1991 and 2010

are for countries like Russia, Belarus, and the Ukraine that have experienced a sharp decline in life expectancy in the 1990s. Except for these few outliers, the RMSEs are less spread for our model than for the other approaches. Overall, the validating forecasts suggest that the rates of mortality improvement strengthen the average forecasting performance of our model in comparison to the other approaches. If we also accounted for mortality trends of *RC*, we could have further strengthened the predictive ability of our model; especially in the presence of long-term trend changes such as in East Germany, Hungary, and Poland after the dissolution of the Soviet Union. We could have assumed that these *COI* would have approached the more favorable mortality conditions in West Germany in the forecast years, resulting in a forecast that accelerates the purely extrapolated mortality decline in the *COI*. Thus, our forecast errors would have even been smaller. However, to guarantee an objective model comparison, we only used the benefits of the rates of mortality improvement; even if we omit the advantage of *RC*, our model performs at least as well as the other approaches, sometimes even better.

The calibration of prediction intervals can be assessed using empirical frequencies (Gneiting et al. 2007; Raftery et al. 2013). They quantify the proportion of actually observed values which are covered in prediction intervals of different levels. Table 5.1 lists such empirical frequencies for 80 % and 95 % prediction

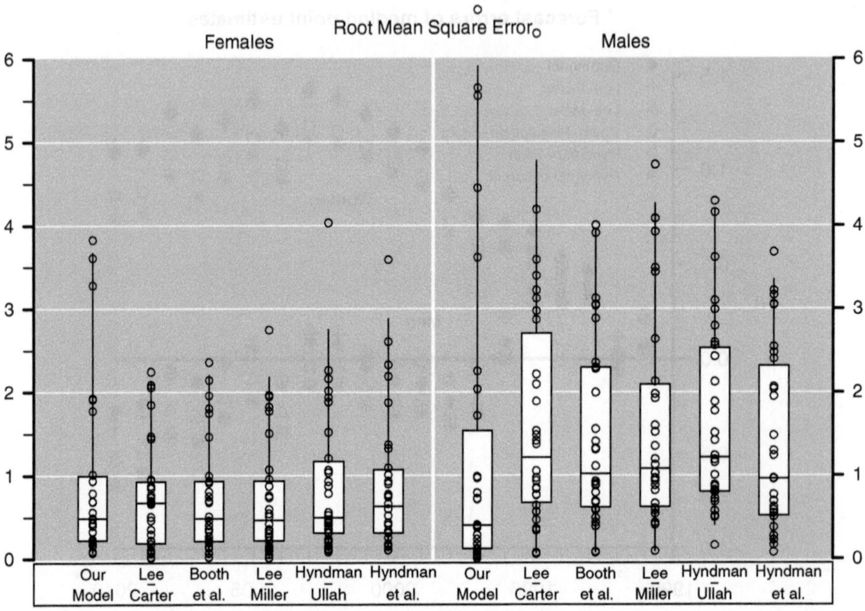

Fig. 5.8 Root mean square errors (RMSE) of validating mortality forecasts from 1991 to 2010 for women (*left*) and men (*right*) in 30 countries of the Human Mortality Database (2014). Box plots depict the level and spread of these country-specific RMSE (*circles*) for all six approaches, i.e. for our model, the Lee-Carter model, and its variants of Lee and Miller, Booth, Maindonald, and Smith, Hyndman and Ullah, and Hyndman and colleagues. The bottom and the top of the boxes represent the first and the third quartile, the inner band represents the median, and the ends of the vertical lines represent the 0.025 and the 0.975 quantiles

Table 5.1 Calibration of the 95 % and 80 % prediction intervals of six approaches

	95 %		80 %	
	Women	Men	Women	Men
Our model (w/o *RC*)	45	100	25	95
Lee-Carter	100	100	40	85
Lee-Miller	60	100	10	100
Booth-Maindonald-Smith	60	100	10	80
Hyndman-Ullah	70	100	25	100
Hyndman et al.	10	100	10	80

Listed are percentages of observed values that are captured in the 95 % and 80 % prediction intervals of the validating U.S. mortality forecasts for women and men from 1991 to 2009 for each approach

intervals of the U.S. validating forecasts. For instance, the 95 % prediction intervals of our model capture 45 % of the female and 100 % of the male observed life expectancies between 1991 and 2010, whereas the 80 % prediction intervals capture 25 % of the female and 95 % of the male observed values. All of the approaches

Table 5.2 Calibration of the 95 % and 80 % prediction intervals of six forecasting approaches

	95 %		80 %	
	Women	Men	Women	Men
Our model (w/o *RC*)	83	87	78	84
Lee-Carter	87	71	75	58
Lee-Miller	76	61	62	46
Booth-Maindonald-Smith	69	62	58	45
Hyndman-Ullah	76	57	67	42
Hyndman et al.	63	48	43	35

Listed are the percentages of observed values that are captured on average in the 95 % and 80 % prediction intervals of 30 validating mortality forecasts for women and men from 1991 to 2009 for each approach

appear to have difficulties capturing the trajectory of female mortality in the forecast years, and calibrating their prediction intervals accordingly. Although 80 % of the observed female values should fall in the 80 % prediction intervals, the fact that each approach catches only a small fraction (i.e., between 10 % and 40 %) indicates that the intervals are too narrow. Conversely, the 95 % prediction intervals of all of the approaches capture 100 % of the observed male values, which indicates that the intervals are slightly too wide. These opposite findings illustrate that it can be difficult to calibrate prediction intervals appropriately. If we compare the mean of the empirical frequencies over the validating forecasts for women and men in 30 countries of the Human Mortality Database (2014) in Table 5.2, we can see that the coverage of the prediction intervals varies among the approaches, and among the sexes. Obviously more observed values fall in the 95 % than in the 80 % prediction intervals, but the coverage of both intervals is for all approaches below the expectations. However, our model and the Lee-Carter model capture on average considerably more observed values than the other models, and the calibration of the prediction intervals appears to be slightly better in the female than in the male forecasts. For example, the 95 % prediction interval of the Lee-Carter model captures on average 87 % of the observed values in the female forecasts, but only 71 % in the male forecasts. These findings also point to the relationship between the accuracy of a forecast, its associated uncertainty, and its level of information. Where is the level of information higher: for a forecast with broad prediction intervals and a median estimate that deviates from the true values, or for a forecast with too narrow prediction intervals and a median estimate that mirrors the actual development quite well? While narrow prediction intervals appear to be precise, broad prediction intervals appear to be more reliable (Levins 1966; Orzack 2012). For instance, although forecasts with narrow prediction intervals appear to be precise, they may be unreliable because of the risk that the actual values will not fall within their closely spaced boundaries. By contrast, forecasts with broad prediction intervals appear to be reliable because the actual values are likely to fall within their widely spaced boundaries. However, these forecasts may also be imprecise if they fail to exclude unlikely pathways that unnecessarily broaden the spectrum of possible developments. Since precision and reliability are both desirable qualities, researchers should weigh how much emphasis to put on each. If a reasonable trade-off is made between precision and reliability, the results should be robust.

5.5 Summary and Concluding Remarks

How long people are likely to live is a question of fundamental interest for individuals, as well as for societies. In recent decades, the life span in the U.S. fell behind the life spans in other highly developed countries like Japan (the current record life expectancy holder), France, and Italy. A forecast of U.S. mortality may be expected to show how likely it is that the U.S. will catch up to international trends. Forecasting the future path of U.S. mortality is challenging due to the irregular developments in that country, particularly among women. Phases of large gains in life expectancy have been followed by phases of slower increases, while survival improvements have shifted from younger to older ages. To account for this dynamic age shift, our model forecasts rates of mortality improvement, rather than death rates. Interruptions in the growth in U.S. life expectancy are mainly due to behavioral factors. For instance, a large number of deaths in the U.S. from lung cancer and diabetes have been attributed to lifestyle factors like cigarette smoking and obesity. But a number of questions about how best to incorporate this knowledge into a forecast remain unsolved. For example, which lifestyle factors will turn out to be associated with substantial mortality risks? How much time will elapse between the initial appearance of these factors and their actual impact on morbidity and mortality in later life? And, how will these developments change in the future? For example, survival improvements due to the declining prevalence of cigarette smoking might be offset by the increasing (or the only very slowly decreasing) prevalence of obesity (Stewart et al. 2009). Hence, it can be beneficial to account for information about potential risk factors in a forecast. However, the choice of risk factors and the estimation of their time-variant effect on mortality as well as of their future trajectory add several sources of uncertainty that are difficult to handle. For that reason, we have chosen to account for such mortality risks implicitly. Our model provides (quasi in exchange for actual covariates) the option to jointly forecast the mortality trends of multiple populations if a long-term trend is expected to change in a country of interest (*COI*) in the forecast years. The choice of reference populations was based on the development of their overall mortality, their risk factor-attributable mortality, and their cultural and political proximity to the *COI*. If a long-term trend is expected to continue, our model simply extrapolates the rates of mortality improvement. Hence, our model provides a flexible forecasting tool employing a basic data-driven method which can be supplemented with additional information via the mortality trends of reference populations. Though the choice of the *RC* is based on subjective expert judgment, it increases the adaptability of our model, making it possible to generate forecasts for populations experiencing regular as well as irregular mortality developments in a single framework.

The out-of-sample forecasts of our model, as well as of the Lee-Carter model (1992) and its variants proposed by Lee and Miller (2001), Booth et al. (2002), Hyndman and Ullah (2007), and Hyndman et al. (2013), predict that further gains in U.S. life expectancy are likely for both sexes between 2011 and 2050. Our model and the model of Hyndman et al. (2013) predict an increasing number of additional years of life, particularly for women. For instance, our model predicts gains of approximately seven and a half years for women and of eight and a half years for men between 2011 and 2050. Moreover, the decennial increase in female life expectancy (1.87 years) is predicted to exceed that of males (1.83 years) from 2041 to 2050. By contrast, the other models predict only moderate increases in life expectancy, of approximately four years for women and six and a half years for men. These results indicate that the gains will be lower for females than for males. The forecast of smaller gains for women than for men implies that the gap between female and male life expectancy will continue to decline, and could even cross over in the long run; though this is considered to be impossible for biological reasons (Luy 2002). Our model and the model of Hyndman et al. (2013) avoid this implausible prediction by generating joint mortality forecasts for multiple (sub)populations: Hyndman et al. (2013) combine female and male mortality, whereas our model combines the mortality trends of U.S.-American, Japanese, French, Italian and Swedish women. Since women in Japan, France, Italy, and Sweden have lower overall mortality, as well as lower mortality attributable to cigarette smoking, our coherent forecast also accounts for mortality risk information via the selection of those reference countries. Combining the mortality trends of (sub)populations affects not only the level and the pace of future life expectancy, but also the degree of uncertainty. The prediction intervals of the coherent approaches are much broader than those of the other approaches. This suggests that these intervals could be more robust and reliable, because the chances are greater that the values which will be actually observed in the future will fall within their boundaries.

Acknowledgments The European Research Council has provided financial support under the European Community's Seventh Framework Programme (FP7/2007-2013) / ERC grant agreement no. 263744.

References

Bohk, C., & R. Rau (2014a). Bayesian mortality forecasts with a flexible age pattern of change for several European countries. In *Proceedings of the sixth Eurostat/Unece work session on demographic projections* (pp. 360–371).

Bohk, C., & Rau, R. (2014b). Probabilistic mortality forecasting with varying age-specific survival improvements. *arXiv:1311.5380v2[stat.AP].*

Booth, H., Maindonald, J., & Smith, L. (2002). Applying Lee-Carter under conditions of variable mortality decline. *Population Studies, 56*(3), 325–336.

Booth, H., Hyndman, R. J., Tickle, L., & de Jong, P. (2006). Lee-Carter mortality forecasting: A multi-country comparison of variants and extensions. *Demographic Research, 15*(1–2), 289–310.

Cairns, A. J. G., Blake, D., Dowd, K., Coughlan, G. D., & Khalaf-Allah, M. (2011). Bayesian stochastic mortality modelling for two populations. *ASTIN Bulletin, 41*, 29–59.

Caselli, G., Vaupel, J. W., & Yashin, A. I. (1985). Mortality in Italy: Contours of a century of evolution. *Genus, 41*(1–2), 39–55.

Crimmins, E. M., Preston, S. H., & Cohen, B. (Eds.). (2011). *Explaining divergent levels of longevity in high-income countries*. Washington, DC: The National Academy of Sciences.

Danaei, G., Ding, E. L., Mozaffarian, D., Taylor, B., Rehm, J., Murray, C. J. L., & Ezzati, M. (2009). The preventable causes of death in the United States: Comparative risk assessment of dietary, lifestyle, and metabolic risk factors. *PLoS Medicine, 6*(4), 1–23.

Davis, K., Stremikis, K., Squires, D., & Schoen, C. (2014). *Mirror, mirror on the wall. 2014 update: How the performance of the U.S. health care system compares internationally*. The Commonwealth Fund.

Ezzati, M., Martin, H., Skjold, S., Hoorn, S. V., & Murray, C. J. L. (2006). Trends in national and state-level obesity in the USA after correction for self-report bias: Analysis of health surveys. *Journal of the Royal Society of Medicine, 99*, 250–257.

Gambill, B. A., & Vaupel, J. W. (1985). *The LEXIS program for creating shaded contour maps of demographic surfaces*. Technical report, International Institute for Applied Systems Analysis (IIASA) (Working Paper WP-85-094).

Geman, S., & Geman, D. (1984). Stochastic relaxation, Gibbs distributions, and the Bayesian restoration of images. *IEEE Transactions on Pattern Analysis and Machine Intelligence, 6*(6), 721–741.

Gneiting, T., Balabdaoui, F., & Raftery, A. E. (2007). Probabilistic forecasts, calibration and sharpness. *Journal of the Royal Statistical Society, Series B, 69*(Part 2), 243–268.

Haberman, S., & Renshaw, A. E. (2012). Parametric mortality improvement rate modelling and projecting. *Insurance: Mathematics and Economics, 50*, 309–333.

Hyndman, R. J. (2014). *Demography: Forecasting mortality, fertility, migration and population data*. https://cran.r-project.org/package=demography

Hyndman, R. J., & Ullah, M. S. (2007). Robust forecasting of mortality and fertility rates: A functional data approach. *Computational Statistics and Data Analysis, 51*(10), 4942–4956.

Hyndman, R. J., Booth, H., & Yasmeen, F. (2013). Coherent mortality forecasting: The product-ratio method with functional time series models. *Demography, 50*(1), 261–283.

Janssen, F., van Wissen, L. J. G., & Kunst, A. E. (2013). Including the smoking epidemic in internationally coherent mortality projections. *Demography, 50*(4), 1341–1362.

King, G., & Soneji, S. (2011). The future of death in America. *Demographic Research, 25*(1), 1–38.

Lee, R. D., & Carter, L. R. (1992). Modeling and forecasting U.S. Mortality. *Journal of the American Statistical Association, 87*(419), 659–671.

Lee, R., & Miller, T. (2001). Evaluating the performance of the Lee-Carter method for forecasting mortality. *Demography, 38*, 537–549.

Levins, R. (1966). The strategy of model building in population biology. *American Scientist, 54*(4), 421–431.

Li, N., & Lee, R. (2005). Coherent mortality forecasts for a group of populations: An extension of the Lee-Carter method. *Demography, 42*(3), 575–594.

Li, N., Lee, R., & Gerland, P. (2013). Extending the Lee-Carter method to model the rotation of age patterns of mortality decline for long-term projections. *Demography, 50*(6), 2037–2051.

Luy, M. (2002). Die geschlechtsspezifischen Sterblichkeitsunterschiede – Zeit für eine Zwischenbilanz. *Zeitschrift für Gerontologie*, Band 35, Heft 5, 412–429.

Mitchell, D., Brockett, P., Mendoza-Arriage, R., & Muthuraman, K. (2013). Modeling and forecasting mortality rates. *Insurance: Mathematics and Economics, 52*(2), 275–285.

National Center for Health Statistics. (2013). *Mortality data – Vital statistics. NCHS's multiple cause of death data*. Available at http://www.nber.org/data/multicause.html

Oeppen, J., & Vaupel, J. W. (2002). Broken limits to life expectancy. *Science, 296*, 1029–1031.

Orzack, S. H. (2012). The philosophy of modelling or does the philosophy of biology have any use? *Philosophical Transactions of the Royal Statistical Society, 367*(1586), 170–180.

Plummer, M. (2011). *JAGS Version 3.1.0 user manual.*
Preston, S. H., & Wang, H. (2006). Sex mortality differences in the United States: The role of cohort smoking patterns. *Demography, 43*(4), 631–646.
Preston, S. H., Glei, D. A., & Wilmoth, J. R. (2010). A new method for estimating smoking-attributable mortality in high-income countries. *International Journal of Epidemiology, 39*(2), 430–438.
R Core Team. (2014). *R: A language and environment for statistical computing*. R Foundation for Statistical Computing, Vienna, Austria, http://www.R-project.org/
Raftery, A. E., & Lewis, S. M. (1992). Comment: One long run with diagnostics: Implementation strategies for Markov Chain Monte Carlo. *Statistical Science, 7*, 493–497.
Raftery, A. E., Chunn, J. L., Gerland, P., & Ševčíková, H. (2013). Bayesian probabilistic projections of life expectancy for all countries. *Demography, 50*(3), 777–801.
Renshaw, A. E., & Haberman, S. (2003). Lee-Carter mortality forecasting with age-specific enhancement. *Insurance: Mathematics and Economics, 33*, 255–272.
Renshaw, A. E., & Haberman, S. (2006). A cohort-based extension to the Lee-Carter model for mortality reduction factors. *Insurance: Mathematics and Economics, 38*, 556–570.
Shang, H. L. (2012). Point and interval forecasts of age-specific life expectancies: A model averaging approach. *Demographic Research, 27*, 593–644.
Shang, H. L., Booth, H., & Hyndman, R. (2011). Point and interval forecasts of mortality rates and life expectancy: A comparison of ten principal component methods. *Demographic Research, 25*, 173–214.
Soneji, S., & King, G. (2010). The future of death in America. *Demographic Research, 25*(1), 1–38.
Stewart, S. T., Cutler, D. M., & Rosen, A. B. (2009). Forecasting the effects of obesity and smoking on U.S. life expectancy. *New England Journal of Medicine, 361*, 2252–2260.
Stoeldraijer, L., van Duin, C., van Wissen, L., & Janssen, F. (2013). Impact of different mortality forecasting methods and explicit assumptions on projected future life expectancy: The case of the Netherlands. *Demographic Research, 29*(13), 323–354.
Torri, T., & Vaupel, J. W. (2012). Forecasting life expectancy in an international context. *International Journal of Forecasting, 28*, 519–531.
University of California, Berkeley (USA), & Max Planck Institute for DemographicResearch, Rostock, (Germany). (2014). *Human mortality database*. Available at www.mortality.org
Vaupel, J. W., Gambill, B. A., & Yashin, A. I. (1985). *Contour maps of population surfaces*. Technical report, International Institute for Applied Systems Analysis (IIASA) (Working Paper WP-85-047).
Wang, H., & Preston, S. H. (2009). Forecasting United States mortality using cohort smoking histories. *PNAS, 106*(2), 393–398.
White, K. M. (2002). Longevity advances in high-income countries, 1955–96. *Population and Development Review, 28*(1), 59–76.
World Health Organization. (2000). *Obesity: preventing and managing the global epidemic* (Report of a WHO Consultation. WHO Technical Report Series 894)

Chapter 6
Modeling the Dynamics of an HIV Epidemic

Jason R. Thomas and Le Bao

6.1 Introduction

HIV epidemics in sub-Saharan Africa grabbed the world's attention by jeopardizing future improvements in life expectancy and threatening to fray the economic and social fabric of these populations. The responses from these populations has been just as drastic, as the focus of the story has shifted to one of behavior change and concomitant declines in HIV incidence, or the rate of new infections. Throughout the maturation of the pandemic there has been a persistent need to understand the population dynamics in this rather unique epidemiological environment. Knowledge of these population dynamics helps policy makers and program planners track the epidemic, design interventions, measure progress in the fight against HIV, and forecast the future impact of the epidemic. However, advancing on this front is difficult because of the paucity of data on the primary engine of change, HIV incidence, particularly as it varies by age and over time.

The lack of direct information on HIV incidence has motivated scholars to adopt a modeling approach to study the population dynamics in an epidemic. A general strategy is to specify a model that takes parameters for HIV incidence and combines them with mortality rates to produce estimates of HIV prevalence, or the proportion of the population that is HIV positive. The primary advantage of this approach is that the model outputs can be compared with data on HIV prevalence that are now widely

J.R. Thomas (✉)
Department of Sociology and Criminology, Penn State University, University Park, PA 16802, USA
e-mail: jrt17@psu.edu

L. Bao
Department of Statistics, Penn State University, University Park, PA 16802, USA
e-mail: lebao@psu.edu

R. Schoen (ed.), *Dynamic Demographic Analysis*,
The Springer Series on Demographic Methods and Population Analysis 39,
DOI 10.1007/978-3-319-26603-9_6

available. Optimization routines are then used to search for the set of HIV incidence rates (and mortality rates) that generates model estimates of HIV prevalence that are as close as possible to the observed data.

The modeling and estimation strategy just described is employed by Heuveline (2003) in a multi-state extension of the classic cohort-component method for population projection (Preston et al. 2001; Keyfitz and Caswell 2005). In keeping with the spirit of the original, the new model classifies the population by age and thus captures an important source of heterogeneity in the risk of infection. Time, an equally important source of variability, is incorporated by assuming a trend in HIV incidence that is not directly estimated using the available data. Despite this weakness, the adaptability of Heuveline's approach and its age-specific estimates and projections make it a promising tool for studying the population dynamics underlying HIV epidemics.

In this chapter we generalize Heuveline's model by parameterizing the trend in HIV incidence with penalized B-splines, a flexible technique that has been successfully implemented in the modeling of HIV epidemics (Hogan et al. 2010; Hogan and Salomon 2012). Our new version of the model is used to examine the HIV epidemic in Tanzania, an East African country where adult HIV prevalence reached 7 % in 2003 before declining to 5 % in 2011 (Tanzania Commission for AIDS (TACAIDS), National Bureau of Statistics (NBS) and ORC Macro 2005; Tanzania Commission for AIDS (TACAIDS), Zanzibar AIDS Commission (ZAC), National Bureau of Statistics (NBS), Office of the Chief Government Statistician (OCGS), and ICF International 2013). Nearly 20 years of data on HIV prevalence from antenatal clinics (ANCs), stretching back to the mid-1980s, help inform our estimate of the trend in HIV incidence, and age-specific HIV prevalence from the nationally representative Demographic and Health Surveys and AIDS Indicator Surveys (DHS/AIS) shape our estimated age patterns of incidence. Next, we extrapolate the B-spline model in two different ways to make forecasts of HIV prevalence, and validate these projections using the most recent DHS/AIS data from Tanzania. An additional version of the model, which employs a random walk during the forecast period, is also examined for comparative purposes. In all of our analyses, we use the B-spline specification to estimate the incidence trend because it is a more parsimonious model that greatly facilitates parameter estimation. Estimating annual incidence up to 2003 using a random walk model more than doubles the number of model parameters. Finally, we conclude by summarizing the results and making recommendations for improving efforts to model the dynamics of HIV epidemics.

6.2 Background

In this section, we discuss several important features of HIV incidence as they pertain to the larger epidemics in Eastern and Southern Africa. Our treatment of this subject is not intended to be exhaustive, but to provide a general context for our analysis of the epidemic in Tanzania. We begin by discussing the literature focusing on changes in HIV incidence and prevalence, and then touch on a few characteristics of the epidemic in Tanzania. Many important topics are neglected, such as the dynamics of HIV incidence across geographic locations or the evolution of sexual networks and the characteristics binding people together. We leave these challenges to be tackled by future research.

6.2.1 Dynamics of HIV Incidence

Early in the evolution of HIV epidemics, infections primarily occur among groups whose behavior exposes them to a high risk of infection (UNAIDS 1998a). These groups, who typically include sex workers and their clients, drug injectors, and men who have sex with men, tend to make up a small proportion of the population. While most epidemics remain concentrated with low levels of HIV incidence among the total population, this is not always the case. In over 30 countries, many of which are located in Eastern and Southern Africa, the HIV epidemic has generalized to the broader population with most infections occurring through sexual contact between men and women (UNAIDS 2013). Bongaarts et al. (2008) note that the probability of infecting a heterosexual partner is quite low[1] and thus argue that at least three of the following conditions must be present for an epidemic to become generalized and reach the levels observed in sub-Saharan Africa: (1) multiple and concurrent sexual partners; (2) absence of male circumcision; (3) other sexually transmitted infections; and (4) low condom use. The confluence of these conditions has resulted in levels of HIV prevalence among adults aged 15–49 that generally exceed 1 %, and have likely surpassed 20 % in Botswana, Lesotho, Swaziland, and Zimbabwe (UNAIDS 2013).

UNAIDS estimates suggest that the relatively high levels of HIV prevalence observed in sub-Saharan Africa are driven by annual incidence rates that started from well below 0.001 in 1980 to a peak of nearly 0.008 in the early 1990s (Bongaarts et al. 2008). Since then, it seems that the rate of new infections has declined throughout most of the region, although there are differences across the countries in the timing and extent of the decline (UNAIDS 2013; Bongaarts et al. 2008; Shelton et al. 2006). Some point out that part of the decline may be due to the natural dynamics of an epidemic. If only a subset of the population faces a non-negligible risk of infection and this susceptible sub-population becomes saturated

[1]Wawer et al. (2005) estimate the average probability of heterosexual HIV transmission per coital act using data from HIV-discordant, monogamous couples. They find that probability of transmission depends on the stage of infection in the index partner, who first became HIV+. The risk of transmission peaks at 0.008 (per coital act) during the first few months after the index partner's infection, then declines and stabilizes around 0.001 over the next several years, and subsequently increases to 0.004 several months before the index partner's death. Hollingsworth et al. (2008) report similar findings.

with infections, then the incidence will decline because most of the individuals at risk are already infected or dead (UNAIDS 1998b).[2]

It is also possible that among HIV-positive individuals the average duration of infection will increase after HIV incidence peaks and the group composition shifts toward people who have been infected for longer periods of time.[3] Furthermore, an individual's infectiousness tends to decline after seroconversion and remains at a suppressed level for several years.[4] Together, these dynamics imply that the average level of infectiousness in the infected population declines as the epidemic matures, thus creating a downward pressure on the HIV incidence rate over time (Nagelkerke et al. 2014). We illustrate this point with a simple population projection using the model developed by Heuveline (2003), which classifies the infected population into four groups defined by the duration of infection (i.e., 0–4 years, 5–9 years, 10–14 years, and 15+ years). For this exercise we assume that once the epidemic begins HIV incidence increases for 10 years and then stabilizes. The resulting distribution across the HIV+ groups is shown in Fig. 6.1. The composition of the infected population continues to change for at least another decade after HIV incidence stabilizes. As the epidemic matures, the population share of recent infections declines to about 50 %, while the average duration of infection increases to over 6 years. If we use the estimates of Wawer et al. (2005) as a rough guide and assume group-specific HIV transmission probabilities per coital act of 0.001 (HIV+ for 0–4 years), 0.0007 (5–9, and 10–14 years), and 0.003 (15+ years), then the average rate of infectiousness declines by nearly 40 % in our hypothetical example.

Declines in HIV incidence may also result from changes in behaviors that protect individuals from or expose them to a greater risk of infection. Behavior change is built upon increases in knowledge about HIV prevention, changes in attitudes, and access to health services for testing, counseling, and treatment. Progress has been made on these fronts, with additional success from interventions (Scott-Sheldon et al. 2011), but not all countries have effective strategies and programs in place at a national scale to realize the full potential of behavior change. The situation may be complicated further by barriers associated with stigma, criminalization, gender inequality, and a lack of economic security (UNAIDS 2013). However, there is mounting evidence across various countries pointing to increases in knowledge of HIV prevention, condom use, male circumcision, and age of sexual debut, as well as decreases in the number of people reporting multiple and concurrent sexual partners (Cheluget et al. 2006; Fylkesnes et al. 2001; The International Group on

[2]This dynamic relies on the assumption that the susceptible sub-population is not replenished with new, uninfected individuals at a rate faster than the exit of uninfected members.

[3]In Eastern and Southern Africa, the median survival time from seroconversion is roughly 11 years in the absence of antiretroviral therapy (Todd et al. 2007).

[4]Using data from Rakai, Uganda, Hollingsworth et al. (2008) estimate that HIV-positive individuals spend about 8 years (80 %) of their remaining life in the asymptomatic stage of HIV, when infectiousness is relatively low. This result applies to a context without antiretroviral therapy that prolongs survival.

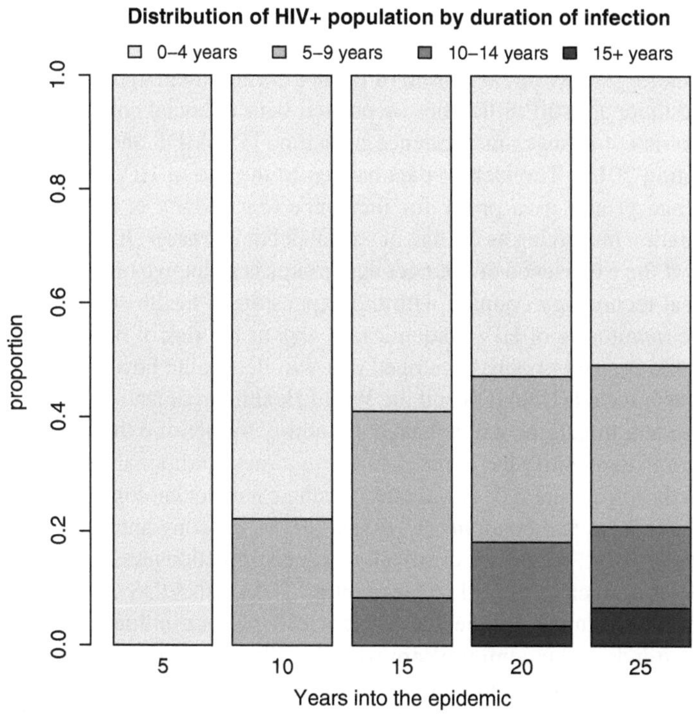

Fig. 6.1 A hypothetical example of change in the composition of the HIV+ population over time. The sub-groups are defined by the time since infection, and the changes are calculating using the cohort-component projection model and an assumed trend in HIV incidence that increases for the first 10 years of the epidemic, and then stabilizes at a constant values for the next 15 years

Analysis of Trends in HIV Prevalence and Behaviours in Young People in Countries most Affected by HIV 2010; Halperin et al. 2011; Stoneburner and Low-Beer 2004; UNAIDS 2013).

In at least two cases, namely, Uganda and Zimbabwe, some have argued that behavior changes are responsible for declines in HIV incidence (Gregson et al. 2006; Halperin et al. 2011; Stoneburner and Low-Beer 2004). The primary point of contention is that a decline in HIV prevalence is the observed outcome, not incidence. While the two are related, with HIV prevalence determined by incidence and mortality (Keiding 1991), a decline in prevalence does not necessarily indicate a decline in incidence. Sampling error, biased data, and changes in mortality among the infected population can lead to either actual or perceived changes in HIV prevalence, holding incidence constant (Ghys et al. 2006; UNAIDS 1998b).[5] There are older studies of longitudinal data on HIV incidence (e.g. Mbulaiteye et al. 2002), but the cost and logistical difficulties with collecting such data make these studies

[5]Brookmeyer and Konikoff (2011) discuss techniques for estimating HIV incidence from data on HIV prevalence at two points in time.

the exception rather than the rule. More recently, HIV incidence assays provide an exciting new source of data on recent infections (e.g. Kimanga et al. 2014). This technology, however, is in need of further development to reduce false recency rates, and there are still difficulties associated with financial cost and sample sizes needed to detect changes in incidence over time (UNAIDS and the World Health Organization 2015). The lack of data has led to the use of HIV prevalence among younger age groups as a proxy for incidence (e.g., Mahy et al. 2012), with the approximation improving as the age of sexual debut increases. It is difficult to know, however, if the experience of younger age groups is reflective of older cohorts.

Medical technology coupled with the expansion of health services will impact the future monitoring of HIV epidemics, as well as the risk of infection itself. New methods are currently being developed that will determine how long ago an HIV+ person was infected (UNAIDS and the World Health Organization 2015). Access to tests for recent infections will enhance our ability to measure the momentum of an epidemic and to quantify the impact of interventions. Additional developments may also directly impact the risk of infection, such as new technologies associated with male circumcision and condoms (UNAIDS 2013). Existing antiretroviral therapies (ART) have been shown to drastically reduce infectiousness and transmission probabilities (Cohen et al. 2011; Tanser et al. 2013; UNAIDS 2013), and thus the expansion of treatment and health services will play an influential role in future incidence trends.[6] The coming years will also likely feature continued challenges to educational and treatment campaigns associated with treatment adherence and retention among the infected population (Kranzer et al. 2012; UNAIDS 2013).

6.2.2 *Tanzania*

The United Republic of Tanzania is a country in East Africa with a population of roughly 48 million people growing at a rate close to 3 %. Life expectancy at birth is just over 60 years, the total fertility rate is about 5.2, and nearly 45 % of the population is below the age of 15. The economy is largely agricultural, yielding a gross domestic product per capita equal to 1.2 % of that of the United States (United Nations, Department of Economic and Social Affairs, Statistics Division 2014). In addition to a heavy burden associated with malaria, Tanzania is battling an HIV epidemic stretching back to the three AIDS cases first documented in 1983 (Ministry of Health 2000). Since then, HIV has spread across the country with adult prevalence in 2011–2012 ranging from a low of 0.1 % in the Kaskazini Unguja region to a high of 14.8 % in Njombe (see Fig. 6.2). On the mainland,[7] there is a

[6]The importance of this point is emphasized by the work of Eyawo et al. (2010), who estimate that the percentage of HIV+ individuals who are in stable heterosexual serodiscordant relationships may exceed 40 %.

[7]The HIV/AIDS Indicator Survey from 2003–2004 only tested for HIV on the mainland, while subsequent surveys also included the island of Zanzibar. In 2007–2008 and 2011–2012, estimates of adult HIV prevalence at the national level (including Zanzibar) stand at 5.7 % and 5.1 %, respectively.

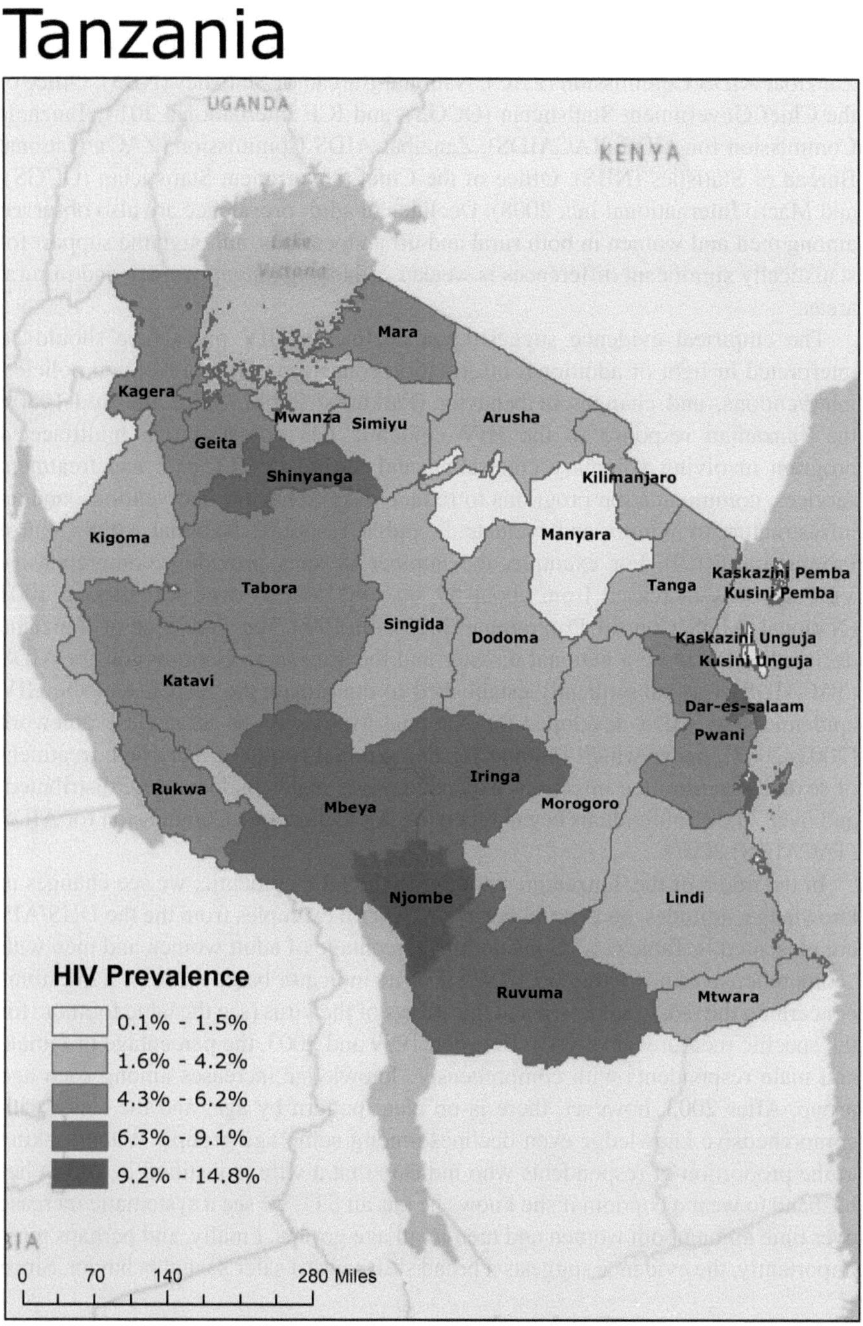

Fig. 6.2 Percent HIV positive among persons age 15–49 who were tested (Source: Tanzania HIV/AIDS and Malaria Indicator Survey 2011–2012)

statistically significant decline ($\alpha = 0.01$) in adult HIV prevalence from 7.0 % in 2003–2004 to 5.3 % in 2011–2012 (Tanzania Commission for AIDS (TACAIDS), Zanzibar AIDS Commission (ZAC), National Bureau of Statistics (NBS), Office of the Chief Government Statistician (OCGS), and ICF International 2013; Tanzania Commission for AIDS (TACAIDS), Zanzibar AIDS Commission (ZAC), National Bureau of Statistics (NBS), Office of the Chief Government Statistician (OCGS), and Macro International Inc. 2008). Declines in adult prevalence are also observed among men and women in both rural and urban locations, although the support for statistically significant differences is weaker or lacking among women and in rural areas.

The empirical evidence suggesting a decline in HIV prevalence should be interpreted in light of additional information concerning HIV prevention policies, interventions, and changes in behavior (Parkhurst 2008). Since the mid-1980s, the Tanzanian response to the HIV epidemic has grown into a multifaceted program involving voluntary counseling and testing (VCT), care and treatment services, communication programs to reduce risky behavior, interventions, and an infrastructure to monitor and evaluate the public response (National AIDS Control Programme 2010). For example, the number of sites providing comprehensive VCT services increased from about 59 sites in 1997 to over 480 sites in 2004 (National AIDS Control Programme 2005). In 1999, the President of Tanzania declared AIDS to be a national disaster and the Tanzanian Commission for AIDS (TACAIDS) was subsequently established to coordinate the response to the HIV epidemic. TACAIDS developed the National Multi-sectoral Strategic Framework (2003–2007), under which funding for the national response increased, treatment of sexually transmitted infections expanded, more male condoms were distributed, and over 70,000 individuals began receiving ART (Tanzania Commission for AIDS (TACAIDS) 2007).

In the midst of the Tanzanian response to the HIV epidemic, we see changes in knowledge, attitudes, and sexual behavior. Several examples from the the DHS/AIS are presented in Table 6.1.[8] Consider the percentage of adult women and men with "comprehensive knowledge" of HIV/AIDS, an indicator based on several questions concerning the risk of infection and the effects of the virus (see the table footnote for the specific measures involved). Between 1999 and 2003, the percentage of female and male respondents with comprehensive knowledge increases among each age group. After 2003, however, there is no clear pattern by age, and the share with comprehensive knowledge even declines among some age groups. When looking at the proportion of respondents who indicate that a wife is justified in asking her husband to wear a condom if she knows he has an STI, we see a systematic increase over time among both women and men in all age groups. Finally, and perhaps most importantly, the evidence suggests a broad shift toward safer sexual behavior. Since

[8]The DHS/AIS from 1999 and 2003–2004 only include information collected from the mainland, and thus we exclude respondents from Zanzibar when calculating the estimates from the 2011–2012 DHS/AIS.

Table 6.1 Age-specific percentages of (mainland) Tanzanian women and men with comprehensive knowledge of HIV, positive attitudes towards protective behavior among wives, and who had sex with two or more partners in the preceding 12 months (and the percentage of this group that reported using a condom at last intercourse) (Source: DHS/AIS surveys)

	Comprehensive knowledge[a]			Wife can request husband use condom[b]			2+ partners (used condoms)[c]		
Age	*1999*	*2003*	*2011*	*1999*	*2003*	*2011*	*1999*	*2003*	*2011*
(a) Women									
15–19	15.8	38.5	39.1	35.5	61.0	69.8	6.5 (22.7)	3.8 (15.3)	3.0 (37.7)
20–24	27.7	50.4	47.5	64.2	72.1	82.1	11.9 (16.1)	5.7 (22.7)	4.7 (30.8)
25–29	34.9	51.4	50.3	68.4	75.3	83.8	6.9 (13.4)	5.1 (18.8)	4.6 (34.2)
30–39	32.5	50.5	48.2	58.7	67.6	82.6	9.7 (16.8)	4.7 (12.9)	3.7 (21.2)
40–49	21.8	38.0	42.6	50.0	64.2	79.6	4.9 (1.5)	4.2 (8.5)	3.3 (11.8)
N	**4,029**	**6,863**	**10,892**	**4,029**	**6,863**	**10,892**	**4,029 (249)**	**6,863 (327)**	**10,892 (364)**
(b) Men									
15–19	18.7	42.6	45.1	40.1	64.0	76.7	18.3 (28.9)	9.2 (37.1)	7.1 (45.2)
20–24	31.4	56.5	55.2	68.1	78.4	88.0	33.9 (27.8)	24.8 (34.9)	23.0 (38.8)
25–29	44.2	59.9	54.0	71.1	78.5	85.3	31.0 (23.8)	27.2 (28.4)	25.5 (31.6)
30–39	43.8	59.3	54.7	67.4	79.7	86.5	28.8 (15.2)	23.0 (17.8)	25.8 (21.2)
40–49	30.2	54.4	53.6	58.6	76.5	84.7	28.8 (9.5)	18.3 (12.2)	25.8 (14.4)
N	**3,542**	**5,659**	**8,317**	**3,542**	**5,659**	**8,317**	**3,542 (842)**	**5,659 (1,114)**	**8,317 (1,584)**

[a]Having comprehensive knowledge is determined by: (1) having heard of AIDS; (2) responding that people can reduce their chances of getting the AIDS virus by having just one sex partner who is not infected and who has no other partners, *and* by using a condom every time they have sex; (3) knowing it is possible for a healthy-looking person to have the AIDS virus; and (4) *rejecting* at least two of the following three misconceptions about AIDS – mosquitoes transmit the AIDS virus; the AIDS virus can be contracted from sharing food with an infected person; and people can get the AIDS virus from witchcraft or other supernatural means

[b]Proportion of respondents who indicate that a wife is justified in asking her husband to wear a condom if she knows he has an STI

[c]The first indicator identifies respondents who report having sexual intercourse with two or more partners in the preceding 12 months. Among this group, the percentage that used a condom during their last sexual intercourse is reported in parentheses

1999, the percentage of respondents who report having two or more sexual partners in the last 12 months shows a steady decline for women and men in most age groups. Furthermore, among those with multiple partners, condom use appears to be more common over time. There are exceptions to this general pattern, notably, older men, but these behavior changes are consistent with a decline in HIV incidence.

In addition to the knowledge, attitudinal, and behavioral mechanisms for change, we can also rely on HIV prevalence among young adults as a proxy for incidence. Age-specific prevalence from the Tanzanian DHS/AIS surveys from both 2003–2004 and 2011–2012 are shown in Fig. 6.3 for women and men. The point estimates for prevalence are indicated by the symbols and the 95 % confidence intervals are denoted by the vertical lines. Focusing on the 15–19 year age group, we do see a decline in prevalence among both women and men, but uncertainty around the point estimate for women suggest that the difference could be due to sampling variability. Among the older age groups, a lack of statistical precision is compounded by survival among the infected population. Improvements in survival associated with

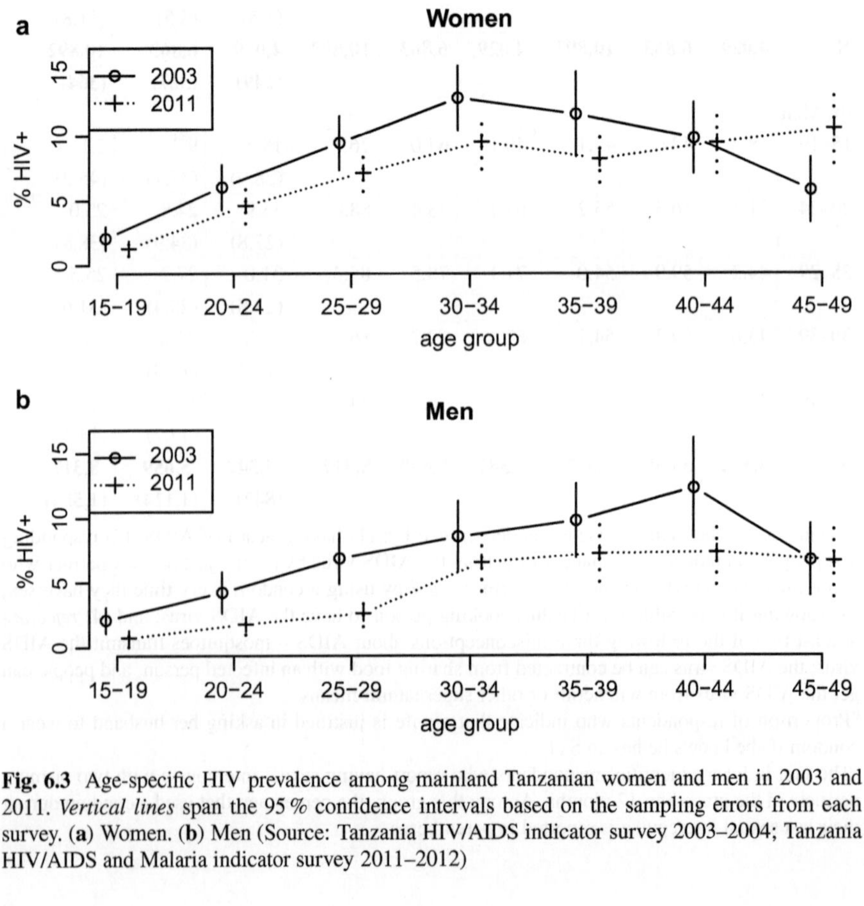

Fig. 6.3 Age-specific HIV prevalence among mainland Tanzanian women and men in 2003 and 2011. *Vertical lines* span the 95 % confidence intervals based on the sampling errors from each survey. (**a**) Women. (**b**) Men (Source: Tanzania HIV/AIDS indicator survey 2003–2004; Tanzania HIV/AIDS and Malaria indicator survey 2011–2012)

ART, creates an upward pressure on HIV prevalence and may account for the stability or potential increase in prevalence observed among women ages forty and over, as well as men who are 45–49 years old. These complications motivate our modeling approach to assessing potential changes in HIV incidence, which we turn to in the next section.

6.3 Methods

Our general approach to investigate the dynamics of HIV incidence is to use a model that projects a population classified by age, sex, and HIV status over time. The composition of the population depends on age- and sex-specific parameters that determine the level and trend in HIV incidence. In order to choose the set of parameter values that most closely reflects reality, we compare the model estimates of HIV prevalence to the observed DHS/AIS data using a Bayesian approach. Furthermore, we assess the model's predictive performance by calibrating the model to data up to 2003–2004, using these parameter estimates to make forecasts of HIV prevalence in 2007–2008 and 2011–2012, and comparing our forecasts to the observed data from those years. In the following sections we describe the data used in our analysis and then turn to a discussion of the projection model, statistical estimation, and forecasting.

6.3.1 Data

Two types of data are used in this analysis, the first being sex- and age-specific HIV prevalence from the Tanzanian DHS/AIS surveys collected in 2003–2004, 2007–2008, and 2011–2012. We combine these data into the standard 5-year age groups presented in Table 6.2. The earliest survey only includes data from mainland Tanzania and thus we exclude information from Zanzibar in the latter surveys. Despite this omission these data are still representative of over 95 % of the Tanzanian population.

The second type of data used in our analysis consists of HIV prevalence observed among female attendees of ANCs. Although these data are not stratified by age, they do stretch back much further in time, relative to the DHS/AIS data, and contribute information about the early years of the Tanzanian epidemic. The time span covered by the ANC data starts in 1986 and extends to 2003 as depicted in Fig. 6.4, which presents the data separately for the urban and rural sites. We restrict our analysis to the rural and urban sites that provide observations across multiple years. Overall, 22 urban ANCs and 21 rural ANCs contribute 118 and 110 observations, respectively, with each observation based on at least 100 (and up to 1,200) women being tested for HIV each year. ANC data are not nationally representative and are subject to at least two sources of bias. More specifically, previous research has shown that

Table 6.2 Sex- and age-specific HIV prevalence and number tested (in parentheses) in mainland Tanzanian (Source: DHS/AIS surveys)

	2003–2004		2007–2008		2011–2012	
Age	% positive	N	% positive	N	% positive	N
(a) Women						
15–19	2.1	1,257	1.4	1,311	1.3	1,967
20–24	6.0	1,185	6.5	1,128	4.6	1,514
25–29	9.4	1,108	8.0	1,061	7.1	1,452
30–34	12.9	872	10.7	886	9.5	1,186
35–39	11.6	685	9.7	735	8.2	1,147
40–44	9.8	482	7.9	524	9.5	830
45–49	5.8	380	7.0	476	10.5	675
(b) Men						
15–19	2.1	1,137	0.7	1,244	0.7	1,610
20–24	4.2	839	1.8	754	1.8	1,126
25–29	6.8	782	5.1	625	2.6	862
30–34	8.6	677	7.6	653	6.5	795
35–39	9.8	582	10.9	562	7.2	783
40–44	12.3	408	6.9	418	7.3	698
45–49	6.7	349	6.2	392	6.7	510

ANC data tend to overstate the level of HIV prevalence in the general population because women who attend ANCs are at higher risk of infection (Marsh et al. 2014; Garcia-Calleja et al. 2004; Gouws et al. 2008; Walker et al. 2001) and because of site-selection bias – where areas with high HIV prevalence are brought into the surveillance system earlier (Cheluget et al. 2006; Diaz et al. 2009; UNAIDS 2006; Walker et al. 2004). Gouws et al. (2008) provide estimates of bias in urban and rural ANC data, which we use to adjust the observations used in our analysis.[9]

6.3.2 *Model*

The projection model used in our analysis originates from Heuveline (2003), who generalizes the classic cohort-component projection technique (Preston et al. 2001; Keyfitz and Caswell 2005) to include additional states for HIV+ sub-populations defined by the duration of infection. Projections are made separately for men and women, although the number of births for both groups are generated from the female population. To simplify the notation we focus our discussion on women and

[9]Data from urban ANCs are deflated by a factor of 0.973, and rural ANC data are deflated by a factor of 0.964 (see Table 1, Gouws et al. 2008).

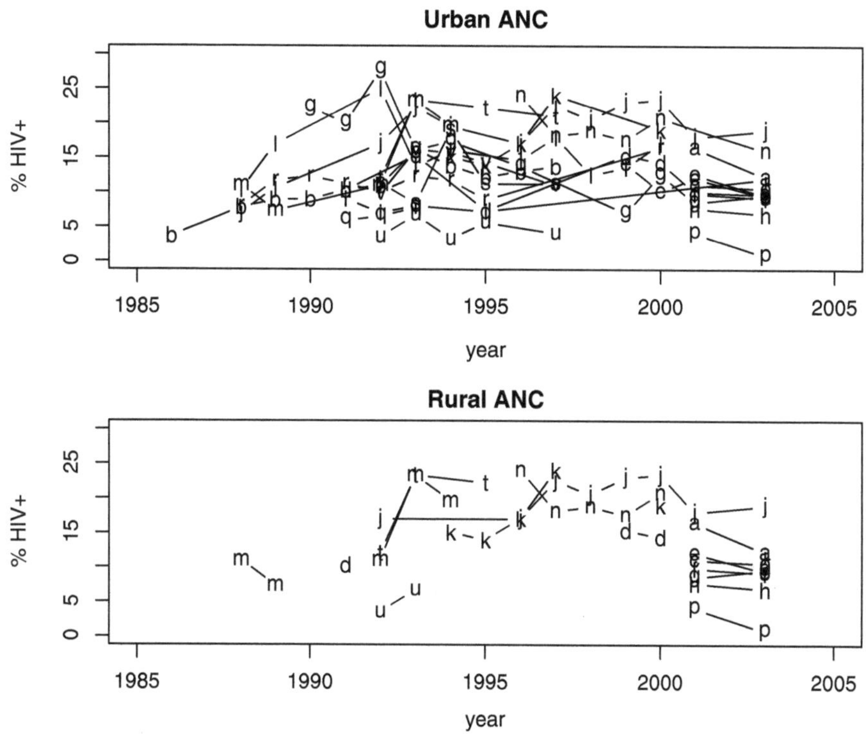

Fig. 6.4 HIV prevalence among female attendees of Tanzanian antenatal clinics in urban and rural locations from 1986 to 2003 (Data from the same clinic are indicated by the same lower case letter with lines connecting consecutive years of data collection)

only point out key sex-specific differences in the text. In the single-state model without HIV, the population projections from time t to time $t + 5$ can be written as $\boldsymbol{n}_{t+5} = \boldsymbol{A}_t \boldsymbol{n}_t$, where $\boldsymbol{n}_t$ is a column vector containing the number of women in each 5-year age group at time t, and $\boldsymbol{A}_t$ is the so-called Leslie matrix. The top row of the Leslie matrix contains the age-specific fertility rates and the sex ratio at birth, and the elements just below the diagonal contain the age-specific probabilities of surviving to the next 5-year age group. Generalizing the classic cohort-component model to include additional states for the infected population involves expanding the population vector, $\boldsymbol{n}_t$, to include the age-specific counts in each HIV duration group. The Leslie matrix also swells to allow transitions across age groups and HIV duration states over time (for more details, see Thomas and Clark 2011).

Consider a population classified into 17 five-year age groups – 0–5, 5–9, ..., 75–79, and a final, open age group 80+ years – and denote the age groups by $a = 1, \ldots, 17$. Furthermore, the population is also divided into five groups, indexed by $d = 0, \ldots, 4$, that include those who are HIV– ($d = 0$), as well as those who have been infected for 0–4 ($d = 1$), 5–9 ($d = 2$), 10–14 ($d = 3$), and 15+ years ($d = 4$). Let the number of women in age group a who are HIV– at time t be

$n_{a,d=0,t}$. Transitions from an age-specific HIV– group to the HIV+ group of the shortest duration, $d = 1$ (i.e., infected for less than 5 years), are defined by

$$n_{a,d=1,t} = n_{a-5,d=0,t-5} \times s_{a,d=0,t} \times i_{a,t} \times v_{a,d=1,t}$$

where $s_{a,d=0,t}$ is the proportion of uninfected women who survive to age group a between $t-5$ and t, $i_{a,t}$ is the proportion of women who would be infected if they had survived to age group a between $t-5$ and t, and $v_{a,d=1,t}$ is an adjustment factor used to lower the survival probability among those who have been HIV+ for less than 5 years. To clarify, each HIV duration group experiences the same baseline survival probability, $s_{a,d=0,t}$, and the survival prospects are lowered by a duration specific adjustment factor, $v_{a,d,t}$ for $d > 0$. The adjustment factor applies to longer duration groups as follows

$$n_{a,d,t} = n_{a-5,d-1,t-5} \times s_{a,d=0,t} \times v_{a,d,t}$$

for $d > 1$. Vital rates for both the uninfected and infected sub-populations are assumed to be known, and the specific values are taken from Heuveline (2003).

We develop this model further by using a new specification for how the age-specific incidence ratios, $i_{a,t}$, change over time. Heuveline (2003) uses the following decomposition

$$i_{a,t} = 1 - \exp\{-\Gamma_{t-t_0} \times H \times j_a\} \tag{6.1}$$

where Γ_{t-t_0} is a parametric curve used to model the temporal change in HIV incidence since the epidemic began (t_0).[10] The parameter H is a population-specific scale parameter which captures the size of the epidemic. The parameter j_a is the age- and sex-specific scaling factor of incidence relative to women aged 20–25, for whom the parameter is constrained to be one in order for the model to be identifiable (i.e., $j_{20-25} = 1$).

We follow the lead of Hogan et al. (2010) and model Γ_{t-t_0}, the trend in HIV incidence, using penalized B-splines (Eilers and Marx 1996). Starting from the assumption that the Tanzanian epidemic becomes generalized to the total adult population in 1980, we use equidistant knots to divide the subsequent 25 years into 5-year intervals. The B-spline is created by linking polynomial pieces together at the knots. In a particular interval the polynomial piece, $f(t)$ (for $t \in$ the interval), is defined as a linear combination of four cubic polynomials, $f(t) = \sum_{i=1}^{4} \alpha_i \mathrm{B}_i(t)$, where α_i are estimated coefficients (as described in the following section) and B_i are the B-spline (cubic) basis functions (see de Boor 1978, for a derivation). The coefficients control the smoothness of the B-spline, with smaller differences corresponding to a

[10]The Γ_{t-t_0} specification is adopted from an earlier model (Chin and Lwanga 1991), and was chosen because it provided a close fit to the data available at that time.

smoother function and thus smaller year-to-year fluctuations in incidence. A penalty term, described in the next section, also enforces smoothness on the B-spline.

6.3.3 Estimation and Forecasting

We estimate our model using a Bayesian approach, which treats unknown parameters as random variables. Beliefs about the parameters held prior to observing the data are expressed as a probability distribution. These beliefs are updated by constructing a likelihood function for the data and model parameters, and multiplying it with the prior distribution. The product, or posterior distribution, is used to carry out inference. The estimated parameters in our model are associated with HIV incidence, as shown in Eq. (6.1). We estimate sex-specific age patterns of infection, which includes two sets of j_a parameters for women and men in 5-year age groups from 15–19 up to 45–49 years old. Recall that j_{20-25} for women is fixed at 1 for identification purposes, resulting in six parameters for women and seven for men. We also estimate two scale parameters, H, which allow the size of the epidemic to vary across rural and urban settings. This specification is useful since HIV prevalence tends to be higher in urban areas, and because the ANC data are separated into rural and urban sites.

Note that the age patterns of infection are assumed to be fixed from one projection period to the next. The variation in HIV incidence over time is captured by Γ_{t-t_0} in Eq. (6.1), which we derive from a penalized B-spline. In exploratory analyses, we experimented with various numbers of B-spline basis functions (e.g., 3, 5, 7, and 9). This work suggests that five B-spline basis functions work well with a second-degree difference penalty for the coefficients. Within a Bayesian framework, this penalty can be implemented with a second-order random walk for the coefficients

$$\alpha_i = \alpha_{i-1} + (\alpha_{i-1} - \alpha_{i-2}) + \varepsilon_i \quad \text{for} \quad i > 2 \tag{6.2}$$

where ε_i follows a normal distribution with mean zero and variance τ^2, and an uninformative prior distribution is assumed for α_1 and α_2 (Lang and Brezger 2004). In our analysis, we use a diffuse joint prior distribution, placing most of the influence with the observed data.[11]

The likelihood function used to update our prior beliefs is developed by Alkema et al. (2007). HIV prevalence data are transformed to the probit scale and a

[11]Uniform prior distributions with a domain from 0 to 2 are used for the sex- and age-specific incidence parameters, j_a (in Heuveline (2003), the upper bounds of the confidence intervals for j_a are all below 1.5). Similarly, the scale parameters, H, have a uniform(0,1) prior, and the B-spline coefficients α_1 and α_2 follow uniform(0,5) distributions. Finally, the error term for the B-spline coefficients follows a normal distribution, $\varepsilon \sim$ N(0, τ^2), and τ^2 is assumed to follow an inverse gamma distribution with shape and scale parameters set equal to 1. We assume these priors are independent and take their product to form the joint prior distribution.

hierarchical normal linear model is used with a random effect that accounts for the correlation across multiple observations from the same ANC over time. We use the same probit transformation for the sex- and age-specific data from the 2003–2004 Tanzanian DHS/AIS, specify a likelihood function using a normal distribution, and assume it is independent of the likelihoods for the urban and rural ANC data. The predicted values are simply the projected values of HIV prevalence generated from the cohort-component model (based on a particular set of parameter values). An analytic solution to the posterior distribution is intractable, and thus inference is carried out by sampling from the posterior distribution using the IMIS algorithm with optimization (Raftery and Bao 2010). This technique has proven useful in previous analyses with a similar version of the multi-state cohort-component model (Clark et al. 2012).

In the second part of our analysis, we assess how well the model predicts future values of HIV prevalence. Data observed up to 2003–2004 are used to estimate the model parameters, and these estimates are then used to project HIV prevalence up through 2011–2012. The predictive performance of the model is evaluated by comparing the projections with the sex- and age-specific HIV prevalence observed in the 2007–2008 and 2011–2012 Tanzanian DHS/AIS. To move beyond 2004 we add two additional basis functions, B_i, that cover the forecast interval from 2003–2004 to 2011–2012, and we extend the coefficients, α_i, using Eq. (6.2) (for a similar approach in a regression context, see Currie et al. 2004). These forecasts are implemented in two different ways, the first of which restricts the errors in Eq. (6.2) to zero. Next, we produce forecasts again using Eq. (6.2), but with error terms drawn from a normal distribution with a mean of zero and the variance equal to our estimate of the hyperparameter τ^2. In summary, seven basis functions are used to span the period from 1980 to 2011–2012, five coefficients are estimated from the data up to 2003–2004, and the forecasts are based on the final two coefficients derived from Eq. (6.2). Again, the last step is implemented first with $\varepsilon_6 = \varepsilon_7 = 0$, and then with $\varepsilon_i \sim N(0, \tau^2)$, for $i = 6$ and 7.

To help assess the predictive performance of the B-spline model, we compare the results to forecasts from a random walk model of HIV incidence, such that $\log(i_{a,t}) - \log(i_{a,t-1}) \sim N(0, \sigma^2)$. The variance in the (log) incidence trend estimated from the B-spline is used for the value of σ^2. Since we are projecting 8 years beyond 2004, we only use the 8 years prior to 2004 to calculate the variance of the random walk. This specification is conservative in the sense that it assumes no systematic drift away from the last estimated value for incidence. Both of these approaches launch off of the incidence trend estimated from the B-spline model. While it is tempting to use a random walk model for both estimation and prediction, such an approach would more than double the number of parameters that need to be estimated. Alternatively, the B-splines offer a much more parsimonious approach (8 parameters versus 25) that is practically feasible with respect to parameter estimation from the available data.

6.4 Results

Our results are summarized by drawing a sample (n = 3,000) from the posterior distribution of the model parameters, and projecting the Tanzanian population for each draw. Figure 6.5 contains our estimate of the posterior distribution of the trend in annual HIV incidence based on the two B-spline specifications[12] and the random walk model. The trend for each draw in our posterior samples is shown in grey, and the posterior medians and 95 % credible intervals are indicated with the black solid and dashed lines, respectively. After 1980, when we assume the epidemic becomes generalized to the total adult population, HIV incidence increases dramatically before peaking around 1992. Furthermore, our results support a decline in HIV incidence that continues for about a decade and overlaps with the national response headed by the Tanzanian government.

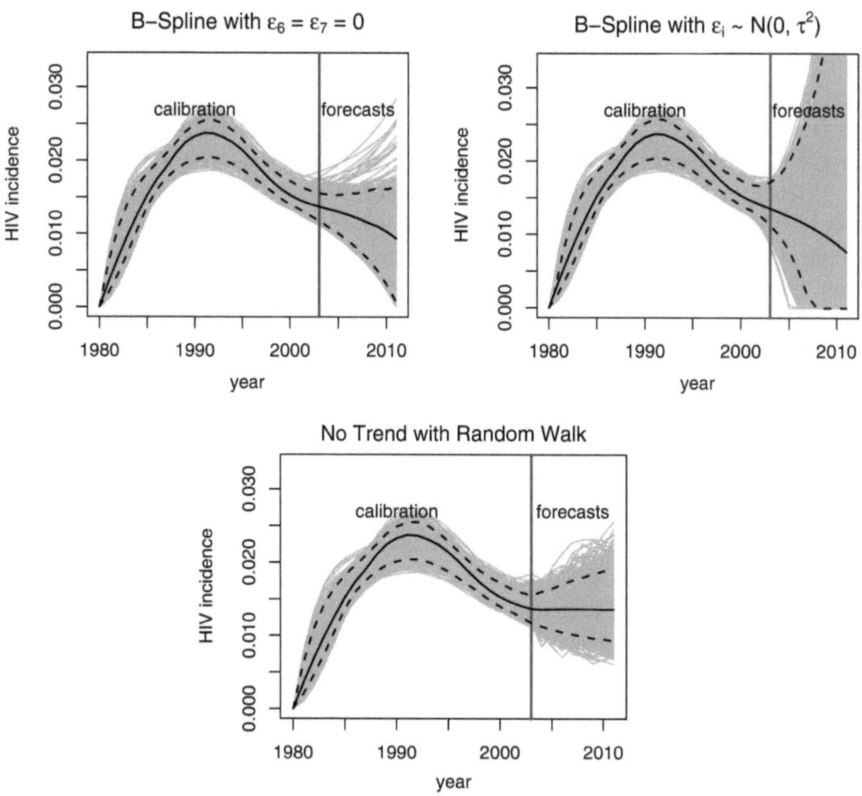

Fig. 6.5 Posterior sample, median, and 95 % credible interval for trends in HIV incidence from the B-spline model, with forecasts based on two extensions of the B-splines and another for the random walk model

[12]The B-spline model produces the annual estimates of HIV incidence, shown in Fig. 6.5, which are then averaged across 5-year periods and used as inputs to the population projection model.

Recall that we use data observed up to 2003 to estimate our model parameters, and then three different techniques are implemented to carry the model forecasts up through 2011 (i.e., the interval to the right of the vertical line demarcating 2003 in Fig. 6.5). Forecasts from the first extension of the B-spline approach, which assumes no additional random variation, tend to continue the downward march beginning in the early 1990s. The uncertainty surrounding the median forecasts increases slightly in the forecast interval, relative to the calibration period, but there are only handful of posterior predictions that suggest a "future" increase in HIV incidence. Conversely, when the additional variation is added, in the form of draws from $\varepsilon_i \sim N(0, \tau^2)$, there is much more support for a potential post-peak increase in incidence. It should be noted, however, that the median forecast still suggest a decline up through 2011. A major drawback from both of these extensions of the B-splines is that the forecasts can descend below zero, implying a negative incidence rate, which is particularly the cases as one moves further into the "future." While this weakness does not seem to plague our third forecasting technique, at least during the interval analyzed here, the random walk model that is assumes no trend is fairly conservative in that it does not make use of the information before 2003 that suggests a subsequent decline in incidence.

Our estimates of the model parameters associated with HIV incidence are used to produce estimates of HIV prevalence in 2003, as well as forecasts for HIV prevalence in 2007 and 2011. Moving forward, we assign a value of zero to any forecasts of HIV incidence that exceeds the lower bound of zero. The assessment of the forecasts for each of the three approaches is presented in Table 6.3. For each of the three approaches, the positive mean bias indicates that the forecasts tend to be higher than the values of HIV prevalence observed in the DHS/AIS. The random walk model with no trend has the largest bias with the forecast being, on average, just under a percentage point higher than the observed value. The bias associated with the forecasts from the B-spline approaches are less than a half of a percentage point too high (on average). A somewhat similar result holds for the mean absolute error, such that the B-spline model without the extra variation has the lowest average error rate at 1.2 % points. Using this metric, however, the the random walk specification and the B-spline with added variation have similar mean absolute errors of about 1.6 across the twenty-eight observed values of HIV prevalence. Finally, we find that the 95 % prediction intervals fail to reach the appropriate amount of coverage, which roughly consists of covering 27 of the 28

Table 6.3 Evaluation of probabilistic forecasts for sex- and age-specific HIV prevalence in Tanzania for 2007 and 2011. Forecasts are made for 28 values of HIV prevalence (seven per sex, for two DHS/AIS). Mean bias and absolute error are measured in percentage points

Model	Average bias	Mean absolute error	Coverage of 95 % prediction interval (%)
B-Spline with $\varepsilon_i = 0$	0.367	1.235	53.57
B-Spline with $\varepsilon_i \sim N(0, \tau^2)$	0.438	1.560	82.14
Random walk no trend	0.926	1.552	25.00

age-specific targets. The B-spline model with added variation is the closest, with roughly 82 % of the observations lying within the 95 % prediction interval. When the added variation is omitted, the B-spline forecasts are only able to achieve 54 % coverage, 41 % points less than expected. Finally, the prediction intervals from the random walk model fail to include 21 of the 28 observed values, or 25 % coverage.

For illustrative purposes, we assess the B-spline model without the added variation further by comparing the observed data to the model outputs for the three different DHS/AIS surveys. These comparisons are shown in Fig. 6.6, with the top (bottom) row of plots containing the results for women (men). Starting with 2003, the year from which data are used to estimate the parameters, the distribution of fitted values from our posterior sample are summarized using box-

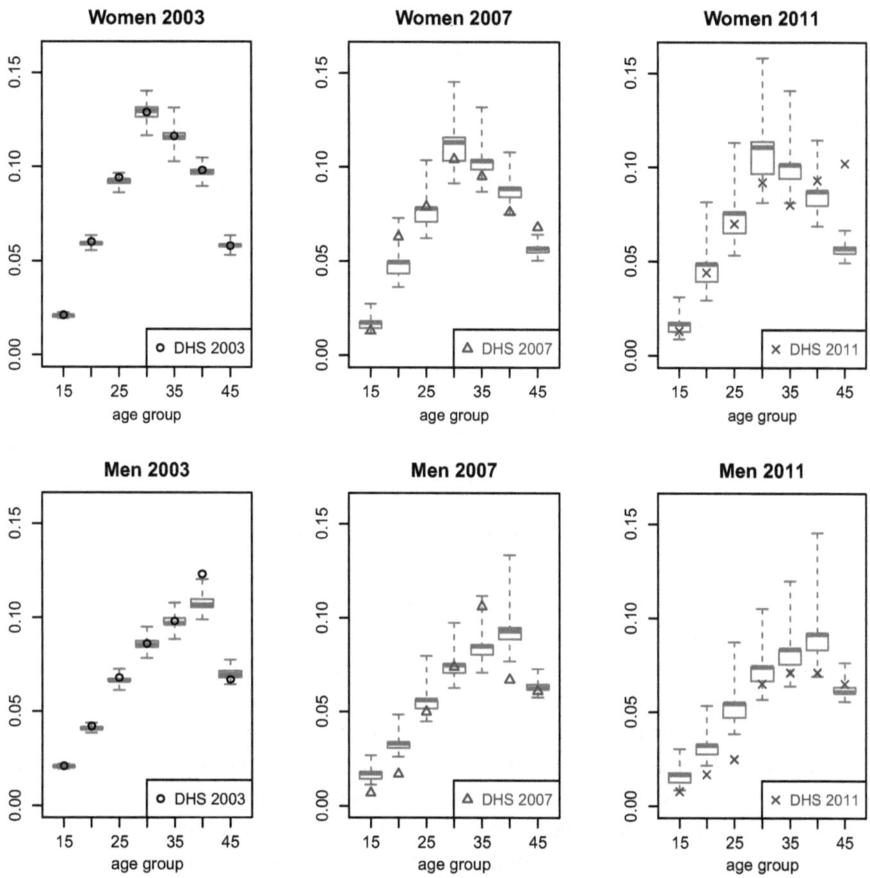

Fig. 6.6 Age- and sex-specific fitted values and forecasts of adult HIV prevalence from B-spline model with $\varepsilon = 0$. Observed HIV prevalence levels from DHS/AIS for 2003 are used to estimate model parameters, and the corresponding observations from DHS/AIS in 2007 and 2011 serve as the forecasting targets

plots with whiskers extending to the extreme values in the sample. The fitted values are generally close to the levels of HIV prevalence observed in the 2003 DHS/AIS, with the means of the age-specific residuals ranging from 0.006 percentage points to 0.184 percentage points for women, and from 0.002 percentage to 1.541 percentage points for men (absolute values are given, and means are taken across the posterior sample). On average, the residuals also tend to be slightly positive, which is true for 11 of the 14 age groups. The most extreme case is men in the 40–44 year age group.

Our forecasts of age-specific HIV prevalence for women and men in 2007 appear in the second column of plots in Fig. 6.6. The observed values from the 2007 DHS/AIS are depicted with triangles. Although the forecasts of HIV prevalence tend to predict a decline in HIV prevalence for each age group between 2003 and 2007, the observed changes in prevalence are more erratic. Observed levels of HIV prevalence either stay roughly the same or increase by 2007 for women ages 20–24 and 45–49, and for men ages 35–39. Among younger men, as well as women in the 40–44 age group, the observed declines exceed the predicted declines from our posterior sample. This finding also holds for four of the age groups (particularly younger men) in the set of forecasts for 2011 (third column in Fig. 6.6). We also see increases in observed prevalence among the two oldest age groups of women, but our posterior predictions are only able to capture the increase experienced in the 40–44 age group.

6.5 Conclusions

In this analysis, we proposed a new model for the trend in HIV incidence and used it to investigate the epidemic in Tanzania. Our results suggest that HIV incidence did indeed decline after an initial exponential rise continuing through the early 1990s. Although the model produced a close fit to the levels of age- and sex-specific HIV prevalence observed in the 2003 DHS/AIS, it was unable to accurately forecast these same targets observed in 2007 and 2011. Across the three techniques examined, the best approach was only able to cover 82 % of the observations with the 95 % prediction interval. Although the forecast errors tend to be within 2 percentage points of the observed values, there is clear room for improvement. One potentially crucial limitation is the assumption of a fixed age-pattern of HIV incidence over time. The population response to the epidemic may not affect all age groups equally. For example, younger individuals may receive greater exposure to educational campaigns, relative to older age groups, through school attendance. Similarly, younger age groups may be more impressionable or accepting of new ideas (e.g., male circumcision) compared to older, more experienced individuals. These potential cohort effects suggest a dynamic age pattern in the risk of infection as the epidemic matures and, more importantly, the population response gains momentum. Parametric models for age patterns could be adopted to help keep the parameter space tractable in terms of statistical estimation. It is also worth noting that these changes may depend on gender to the extent that women experience social barriers to economic opportunities and self-sufficiency. Thus, discriminatory contours in the social structure may limit behavioral responses from women relative to men.

Future work would benefit greatly from more information concerning the trend in the epidemic. More specifically, if ANC data were available for 5-year age groups over time, then modelers would be better equipped to explore potential age dynamics associated with HIV incidence. That said, ANC data have been shown to be biased (Gouws et al. 2008) and the extent of the bias may be changing over time as the average duration of infection increases and associated declines in fertility select older women out of the population of ANC attendees. To the extent that these issues hamper our ability to forecast the epidemic, future efforts to improve the quality and coverage of surveillance data will facilitate knowledge and improve our capacity to mobilize an effective response. Continuing with the idea of leveraging additional information, it may prove useful to develop the cohort-component model to take advantage of data on knowledge, attitudes, and behavior change associated with the risk of infection. Embedding a simple model with covariates that are predictive of risky behavior may be an economical way forward. Finally, as more infected individuals gain access to antiretroviral therapies, it will become imperative to expand the multi-state model to include treatment groups. Feedback effects on the risk of infection may also improve the model's predictive capabilities. In closing, we feel that this model has potential to help inform policy makers and program planners to address the future dynamics of HIV epidemics epidemic.

Acknowledgements This research was supported by grants from the NICHD (R24HD041025) and the NIA (P30AG17266). The authors thank Robert Schoen and an anonymous reviewer for helpful comments on this manuscript.

References

Alkema, L., Raftery, A. E., & Clark, S. J. (2007). Probabilistic projections of HIV prevalence using Bayesian melding. *Annals of Applied Statistics, 1*(1), 229–248. doi:10.1214/07-AOAS111.

Bongaarts, J., Buettner, T., Heilig, G., & Pelletier, F. (2008). Has the HIV epidemic peaked? *Population and Development Review, 34*(2), 199–224. doi:10.1111/j.1728-4457.2008.00217.x.

Brookmeyer, R., & Konikoff, J. (2011). Statistical considerations in determining HIV incidence from changes in HIV prevalence. *Statistical Communications in Infectious Diseases, 3*(9). doi:10.2202/1948-4690.1044.

Cheluget, B., Baltazar, G., Orege, P., Ibrahim, M., Marum, L., & Stover, J. (2006). Evidence for population level declines in adult HIV prevalence in Kenya. *Sexually Transmitted Infections, 82*, i21–i26. doi:10.1136/sti.2005.015990.

Chin, J., & Lwanga, S. (1991). Estimation and projection of adult AIDS cases. *A simple epidemiological model, 69*(4), 399–406.

Clark, S. J., Thomas, J., & Bao, L. (2012). Estimates of age-specific reductions in HIV prevalence in Uganda: Bayesian melding estimation and probabilistic population forecast with an HIV-enabled cohort component projection model. *Demographic Research, 27*(26), 743–774. doi:10.4054/DemRes.2012.27.26.

Cohen, M. S., Chen, Y. Q., McCauley, M., Gamble, T., Hosseinipour, M. C., Kumarasamy, N., Hakim, J. G., Kumwenda, J., Grinsztejn, B., Pilotto, J. H., Godbole, S. V., Mehendale, S., Chariyalertsak, S., Santos, B. R., Mayer, K. H., Hoffman, I. F., Eshleman, S. H., Piwowar-Manning, E., Wang, L., Makhema, J., Mills, L. A., de Bruyn, G., Sanne, I., Eron, J.,

Gallant, J., Havlir, D., Swindells, S., Ribaudo, H., Elharrar, V., Burns, D., Taha, T. E., Nielsen-Saines, K., Celentano, D., Essex, M., & Fleming, T. R. (2011). Prevention of HIV-1 infection with early antiretroviral therapy. *New England Journal of Medicine, 365*(6), 493–505. doi:10.1056/NEJMoa1105243. pMID: 21767103.

Currie, I. D., Durban, M., & Eilers, P. H. C. (2004). Smoothing and forecasting mortality rates. *Statistical Modelling, 4*(4), 279–298. doi:10.1191/1471082X04st080oa.

de Boor, C. (1978). *A practical guide to splines*. Berlin: Springer.

Diaz, T., Garcia-Calleja, J., Ghys, P., & Sabin, K. (2009). Advances and future directions in HIV surveillance in low-and middle-income countries. *Current Opinion in HIV and AIDS, 4*, 253–259. doi:10.1097/COH.0b013e32832c1898.

Eilers, P. H., & Marx, B. D. (1996). Flexible smoothing with b-splines and penalties. *Statistical Science, 11*, 89–121. doi:10.1214/ss/1038425655.

Eyawo, O., de Walque, D., Ford, N., Gakii, G., Lester, R. T., & Mills, E. J. (2010). HIV status in discordant couples in sub-Saharan Africa: A systematic review and meta-analyis. *Lancet, 10*, 770–777. doi:10.1016/S1473-3099(10)70189-4.

Fylkesnes, K., Musonda, R., Sichone, M., Ndhlovu, Z., Tembo, F., & Monze, M. (2001). Declining HIV prevalence and risk behaviours in Zambia: Evidence from surveillance and population-based surveys. *AIDS, 15*(7), 907–916. doi:10.1097/00002030-200105040-00011.

Garcia-Calleja, J., Zaniewski, E., Ghys, P., Stanecki, K., & Walker, N. (2004). A global analysis of trends in the quality of HIV sero-surveillance. *Sex Transm Infect, 80*, i25–i30. doi:10.1136/sti.2004.010298.

Ghys, P. D., Kufa, E., & George, M. V. (2006). Measuring trends in prevalence and incidence of HIV infection in countries with generalised epidemics. *Sexually Transmitted Infections, 82*(suppl 1), i52–i56. doi:10.1136/sti.2005.016428.

Gouws, E., Mishra, V., & Fowler, T. (2008). Comparison of adult HIV prevalence from national population-based surveys and antenatal clinic surveillance in countries with generalised epidemics: Implications for calibrating surveillance data. *Sexually Transmitted Infections, 84*(Supplement 1), i17–i23. doi:10.1136/sti.2008.030452.

Gregson, S., Garnett, G. P., Nyamukapa, C. A., Hallett, T. B., Lewis, J. J. C., Mason, P. R., Chandiwana, S. K., & Anderson, R. M. (2006). HIV decline associated with behavior change in Eastern Zimbabwe. *Science, 311*(5761), 664–666. doi:10.1126/science.1121054.

Halperin, D. T., Mugurungi, O., Hallett, T. B., Muchini, B., Campbell, B., Magure, T., Benedikt, C., & Gregson, S. (2011). A surprising prevention success: Why did the HIV epidemic decline in Zimbabwe? *PLoS Medicine, 8*(2), e1000,414. doi:10.1371/journal.pmed.1000414.

Heuveline, P. (2003). HIV and population dynamics: A general model and maximum-likelihood standards for East Africa. *Demography, 40*(2), 217–245. doi:10.1353/dem.2003.0013.

Hogan, D., & Salomon, J. (2012). Spline-based modelling of trends in the force of HIV infection, with application to the UNAIDS estimation and projection package. *Sexually Transmitted Infections*. doi:10.1136/sextrans-2012-050652.

Hogan, D., Zaslavsky, A., Hammitt, J., & Salomon, J. (2010). Flexible epidemiological model for estimates and short-term projections in generalised HIV/AIDS epidemics. *Sexually Transmitted Infections*. doi:10.1136/sti.2010.045104.

Hollingsworth, T., Anderson, R., & Fraser, C. (2008). HIV-1 transmission, by stage of infection. *Journal of Infectious Diseases, 198*, 687–693. doi:10.1086/590501.

Keiding, N. (1991). Age-specific incidence and prevalence: A statistical perspective. *Journal of the Royal Statistical Society Series A (Statistics in Society), 154*(3), 371–412. doi:10.2307/2983150.

Keyfitz, N., & Caswell, H. (2005). *Applied mathematical demograph* (3rd ed.). New York: Springer.

Kimanga, D. O., Ogola, S., Umuro, M., Ng'ang'a, A., Kimondo, L., Murithi, P., Muttunga, J., Waruiru, W., Mohammed, I., Sharrif, S., De Cock, K. M., & Kim, A. A. (2014). Prevalence and incidence of HIV infection, trends, and risk factors among persons aged 15–64 years in Kenya: Results from a nationally representative study. *Journal of Acquired Immune Deficiency Syndromes, 66*, S13–S26. doi:10.1097/QAI.0000000000000124.

Kranzer, K., Govindasamy, D., Ford, N., Johnston, V., & Lawn, S. (2012). Quantifying and addressing losses along the continuum of care for people living with HIV infection in sub-Saharan Africa: A systematic review. *Journal of the International AIDS Society, 15*(2), 17,838. doi:10.7448/IAS.15.2.17383.

Lang, S., & Brezger, A. (2004). Bayesian p-splines. *Journal of Computational and Graphical Statistics, 13*, 183–212. doi:10.1198/1061860043010.

Mahy, M., Garcia-Calleja, J. M., & Marsh, K. A. (2012). Trends in HIV prevalence among young people in generalised epidemics: Implications for monitoring the HIV epidemic. *Sexually Transmitted Infections, 88*(Suppl 2), i65–i75. doi:10.1136/sextrans-2012-050789.

Marsh, K., Mahy, M., Salomon, J. A., & Hogan, D. R. (2014). Assessing and adjusting for differences between HIV prevalence estimates derived from national population-based surveys and antenatal care surveillance, with applications for Spectrum 2013. *AIDS, 28*, S497–S505. doi:10.1097/QAD.0000000000000453.

Mbulaiteye, S., Mahe, C., Whitworth, J., Ruberantwari, A., Nakiyingi, J., Ojwiya, A., & Kamali, A. (2002). Declining HIV-1 incidence and associated prevalence over 10 years in a rural population in South-West Uganda: A cohort study. *Lancet, 360*. doi:10.1016/S0140-6736(02)09331-5.

Ministry of Health. (2000). *National package of essential health interventions in Tanzania* (Technical report). The United Republic of Tanzania.

Nagelkerke, N. J. D., Arora, P., Jha, P., Williams, B., McKinnon, L., & de Vlas, S. J. (2014). The rise and fall of HIV in high-prevalence countries: A challenge for mathematical modeling. *PLoS Computational Biology, 10*(3), e1003,459. doi:10.1371/journal.pcbi.1003459.

National AIDS Control Programme. (2005). *HIV/AIDS/STI surveillance report*. United Republic of Tanzania, Ministry of Health and Social Welfare: Dar es Salaam.

National AIDS Control Programme. (2010). *National guidelines for quality improvement of HIV and AIDS services*. United Republic of Tanzania. Ministry of Health and Social Welfare: Dar es Salaam.

Parkhurst, J. O. (2008). "What worked?": The evidence challenges in determining the causes of HIV prevalence decline. *AIDS Education and Prevention, 20*(3), 275–283. doi:10.1521/aeap.2008.20.3.275.

Preston, S. H., Heuveline, P., & Guillot, M. (2001). *Demography: Measuring and modeling population processes*. Malden, Massachusetts: Blackwell.

Raftery, A. E., & Bao, L. (2010). Estimating and projecting trends in HIV/AIDS generalized epidemics using incremental mixture importance sampling. *Biometrics, 66*(4), 1162–1173. doi:10.1111/j.1541-0420.2010.01399.x.

Scott-Sheldon, L. A., Heudo-Medina, T. B., Warren, M. R., Johnson, B. T., & Carey, M. P. (2011). Efficacy of behavioral interventions to increase condom use and reduce sexually transmitted infections: A meta-analysis, 19991 to 2010. *Journal of Acquired Immune Deficiency Syndrome*. doi:10.1097/QAI.0b013e31823554d7.

Shelton, J., Halperin, D., & Wilson, D. (2006). Has global HIV incidence peaked? *Lancet, 367*, 1120–1122. doi:10.1016/S0140-6736(06)68436-5.

Stoneburner, R. L., & Low-Beer, D. (2004). Population-level HIV declines and behavioral risk avoidance in Uganda. *Science, 304*(5671), 714–718. doi:10.1126/science.1093166.

Tanser, F., Bärnighausen, T., Grapsa, E., Zaidi, J., & Newell, M. L. (2013). High coverage of ART associated with decline in risk of HIV acquisition in rural KwaZulu-Natal, South Africa. *Science, 339*(6122), 966–971. doi:10.1126/science.1228160.

Tanzania Commission for AIDS (TACAIDS), National nStatistics (NBS) and ORC Macro. (2005). *Tanzania HIV/AIDS indicator survey 2003–2004*. Calverton: TACAIDS and ORC Macro.

Tanzania Commission for AIDS (TACAIDS). (2007). The second national multi-sectoral strategic framework on HIV and AIDS (2008–2012). TACAIDS, United Republic of Tanzania, Prime Minister's Office: nnSalaam.

Tanzania Commission for AIDS (TACAIDS), Zanzibar AIDS Commission (ZAC), National Bureau of Statistics (NBS), Office of the Chief Government Statistician (OCGS), & Macro International Inc. (2008). Tanzania HIV/AIDS and Malaria indicator survey 2007–2008. Dar es Salaam: TACAIDS, ZAC, NBS, OCGS, and Macro International Inc.

Tanzania Commission for AIDS (TACAIDS), Zanzibar AIDS Commission (ZAC), National nStatistics (NBS), nthe Chief Government Statistician (OCGS), & ICF International. (2013). Tanzania HIV/AIDS and Malaria indicator survey 2011–2012. Dar es Salaam: TACAIDS, ZAC, NBS, OCGS, and ICF.

The International Group on Analysis of Trends in HIV Prevalence and Behaviours in Young People in Countries most Affected by HIV. (2010). Trends in HIV prevalence and sexual behaviour among young people aged 15–24 years in countries most affected by HIV. *Sexually Transmitted Infections, 86*(Suppl 2), ii72–ii83. doi:10.1136/sti.2010.044933.

Thomas, J. R., & Clark, S. J. (2011). More on the cohort-component model of population projection in the context of HIV/AIDS: A Leslie matrix representation and new estimates. *Demographic Research, 25*(2), 39–102. doi:10.4054/DemRes.2011.25.2.

Todd, J., Glynn, J. R., Marston, M., Lutalo, T., Biraro, S., Mwita, W., Suriyanon, V., Rangsin, R., Nelson, K. E., Sonnenberg, P., Fitzgerald, D., Karita, E., & Zaba, B. (2007). Time from HIV seroconversion to death: A collaborative analysis of eight studies in six low and middle-income countries before highly active antiretroviral therapy. *AIDS, 21*(Suppl 6), S55–S63. doi:10.1097/01.aids.0000299411.75269.e8.

UNAIDS. (1998a). *Report on the global HIV/AIDS epidemic*. Geneva: Joint United Nations Programme on HIV/AIDS (UNAIDS).

UNAIDS. (1998b). *Trends in HIV incidence and prevalence: Natural course of the epidemic or results of behavioural change?* Geneva: Joint United Nations Programme on HIV/AIDS (UNAIDS).

UNAIDS. (2006). *Improving parameter estimation, projection methods, uncertainty estimation, and epidemic classification*. London: Imperial College.

UNAIDS. (2013). *Global report: UNAIDS report on the global AIDS epidemic 2013*. Geneva: Joint United Nations Programme on HIV/AIDS (UNAIDS).

UNAIDS and the World Health Organization. (2015). *Technical update on HIV incidence assays for surveillance and monitoring purposes*. Geneva: Joint United Nations Programme on HIV/AIDS (UNAIDS).

United Nations, Department of Economic and Social Affairs, Statistics Division. (2014). *World statistics pocketbook 2014 edition*. New York: United Nations.

Walker, N., Garcia-Calleja, J., Heaton, L., Asamoah-Odei, E., Poumerol, G., Lazzari, S., Ghys, P., Schwartlander, B., & Stanecki, K. (2001). Epidemiological analysis of the quality of HIV sero-surveillance in the world: How well do we track the epidemic? *AIDS, 15*, 1545–1554. doi:10.1097/00002030-200108170-00012.

Walker, N., Grassly, N., Garnett, G., Stanecki, K., & Ghys, P. (2004). Estimating the global burden of HIV/AIDS: What do we really know about the HIV pandemic? *Lancet, 363*, 2180–2185. doi:10.1016/S0140-6736(04)16511-2.

Wawer, M. J., Gray, R. H., Sewankambo, N. K., Serwadda, D., Li, X., Laeyendecker, O., Kiwanuka, N., Kigozi, G., Kiddugavu, M., Lutalo, T., Nalugoda, F., Wabwire-Mangen, F., Meehan, M. P., & Quinn, T. C. (2005). Rates of HIV-1 transmission per coital act, by stage of HIV-1 infection, in Rakai, Uganda. *Journal of Infectious Diseases, 191*(9), 1403–1409. doi:10.1086/429411.

Part III
Analyzing Heterogeneity

Chapter 7
Revisiting Mortality Deceleration Patterns in a Gamma-Gompertz-Makeham Framework

Filipe Ribeiro and Trifon I. Missov

7.1 Introduction

Horiuchi and Coale (1990) propose a mortality measure, designated later (Horiuchi and Wilmoth 1997, 1998; Carey and Liedo 1995) as the life-table aging rate (LAR), which captures the age-specific rate of mortality change for a given population. LAR, denoted by $\bar{b}(x)$ (Vaupel and Zhang 2010), is defined as

$$\bar{b}(x) = \frac{1}{\mu(x)} \frac{d\mu(x)}{dx} = \frac{d \ln \mu(x)}{dx}. \tag{7.1}$$

The rate of *individual* aging, defined as the relative derivative of the baseline hazard of death from senescent causes, is a different characteristic which is constant whenever the aging process is captured by a Gompertz curve (for further discussion on the rate of individual aging, see Missov and Vaupel (2015) and Missov and Ribeiro (2016)).

Gampe (2010) finds partial evidence for a leveling-off of human death rates at ages 110–114. If a mortality plateau actually takes place, then it speaks in favor of a relative-risk model for adult mortality (Missov and Vaupel 2015) with a Gompertz-Makeham baseline $\mu(x) = ae^{bx} + c$ (Gompertz 1825; Makeham 1860),

F. Ribeiro (✉)
CIDEHUS.UE – University of Évora, Palácio do Vimioso – Largo do Marquês de Marialva 8, Apartado 94, 7000-809 Évora, Portugal
e-mail: fribeiro@uevora.pt

T.I. Missov
Max Planck Institute for Demographic Research, Konrad-Zuse-Str. 1, 18057 Rostock, Germany

Mathematical Demography, University of Rostock, Ulmenstr. 69, 18057 Rostock, Germany
e-mail: missov@demogr.mpg.de

R. Schoen (ed.), *Dynamic Demographic Analysis*,
The Springer Series on Demographic Methods and Population Analysis 39,
DOI 10.1007/978-3-319-26603-9_7

where a measures the mortality level at the starting adult age, b is the rate of individual aging, and c captures the risk of dying that is not associated with the aging process. Unobserved heterogeneity (frailty) can be captured by a gamma distribution with a unit mean and γ variance at the starting age (Vaupel et al. 1979; Missov and Finkelstein 2011; Missov and Vaupel 2015). This leads to the gamma-Gompertz-Makeham frailty model (Vaupel et al. 1979), which we will shortly address as ΓGM, whose force of mortality is given by

$$\mu(x) = \frac{ae^{bx}}{1 + \frac{\gamma a}{b}\left(e^{bx} - 1\right)} + c\,. \tag{7.2}$$

If we estimate its parameters a, b, c and γ, we can take advantage of the following formula by Vaupel and Zhang (2010) to estimate the ΓGM model-based rate of aging for populations:

$$\bar{b}(x) = b\left(1 - \frac{c}{\mu(x)}\right) - \gamma\left(1 - \frac{c}{\mu(x)}\right)(\mu(x) - c)\,. \tag{7.3}$$

Horiuchi et al. (2012) reconstructed model-based LARs by fitting a three-parameter logistic model, the Kannisto model (Thatcher et al. 1998), for homogeneous populations. In this study we focus on a ΓGM heterogeneous model to reflect the perception that populations consist of individuals that share the same baseline risk of death associated with age-related deterioration of physiological functions, by *senescent* mortality (age-related deterioration of physiological functions), but have different (random) susceptibility to it (Vaupel et al. 1979). We also incorporate a Makeham term c to account for non-zero *background* mortality (Golubev 2004), i.e., mortality that cannot be directly attributed to the aging process. If c is neglected, the estimates of all other ΓGM parameters will be biased (Nemeth and Missov 2014). Moreover, if c is left out of the model, LAR will also be distorted, and it will be impossible to capture its bell-shaped pattern (Horiuchi and Coale 1990). The latter implies that, although age-specific mortality rates seem to follow a linear age pattern on a logarithmic scale, the rate of mortality increase is slowing down at older ages. Mortality deceleration is explained traditionally by either the heterogeneity or the individual-risk hypothesis (Horiuchi and Wilmoth 1998). The former explains observed mortality deceleration as a consequence of the existence of frailer individuals that die out at younger ages leaving a selected subpopulation of robust individuals at the older ages. This assumption implies that at the oldest ages mortality rates register a slower increase even though the hazard for individuals keeps growing exponentially (Vaupel et al. 1979; Vaupel and Yashin 1985). The individual-risk hypothesis assumes a slowdown in the rate of *individual* aging at older ages. In this article we assume that the heterogeneity hypothesis holds.

Ribeiro and Missov (2014) show that ΓGM model-based LAR not only fits well the observed LAR patterns after age 65, but it also captures the shift in x^*, the age of mortality deceleration, with time. The latter might suggest a relationship between the rate of increase in e_0, the life expectancy at birth, and the estimated corresponding x^*: statistically significant changes in the slope of life-expectancy growth over time are accompanied by changes in the age of mortality deceleration. Life expectancy at birth, though, is a mortality measure that is influenced by early-life mortality. The modal age at death M, on the other hand, is a characteristic of the distribution of adult deaths Pollard (1998a,b); Kannisto (2001); Cheung et al. (2005); Cheung and Robine (2007); Canudas-Romo (2008); Missov et al. (2015) and captures mortality shifts more accurately than remaining life expectancies such as e_{65} (Horiuchi et al. 2013). In a ΓGM setting, getting an exact expression (see Sect. 7.2.4, Eq. (7.10)) for the age of mortality deceleration x^*, i.e., the age at which LAR reaches its maximum, facilitates the study of the relationship between x^* other longevity measures.

We fit the ΓGM model to mortality data from six selected countries to reflect different types of mortality experience: steady increase (France, Sweden and Japan), increase at a changing pace (USA) and fluctuation (Russia and Ukraine) of age-specific death rates over time. The ΓGM fits the data with high accuracy, and for the chosen countries we find evidence for point *(b)* of the heterogeneity hypothesis[1] by Horiuchi and Wilmoth (1998). Using segmented regression to identify the curvilinear structure of the three longevity measures over time, we find evidence for a plausible relationship between LAR patterns and the changes in the rate of life-expectancy increase over time, also reflected in x^*. Finally, the age of mortality deceleration x^* seems to be strongly influenced by changes in the overall pattern of mortality.

This study not only revisits recent mortality deceleration patterns by calculating LARs for overall mortality from a ΓGM model, but also evaluates mortality dynamics by focusing on longevity development across time as a result of possible improvement or deterioration at different ages.

7.2 Data and Methods

This section presents the methodological foundations of the analysis carried out in Sect. 7.3. Section 7.2.1 identifies the source and format of analyzed data as well as defines the time horizon of the study. Section 7.2.2 revisits formal definitions of remaining life expectancy e_x, the median age at death Md, and the modal age at death M. Section 7.2.3 presents an empirical approximation of LAR, while Sect. 7.2.4 introduces the LAR formula in a ΓGM setting. Section 7.2.5 presents a modified version of the ΓGM LAR by re-parameterizing the Gompertz force of mortality in

[1]The hypothesis states that *(a)* deceleration occurs for the most major causes of death (COD), being less pronounced for CODs with lower death rates and should start at earlier ages for CODs with higher death rates; and *(b)* mortality deceleration, due to selection effects, should shift to older ages as the level of total adult mortality declines.

terms of M. Finally, Sect. 7.2.6 provides an overview of segmented regression, a tool that we use to single out different time intervals of linear increase in M over time.

7.2.1 Data

We use overall death counts $D(x, y)$ and exposures $E(x, y)$ from the Human Mortality Database (http://www.mortality.org/, HMD 2015). We focus on the period from 1970 to the last available year for each country to have, on the one hand, a common study period and, on the other hand, to avoid potential data quality problems in one or more of the selected countries (e.g., data for Ukraine (Pyrozhkov et al. 2011) need to be handled carefully before 1970).

Figure 7.1 presents period female life expectancy at birth e_0 for the six selected countries: France, Japan, Sweden, Russia, Ukraine, and USA. Qualitatively we can identify three different patterns of e_0 over time: steady increase at almost constant pace (Sweden, Japan, France), steady increase at a changing pace (USA), and stagnation with periods of strong fluctuation (Russia, Ukraine). In the 1970s life expectancy at birth varies from 73.4 in Russia to 77.2 in Sweden. France registers the second highest value from the group with 75.8, while Japan, Ukraine, and USA share an almost identical life expectancy at birth (74.7, 74.4 and 74.7, respectively). In the end of the observed period, Japan is leading with 86.5, while Russia and

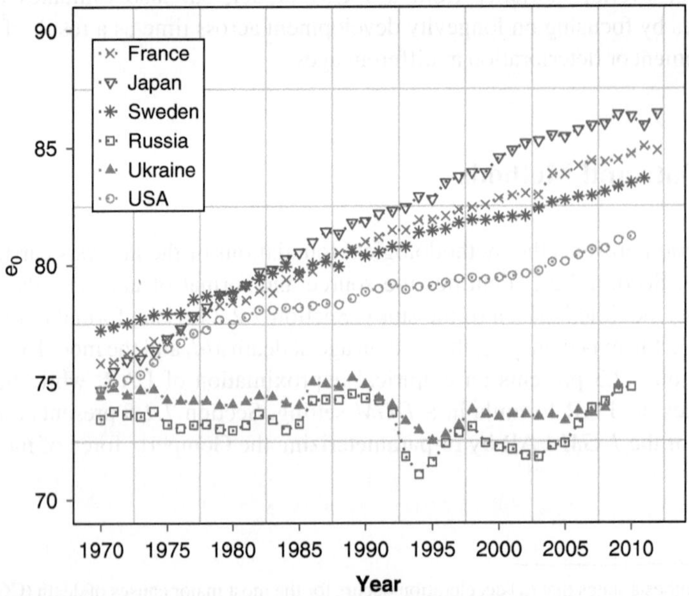

Fig. 7.1 Female life expectancy at birth for the studied countries from 1970 to the last available year (Source: HMD 2015)

Ukraine are still on the bottom (both with 74.8 years of life expectancy at birth in the end of each correspondent series). However, USA's e_0 lags further behind the Swedish and French, and the latter surpasses the former. The selection of countries with different patterns of e_0 increase reflects our aim to understand how LAR patterns may explain (at least partially) the observed changes in the "longevity hierarchy".

For completeness of the study, we also consider cohort mortality data. However, we focus on Sweden and France, cohorts 1800–1900 only as data for these countries are characterized by both high quality and availability over a long time horizon.

7.2.2 *Measuring Longevity*

The age distribution of deaths is usually bimodal, representing the distinctive patterns of early and adult mortality. The distribution of adult deaths is skewed to the left and this results in $e_0 < Md < M$. In most developed countries the gap among these three longevity indicators narrows down, and Md is located approximately at the midpoint between M and e_0 (Horiuchi et al. 2013).

Remaining life expectancy e_x at age x can be calculated from a single-decrement life table by Preston et al. (2001)

$$e_x = \frac{T_x}{l_x}, \tag{7.4}$$

where l_x is the proportion of individuals alive at age x and T_x denotes the person-years lived at age x and above.

The median age at death Md is the age by which half of the population is dead, i.e., when the survival function equals 0.5: $l_{Md} = 0.5$. If $l_{Md} = 0.5$ in $[x, x+1)$, then the median age at death is calculated by

$$Md = x + \frac{0.5 - l_x}{l_{x+1} - l_x}. \tag{7.5}$$

The modal age at death M is the age when most adult deaths occur:

$$M = \left\{ \max_x d_x, x > 5 \right\}, \tag{7.6}$$

where d_x is the life-table density function of the distribution of deaths. Kannisto (2001) proposed a formula that uses life-table input to calculate the modal age at death with decimal precision. If x is the age with the highest number of deaths in the life table, then

$$M = x + \frac{d_x - d_{x-1}}{(d_x - d_{x-1}) + (d_x - d_{x+1})}. \tag{7.7}$$

7.2.3 *The Life-Table Aging Rate (LAR)*

If data are available for single-year age groups, the empirical rate of population aging (LAR) can be obtained by the formula (Horiuchi and Coale 1990)

$$\bar{b}^*(x) = \ln(M(x)) - \ln(M(x-1)) \ , \tag{7.8}$$

where $M(x)$ is the central death rate at age x. The small number of deaths at very old ages, though, leads to large stochastic variation of death rates and, consequently, a two-step procedure needs to be employed (Horiuchi and Coale 1990):

1. applying a 5-year moving average to the central death rates $M(x)$ and then calculating LAR using (7.8);
2. taking subsequently a triangularly weighted 9-year moving average, i.e, being the weight distributed triangularly over nine values:

$$\bar{b}_{emp}(x) = \sum_{n=-4}^{4} \frac{5-|n|}{25} * \bar{b}^*(x+n) \ . \tag{7.9}$$

7.2.4 *LAR in a ΓGM Framework*

Human populations are comprised of different heterogeneous subpopulations in which individuals, despite sharing the same rate of increase in the hazard of death at adult ages, are described by different levels of susceptibility (Vaupel et al. 1979). Within this framework, we assume that the baseline hazard follows a Gompertz-Makeham pattern (Gompertz 1825; Makeham 1860): $ae^{bx} + c$.

The rate of individual aging in a ΓGM setting is captured by the relative derivative of the Gompertz part ae^{bx}, while the rate of aging for the entire population, i.e., LAR, equals the relative derivative of $\mu(x)$ in (7.2). If the former is constant over age, corresponding to a constant b in (7.2), the latter varies age-wise and its pattern is bell-shaped (Horiuchi and Coale 1990; Horiuchi and Wilmoth 1997, 1998; Vaupel and Zhang 2010). LAR has been widely studied, both for human (Horiuchi and Wilmoth 1997, 1998) and non-human populations (Carey and Liedo 1995). In addition, Vaupel and Zhang (2010) derive in a ΓGM setting an explicit relationship between b, the rate of individual mortality increase, and the LAR (see Eq. (7.3)).

The age at which the relative derivative of $\mu(x)$ reaches its maximum is the age when mortality starts decelerating. In a ΓGM framework we can derive a closed-form expression for the age of mortality deceleration:

$$x^* = \frac{1}{b} \ln\left(\frac{(b+c\gamma)\,c}{2ab} + \frac{\sqrt{(b+c\gamma)\,c\gamma\,[(b+c\gamma)\,c - 4b\,(a\gamma - b)]}}{2ab\gamma}\right). \tag{7.10}$$

To capture accurately mortality dynamics across different periods we fit model (7.2). As a result, in year y, we capture the overall force of mortality by Vaupel and Missov (2014):

$$\mu(x, y) = \frac{a(y)e^{b(y)x}}{1 + \frac{\gamma(y-x)a(y)}{b(y)}\left(e^{b(y)x} - 1\right)} + c(y) , \tag{7.11}$$

where, in year y, $a(y)$ is the starting level of mortality, $b(y)$ is the rate of individual aging, $c(y)$ is the Makeham term, $\gamma(y - x)$ the variance of frailty at the initial age of analysis x_0 ($x_0 < x$) among survivors from cohort $y - x$.

The fitting procedure is based on the assumption that death counts *D(x,y)* are Poisson-distributed: $D(x, y) \sim Poisson(E(x, y) \cdot \mu(x, y))$ (Brillinger 1986), where *E(x,y)* denotes the corresponding exposure. For each year y we maximize a Poisson log-likelihood:

$$\ln L\left(a(y), b(y), \gamma(y), c(y)\right) = \sum_x \left[D(x, y) \ln \mu(x, y) - E(x, y)\mu(x, y)\right] . \tag{7.12}$$

LAR follows a bell-shaped pattern at older ages (Horiuchi and Coale 1990; Horiuchi and Wilmoth 1997, 1998) which is well-captured by a ΓGM model (Vaupel and Zhang 2010). If the LAR pattern is not bell-shaped, the ΓGM approximation is not accurate (see Fig. 7.2). As a result, it is important to identify the onset of this pattern and select it as the *starting age* to fit the ΓGM from. Figure 7.2 presents observed and ΓGM model-based (with four different starting ages) LARs for France. For both sexes, the higher the starting age of fitting the ΓGM, the better the accuracy of approximating the empirical LAR. As we aim at the best fit for both sexes and across all the selected countries, we decide to start all our fitting procedures at age 65. Note that ΓGM estimates are less accurate if a substantial proportion of the deaths in the study population occurs prior to age 65 (Russia, Ukraine and males in comparison to females, in general).

7.2.5 *Model-Based* Modal Age at Death

Death rates at young ages decline substantially during the first half of the twentieth century (Oeppen and Vaupel 2002). This results in a steeply increasing life expectancy at birth. From that point in time on, "the extension of length of human life in low-mortality countries is primarily due to improvements in old-age survival" (Horiuchi et al. 2013). Consequently, the modal age at death becomes a convenient lifespan indicator as it is not influenced by mortality at younger ages (Horiuchi et al. 2013; Kannisto 2001).

Different stochastic models, e.g., the Gompertz, logistic and Weibull models, as well as their extensions accounting for the Makeham term (Horiuchi et al. 2013),

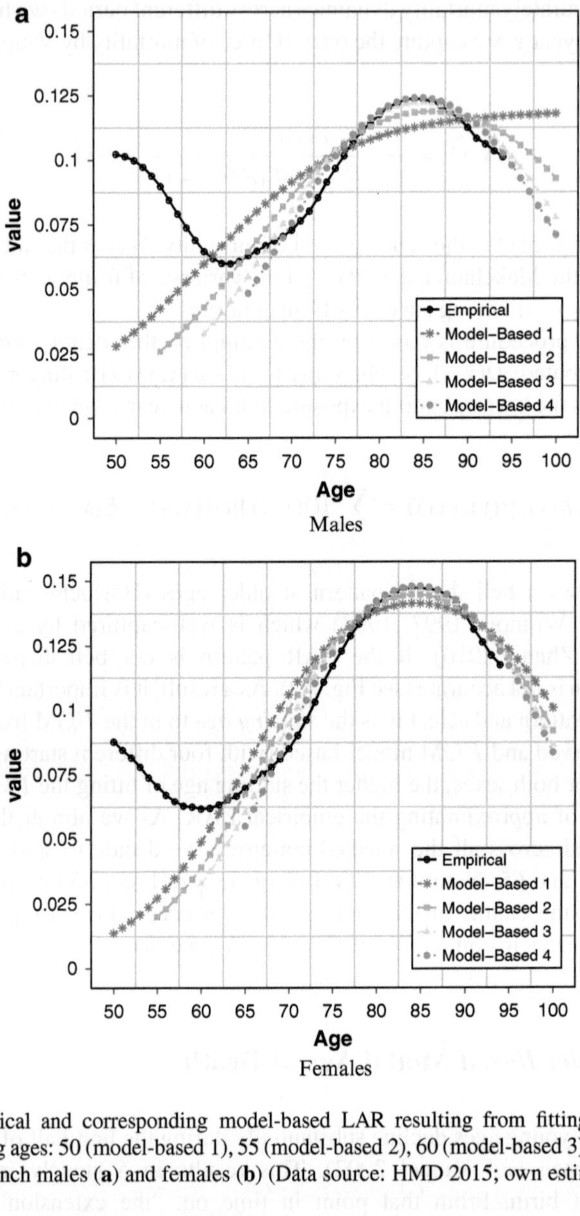

Fig. 7.2 Empirical and corresponding model-based LAR resulting from fitting a ΓGM with different starting ages: 50 (model-based 1), 55 (model-based 2), 60 (model-based 3) and 65 (model-based 4) for French males (**a**) and females (**b**) (Data source: HMD 2015; own estimation)

explain well the variation in observed adult lifespans. The Gompertz part ae^{bx} can be expressed using M as[2] $be^{b(x-M)}$, and the Gompertz-Makeham force of mortality can also be re-parameterized via the Gompertz old-age mode M: $be^{b(x-M)} + c$.

[2]For a detailed description please see Horiuchi et al. (2013), Appendix D.

The difference between *overall M* and the Gompertz-Makeham mode (*senescent M*), is very small in practice. Due to the fact that at old ages the estimated level of senescent mortality is considerably higher than background mortality, the latter registers slightly higher values than the former (especially for males).

In a ΓGM framework the force of mortality can be expressed via M as

$$\mu(x) = \frac{be^{b(x-M)}}{1 + \gamma\left(e^{b(x-M)} - e^{-bM}\right)} + c\,, \tag{7.13}$$

where

$$M = \frac{1}{b}\ln\frac{b}{a}\,. \tag{7.14}$$

Representing the Gompertz force of mortality in terms of M rather than a has also statistical advantages as the maximum-likelihood estimators of M and b are much less correlated than the ones of a and b (for broader discussion see Missov et al. 2015).

7.2.6 *Identifying the Curvilinear Structure of Longevity Measures over Time*

To elaborate on the dynamics of the relationship between the estimated x^* and the three longevity measures (M, Md, and e_0) we differentiate between different period segments, i.e., we estimate a piece-wise linear model for M over calendar time (Muggeo 2003). If we denote a break-point, i.e., a point at which the longevity measure LM changes its slope, by ψ_i, $i = 1, 2, \ldots$, then

$$LM = \alpha + \beta_0 y_0 + \beta_i (y_i - \psi_i)_+\,, \tag{7.15}$$

where α is the intercept, β_0 is the first segment slope, β_i measures the difference in slopes between the first and the i-th segment, ψ_i denotes the breakpoint, and $(y_i - \psi)_+ = (y_i - \psi_i) \cdot I(y_i > \psi_i)$ (Muggeo 2003). The indicator function $I(\cdot)$ equals one when the condition in its argument is true. When the model does not detect a breakpoint, we end up with a simple linear regression, i.e., ψ_i do not exist and β_i are statistical zeroes.

7.3 Results

In this section we use the methodological framework presented in Sect. 7.2 to reconstruct model-based LAR patterns and to elaborate on the link between the evolution of x^* and the three longevity measures (e_0, M and Md) over time for the selected countries. Section 7.3.1 revisits and discusses the shape of empirical LAR patterns on human mortality surfaces. Section 7.3.2 studies the correlations between e_0, M, Md, and the age of mortality deceleration x^*. Section 7.3.3 focuses on ΓGM model-based LAR patterns. Section 7.3.4 searches for possible relationships between the curvilinear structure of longevity measures and the corresponding old-age period mortality deceleration patterns. Finally, Sect. 7.3.5 studies the connection between x^* and e_0 on a cohort perspective.

7.3.1 Revisiting Empirical LAR Patterns

Previous studies (Horiuchi and Coale 1990; Horiuchi and Wilmoth 1997, 1998) acknowledge the fact that LAR follows an approximately bell-shaped pattern at ages 40+. We take advantage of LAR's ΓGM representation (Vaupel and Zhang 2010) (see Eq. (7.3)) to obtain a smooth parametric approximation of Eq. (7.9). Our main goal is to reexamine empirical LAR patterns on human mortality surfaces to get better understanding about their structure and evolution over time.

According to the heterogeneity hypothesis (Horiuchi and Wilmoth 1998), high mortality rates should yield LAR patterns with weaker curvature, and more pronounced LAR peaks should shift to later ages as mortality becomes almost exclusively concentrated at these ages (mortality compression).

Figures 7.3 and 7.4 present empirical LAR patterns at ages 30–94 for USA and Ukraine (see **Appendix** for LAR patterns of the other four study countries). Especially for females (USA Fig. 7.3a and Ukraine Fig. 7.3b), two LAR peaks can be identified – one at younger and another at older adult ages. Increases in LAR at young-adult ages is perhaps related to the contribution of *background* mortality at that stage of human life. If the first observed peak for young adults can be associated with external risks of mortality, the one at older ages reflects the increase of age-related (*senescent*) mortality.

In Sect. 7.2.4, which focuses on selecting the best starting age, we show that it is even harder to construct accurate ΓGM model-based LARs in the absence of a "well-defined" bell-shaped pattern, i.e., when LAR patterns are fluctuating. Consequently, the analysis of empirical LAR already anticipates possible fitting accuracy issues.

Figures 7.3 and 7.4 show that Ukrainian mortality is characterized by higher fluctuation in its LAR patterns when compared with the ones of the USA. In fact these patterns are not exclusive of Ukraine only, but are also observed in Russia (see **Appendix**), while the differences between males and females observed in these two figures are representative for all countries.

Figure 7.4 presents empirical LAR patterns for males in USA and Ukraine. In comparison with females (Fig. 7.3), males register a much flatter and fluctuating LAR pattern in early years, while in most recent decades the peak at older ages becomes more easily distinguishable. These patterns are a consequence of higher male-associated mortality rates that, like in the female case, decrease over time as a result of mortality improvements.

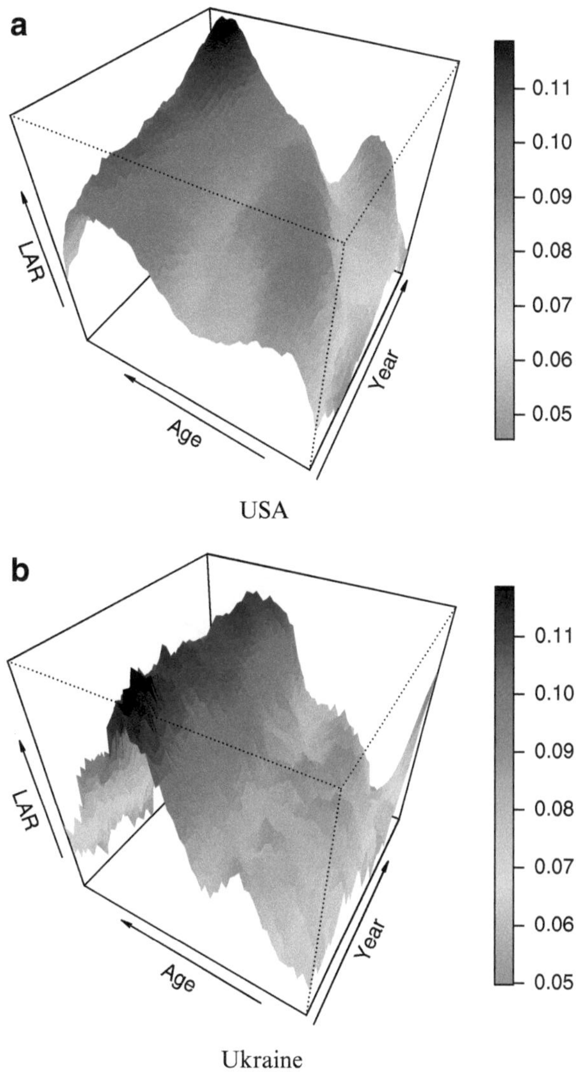

Fig. 7.3 Empirical LAR for females aged 30–94 in USA 1970–2010 (**a**) and Ukraine 1970–2009 (**b**) (Source: HMD 2015, own calculation)

Flatter LAR patterns for males in comparison with females suggests that the ΓGM model accuracy might be higher for females. As a result, the empirical LAR pattern may foresee the existence of possible issues when LAR is specified in a ΓGM framework, not only due to empirical fluctuations, but also because there is more than one peak (resulting from recent mortality improvements). One can expect, though, that in more recent years the concentration of mortality at older ages results in more accurate fits.

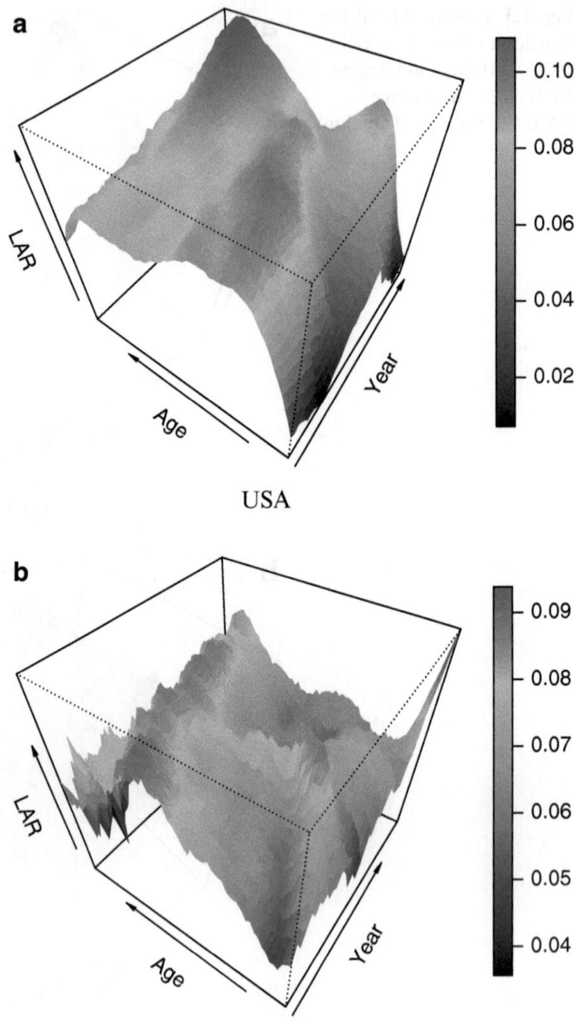

Fig. 7.4 Empirical LAR for males aged 30–94 in USA 1970–2010 (**a**) and Ukraine 1970–2009 (**b**) (Source: HMD 2015, own calculation)

7.3.2 Correlations of Longevity Measures

Rather than explore all possible relationships between statistically significant changes in the rate of increase across the selected longevity measures identified by the employment of (7.15) and the age at which mortality starts to decelerate x^* we perform a preliminary and exploratory evaluation of all possible relationships by calculating the correlation coefficient between each pair of measures (Table 7.1).

The correlation coefficient can be defined as a normalized measurement to evaluate possible linear relationships between two variables. A correlation coefficient

Table 7.1 Correlation coefficients between the different considered longevity measures: life expectancy at birth e_0, median Md and modal M age at death and the age at which mortality starts to decelerate x^* (Source: HMD 2015, own calculation)

Country	Sex	e_0 *Vs* Md	e_0 *Vs* M	e_0 *Vs* x^*	Md *Vs* M	Md *Vs* x^*	M *Vs* x^*
France	Male	1.00	0.95	0.84	0.95	0.84	0.80
	Female	1.00	0.97	0.98	0.97	0.98	0.96
Japan	Male	1.00	0.97	0.65	0.97	0.65	0.70
	Female	1.00	0.98	0.99	0.98	0.99	0.96
Russia	Male	0.97	0.79	−0.24	0.85	−0.40	−0.39
	Female	0.98	0.34	−0.19	0.26	−0.20	0.11
Sweden	Male	1.00	0.90	0.66	0.90	0.65	0.60
	Female	1.00	0.92	0.84	0.92	0.84	0.79
Ukraine	Male	0.99	0.71	−0.61	0.74	−0.65	−0.25
	Female	0.98	0.05	−0.36	0.06	−0.34	0.11
USA	Male	1.00	0.96	0.72	0.96	0.75	0.68
	Female	1.00	0.93	0.97	0.94	0.95	0.90

close to 1 indicates a clear positive linear relationship between the two variables under analysis, while for values closer to -1, it indicates a presence of a negative slope, i.e., a negative correlation. Consequently, correlation values closer to 0 indicates a weak or almost null linear relationship.

The obtained correlation coefficients presented in Table 7.1 indicate that in all the countries under study, exists a strictly positive relationship between e_0 and Md. Nevertheless, concerning the pairwise relationship e_0 *Vs* M and Md *Vs* M, except the generally high positive correlation coefficients, we find a very weak linear relationship in what concerns Russian and Ukrainian females. This weak pairwise relationship foresees the existence of an almost parallel evolution with time of the three widely used longevity measures within each relationship. Nevertheless, the small positive outcomes also suggest that in the near future longevity will possibly increase.

The statistical relationship between e_0 *Vs* x^*, Md *Vs* x^* and M *Vs* x^* in Russia and Ukraine is captured by a negative correlation, especially in the male case. As it could be seen in the previous subsection, LAR for males in Russia and Ukraine registered high fluctuation peaks at young adult ages as a result of higher mortality rates registered at those ages. This influences all three longevity measures as they decrease while x^* increases due to more distinctive old-age LAR peaks. Despite being negative, the female correlation coefficients associated with those two pairwise comparisons are smaller. This situation can be explained by the male-female longevity gaps and sex-related pace of mortality improvements. Nevertheless, a weak positive relationship can be found between M and x^* for females, suggesting that female mortality improvements in Russia and Ukraine occur faster than the ones for males.

After the examination of possible pairwise relationships between the trends over time of "typical" longevity measures (e_0, Md and M) we focus on the possible relationship between the age at which LAR reaches its maximum x^* and the three considered longevity measures. In a broader perspective, it can be seen that the relationship between M and x^* follows a weaker linear pattern. However, females living in France, Japan, Sweden and USA show high positive linear relationships for e_0 *Vs* x^*, *Md Vs* x^* and M *Vs* x^*. In Russia and Ukraine, independently of sex and considering the same two last pairwise relationships, the obtained results suggest a weak negative linear relationship.

The strictly positive and highly significant linear relationship between e_0 and *Md* in Table 7.1 may suggest that the obtained correlation coefficients for e_0 *Vs* x^* and *Md Vs* x^* might not differ significantly or add significant information. A closer look at these two pairwise relationships reveals, though, that for e_0 *Vs* x^*, in presence of significant correlation values, they are slightly higher than the ones for *Md Vs* x^*. Thus, in order to evaluate the presence of a possible connection between LAR peaks (x^*) and longevity measures by applying (7.15), our choice fell on e_0.

7.3.3 Model-Based Patterns

By fitting a ΓGM model from age 65 onwards we estimate the population's rate of aging $\bar{b}(x)$ (see Eq. (7.3)) by country and gender. The onset of mortality deceleration observed when age-specific death rates are plotted on a logarithmic scale is reflected in the peak of the corresponding LAR pattern at age x^*. As postulated by the heterogeneity hypothesis, a fast drop-off of frailer individuals leads to a lower corresponding age of mortality deceleration. Reversely, if deaths occur later in time, x^* increases, while the variability in the ages at death decreases. Due to this concentration of deaths in a narrow age range, the survival curve of the population becomes steeper and rectangular (Wilmoth and Horiuchi 1999). Like e_0 and *Md*, the age of mortality deceleration is strongly influenced by changes in the overall pattern of mortality, while M is highly affected by old-age mortality.

Figure 7.5 presents empirical and model-based ΓGM LAR patterns for females in the six studied countries. As already presented in Fig. 7.2, model-based LAR captures well empirical values. However, not so well-pronounced bell-shaped patterns are associated with less accurate estimates of the ΓGM LAR. The ΓGM captures well the evolution of LAR over time (Fig. 7.5): the flatter patterns at the beginning of the study period and the shifted patterns with stronger curvature after the 1970s. The captured shift of the age of mortality deceleration to older ages reflects point *b)* of the heterogeneity hypothesis: as lifespans increase, mortality deceleration occurs at older ages. Figure 7.5 also shows that countries with higher life expectancy register not only later mortality deceleration, but also more pronounced bell-shaped patterns (i.e., France (a), Japan (b), Sweden (c) and USA (f)). In addition, the observed (empirical) LAR values plotted after age 65 indicate that smaller population countries (such as Sweden) or countries with smaller populations after age 65 due to higher mortality rates at younger ages, present higher fluctuation patterns.

Figure 7.6 shows the empirical and model-based ΓGM LAR patterns for males. In comparison to females (Fig. 7.5), males are characterized by flatter LAR curves. Nevertheless, for almost all countries and for both sexes the age of mortality

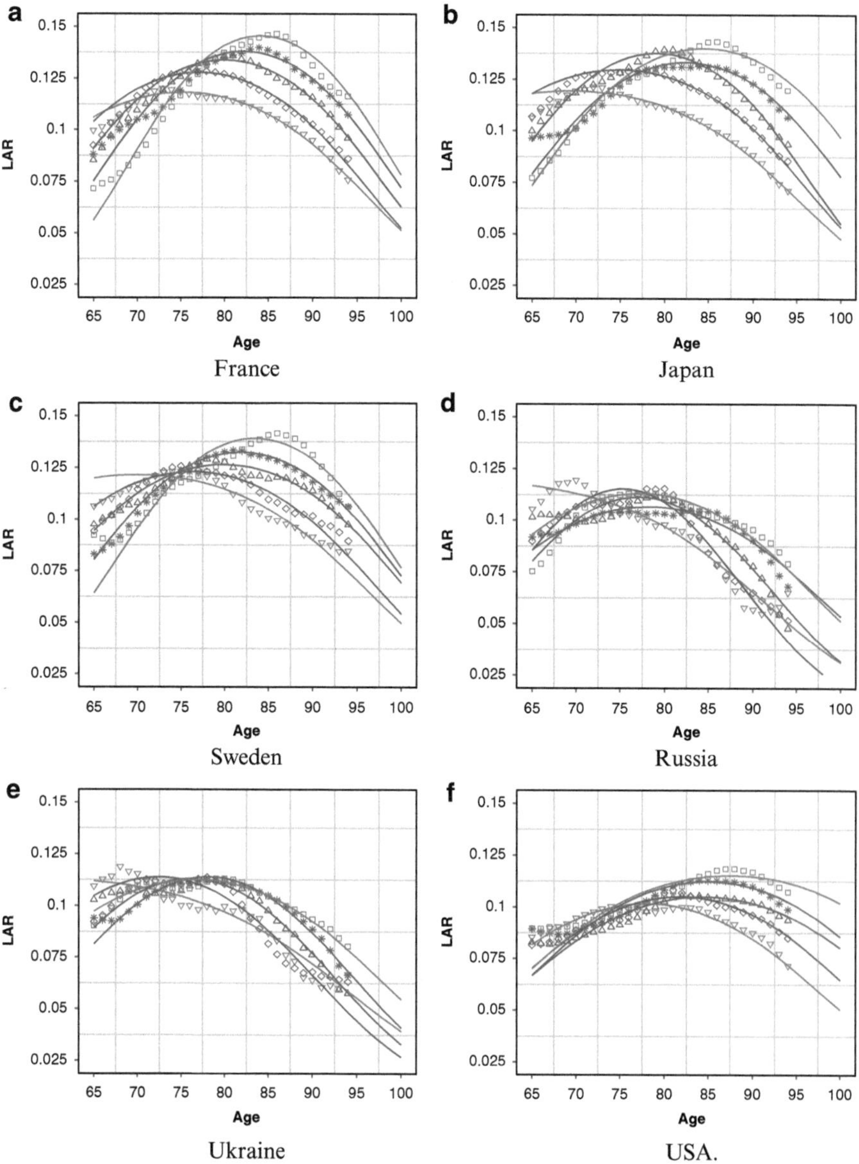

Fig. 7.5 Model-based LAR patterns and goodness of fit for females in France (**a**), Japan (**b**), Sweden (**c**), Russia (**d**), Ukraine (**e**) and USA (**f**) (Data source: HMD 2015; own estimation). Empirical patterns are represented by shapes – 1970: *inverted triangles*, 1980: *diamonds*, 1990: *regular triangles*, 2000: *asterisks* and 2010 (2009 for Ukraine): *squares* – and estimates by *solid lines*

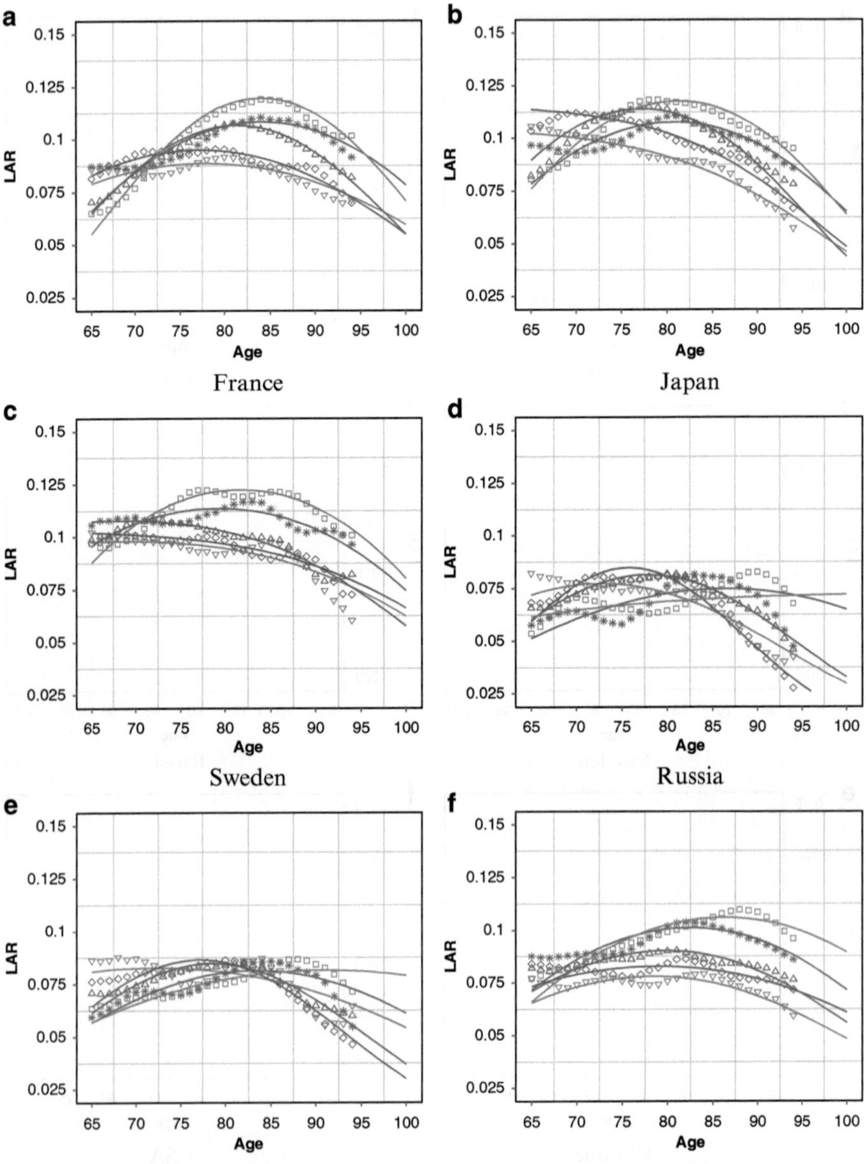

Fig. 7.6 Model-based LAR patterns and goodness of fit for males in France (**a**), Japan (**b**), Sweden (**c**), Russia (**d**), Ukraine (**e**) and USA (**f**) (Source: HMD 2015, own calculation). Empirical patterns are represented by shapes – 1970: *inverted triangles*, 1980: *diamonds*, 1990: *regular triangles*, 2000: *asterisks* and 2010 (2009 for Ukraine): *squares* – and estimates by *solid lines*

deceleration shifts to older ages with time. Deceleration starts at older ages for females, which might be connected with the longevity gap between the sexes: while female mortality is mainly concentrated at older ages, there is still *"excessive"* male mortality at younger ages. Estimated flatter LAR patterns, almost exclusively for males, can also be explained by higher turbulence in death rates. This is strongly pronounced in Russia and Ukraine where life expectancy has been fluctuating in the last couple of decades: Russian and Ukrainian males (Fig. 7.6d, e, respectively) are subjected to substantial turbulence in their (fairly high) mortality rates in comparison to females (Fig. 7.5d, e, respectively). As a result, LAR patterns for Russian and Ukrainian males are flat, the corresponding ΓGM approximation is quite inaccurate, and the associated age of mortality deceleration x^* is poorly captured by Eq. (7.10). Nevertheless, ΓGM LARs capture well the shift of mortality deceleration to older ages, as well as the observed steeper LAR curves as mortality rates decrease with time.

As described previously, flatter LAR patterns impose additional difficulties to estimate the age of mortality deceleration x^*. Figure 7.7 shows model-based LARs and the corresponding x^* estimates (black circles). Although the ΓGM model captures in general x^* accurately, it is less problematic to identify x^* for females than for males mainly due to the higher number of female survivors at older ages

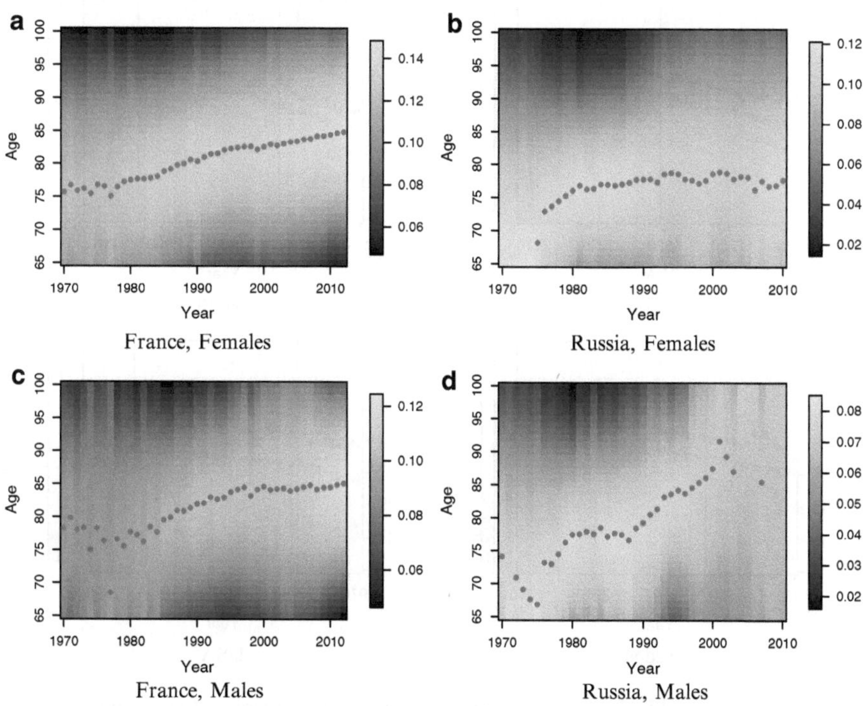

Fig. 7.7 Model-based LAR and the age of mortality deceleration (x^*) for Females and Males in France ((**a**) and (**b**)) and Russia ((**c**) and (**d**)) (Source: HMD 2015, own calculation). Image surface denotes LAR over age and year, and *black circles* x^*

and country mortality dynamics. The estimates for Russia (Fig. 7.7b, d) in Fig. 7.7 illustrate well the estimation difficulties associated with highly fluctuating empirical LARs (see Fig. 7.14 in the Appendix).

7.3.4 LAR and Longevity for Periods

While in Sect. 7.3.1 we presented the empirical patterns of mortality deceleration, in Sect. 7.3.2 we evaluate possible correlations between three of the most employed longevity measures and the age of mortality deceleration x^*, and in Sect. 7.3.3 one can find the ΓGM parametric estimates of LAR. To provide a complementary perspective, in this section we study the evolution of longevity in the selected countries and investigate possible connections between old-age mortality deceleration patterns and life expectancy at birth e_0 (identified previously as the longevity measure with higher correlation coefficients regarding x^*).

In comparison to other widely-used measures of longevity like life expectancy at birth e_0, the median Md and the modal age at death M, the age of mortality deceleration x^* undergoes greater fluctuations: for females (Fig. 7.8) x^* is closer to

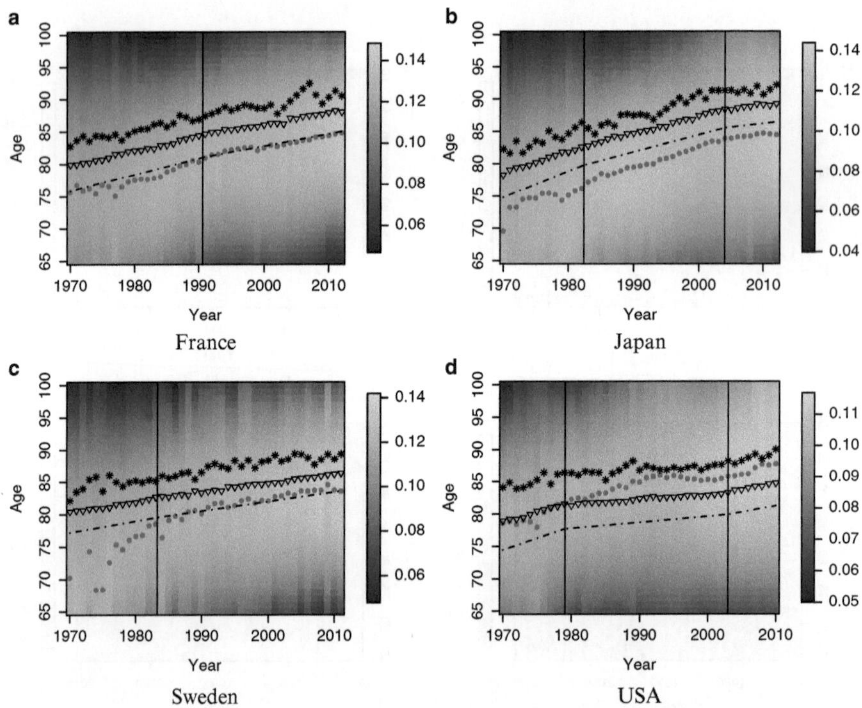

Fig. 7.8 Model-based LAR surface, segmented life expectancy at birth e_0 (*dashed lines*), median Md (*inverse triangles*) and modal age at death M (*asterisks*) and the age of mortality deceleration x^* (*circles*) for Females in France (**a**), Japan (**b**), Sweden (**c**) and USA (**d**). *Vertical solid lines* refers to statistically significant breaks in e_0 (Source: HMD 2015, own calculation)

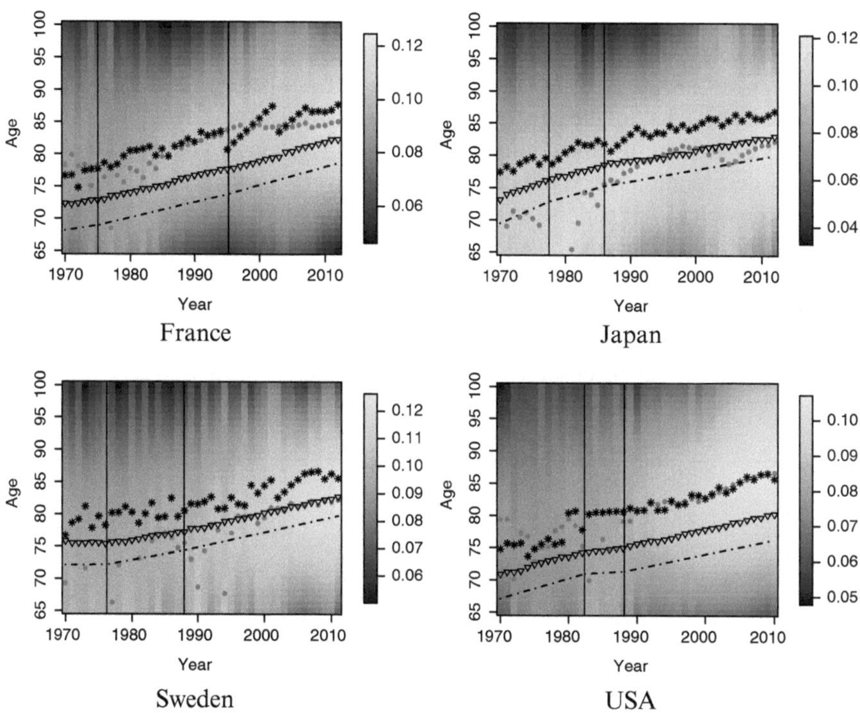

Fig. 7.9 Model-based LAR surface, segmented life expectancy at birth e_0 (*dashed lines*), median *Md* (*inverse triangles*) and modal age at death *M* (*asterisks*) and the age of mortality deceleration x^* (*circles*) for Males in France (**a**), Japan (**b**), Sweden (**c**) and USA (**d**). *Vertical solid lines* refers to statistically significant breaks in e_0 (Source: HMD 2015, own calculation)

e_0 (with the exception of USA), while in the male (Fig. 7.9) case x^* is closer to *Md* for Sweden and Japan, and to *M* in France and USA.

Figures 7.8 and 7.9 not only present the age at which mortality starts to decelerate x^* together with three longevity measures (e_0, *Md* and *M*), but also the results from fitting a segmented regression to life expectancy at birth. It seems that estimated LAR's and the segmented dynamics of e_0 share quite high connection. Statistically significant breaks obtained for e_0 are almost always accompanied by changes in the rate of increase in x^*. Even without a strong association between significant changes in the pace of increase of e_0 and x^*, it seems that whenever a break occurs, the characteristic bell-shaped pattern curvature becomes more pronounced.

7.3.5 LAR and Longevity for Cohorts

For period mortality the age of mortality deceleration shifts to older ages with calendar time. This reflects the improvement of age-specific death rates and the associated selection of frail individuals at later ages in each participating cohort. For a complete overview of (empirical and model-based) LAR patterns and their evolution over time, this section focuses on cohort LAR patterns in a ΓGM setting. It aims to check whether the latter are characterized by the same evolution over time as period LAR patterns. In addition, it studies the relationship between the age of mortality deceleration for cohorts and cohort life expectancy.

To reconstruct LAR patters over a longer period we focus on single cohorts from 1800 to 1900 in France and Sweden. In comparison to empirical LAR patterns for periods, the ones for cohorts are characterized by greater fluctuation, especially for earlier cohorts in which the number of survivors to age 65 and above is small. This is associated with lower accuracy of the ΓGM fit (see Fig. 7.10). LAR patterns with stronger curvature can be observed for females, but unlike the period case for earlier

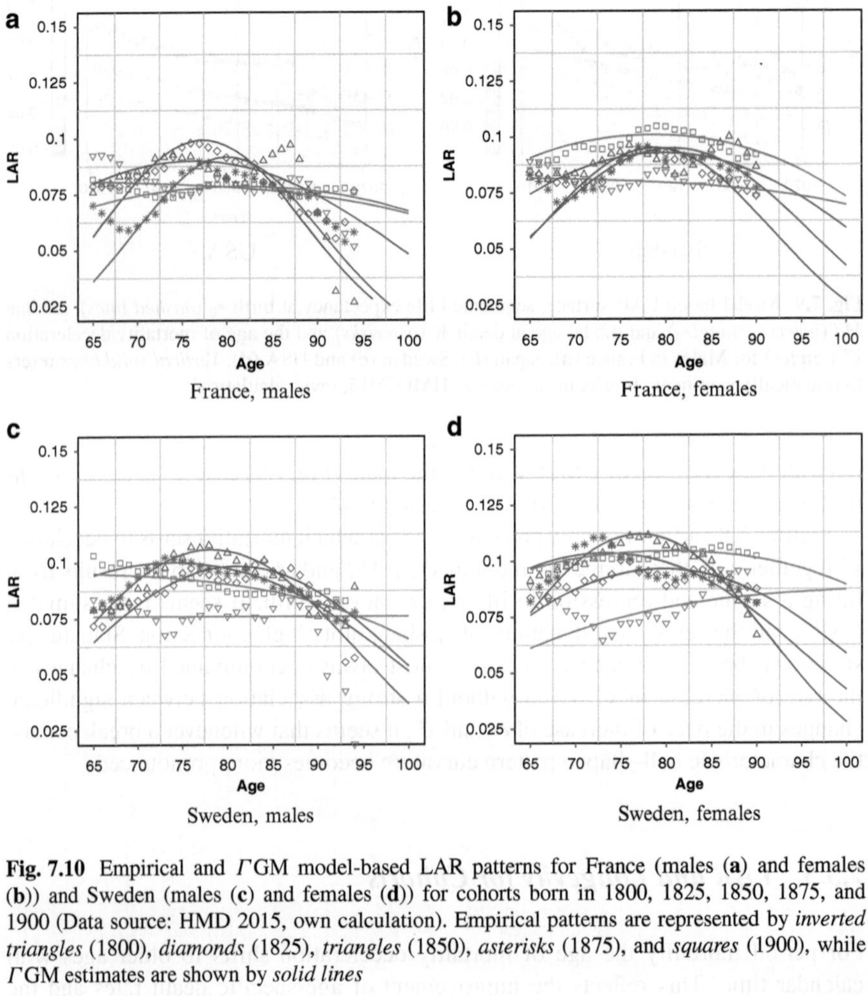

Fig. 7.10 Empirical and ΓGM model-based LAR patterns for France (males (**a**) and females (**b**)) and Sweden (males (**c**) and females (**d**)) for cohorts born in 1800, 1825, 1850, 1875, and 1900 (Data source: HMD 2015, own calculation). Empirical patterns are represented by *inverted triangles* (1800), *diamonds* (1825), *triangles* (1850), *asterisks* (1875), and *squares* (1900), while ΓGM estimates are shown by *solid lines*

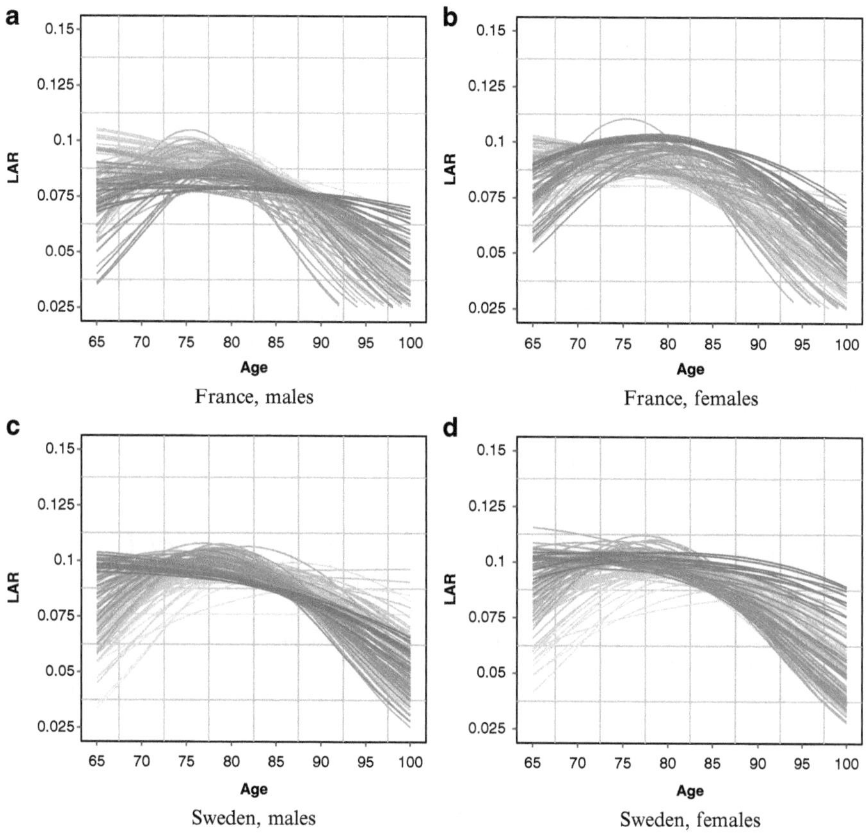

Fig. 7.11 Model-based ΓGM LAR patterns for France (males (**a**) and females (**b**)) and Sweden (males (**c**) and females (**d**)) for all cohorts born between 1800 and 1900. *Lighter colors* represent earlier cohorts (Data source: HMD 2015, own calculation)

cohorts, the only exception being French males (see Figs. 7.10 and 7.11). However, the age of mortality deceleration follows an erratic and, in the case of Sweden, even decreasing pattern (Figs. 7.11 and 7.12) unlike x^* for periods that increases steadily with calendar time. This might be due to the aforementioned greater fluctuation of cohort LAR patterns in comparison to the period ones and, on the other hand, the improvements in age-specific death rates that an aging cohort is exposed to.

Figure 7.12 presents the age of mortality deceleration x^* for cohorts along with cohort life expectancy at birth e_0 and the breaks identifying the segments with different rates of linear increase in cohort life expectancy. The breaks for Sweden are similar between sexes, while French males and females share a common change in the rate of life expectancy increase in 1875. Like in the period case, estimated LAR pattern and the corresponding x^* for cohorts seem to be linked to the segmented linear dynamics of e_0: statistically significant breaks for cohort e_0 are always accompanied by changes in the rate of increase in cohort x^*, the only

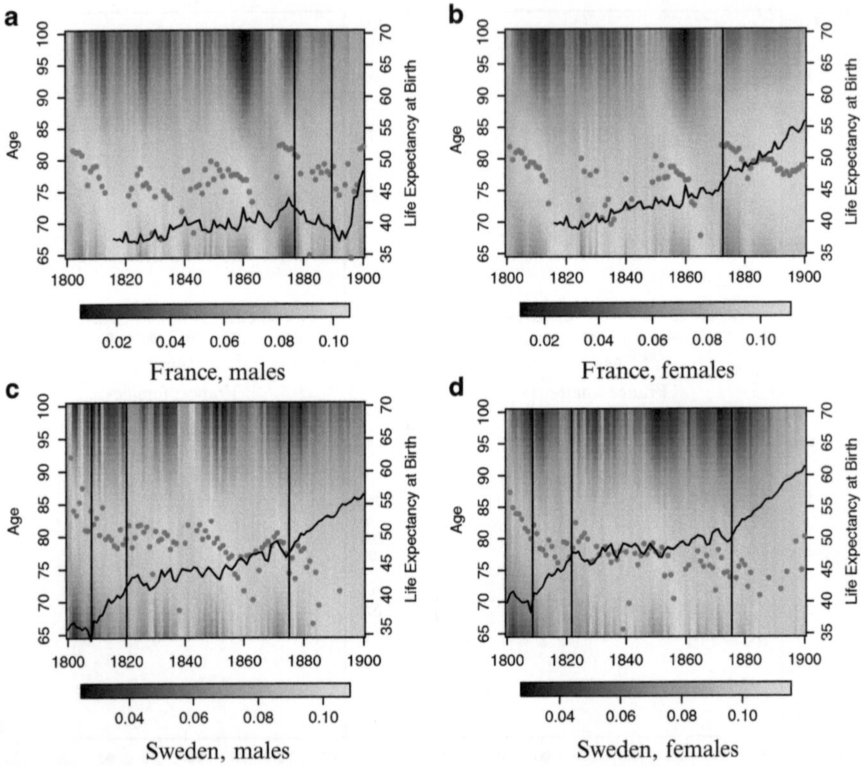

Fig. 7.12 Cohort ΓGM model-based LAR surface, segmented cohort life expectancy at birth e_0 (*solid line*) and the cohort age of mortality deceleration x^* (*circles*) for France (males (**a**) and females (**b**)) and Sweden (males (**c**) and females (**d**)). *Vertical solid lines* refers to statistically significant breaks in cohort e_0 (Source: HMD 2015, own calculation)

exception being perhaps French males (Fig. 7.12a). Figure 7.12d clearly illustrates this connection: in the first segment, the slight decrease in e_0 is accompanied by decreasing x^*; in the second segment, e_0 starts to steeply increase, while x^* seems to stay stable; in the third segment, as the pace of increase in e_0 lowers, x^* starts to decline slowly; and in the last segment, e_0 starts to increase steeply again, while x^* starts following the inverse pattern. Although the evolution of the two measures might be different in direction and scale, the breaks of the applied segmented regression seem to identify periods of different behavior for both measures simultaneously.

7.4 Discussion

A constant life-table aging rate for a population corresponds to a linear mortality increase on a logarithmic scale. LAR patterns, however, detect mortality deceleration resulting from selection of frailer individuals: in agreement with previous studies (Horiuchi and Coale 1990; Horiuchi and Wilmoth 1997, 1998; Vaupel and Zhang 2010), estimated and observed empirical LAR follow a bell-shaped pattern, even though the rate of individual aging, the relative derivative b of the Gompertz function, is constant. An explicit relationship between the rates of individual and population aging is presented in Vaupel et al. (1979) and Vaupel and Zhang (2010).

We choose to work in a gamma-Gompertz-Makeham setting for two reasons: first, the ΓGM model captures with high accuracy both extrinsic mortality at younger ages (given that is convexly increasing) and mortality deceleration at older ages (both "tails" of the S-shaped adult-mortality curve), and, second, the ΓGM provides adequate smooth parametric approximation to the life-table aging rate (Vaupel and Zhang 2010).

The selection of countries reflects three types of mortality experience over the last decades: steady increase, increase at a varying pace, and fluctuation of life expectancy. The inclusion of Japan in our study is important as Japan has not only registered the highest rates of mortality improvement for the last 60 years, but is also well-known for the different mortality risk factors in comparison to the ones in the European and North American countries (Horiuchi and Wilmoth 1997). Russia and Ukraine also deserve special attention as mortality trajectories have undergone major fluctuations for the last decades.

Stochastic variation in LAR estimates increases when the number of deaths is small (Horiuchi and Wilmoth 1997). The latter occurs in our study when either the country of interest has a small population size, e.g., Sweden, or a substantial proportion of deaths in a given country takes place prior to age 65, e.g., Russia. In general, the ΓGM fits for females are better than the ones for males as at ages 65+ the number of women exceeds the number of men. The ΓGM also captures better LAR patterns with stronger curvature.

The estimates of the ΓGM parameters $a(y)$, $b(y)$, $\gamma(y)$ and $c(y)$ in each year y aid understanding the evolution of senescent and background mortality over calendar time. The estimated starting level of mortality $\hat{a}(y)$, i.e., mortality at age 65, declines over time, while estimated background mortality $\hat{c}(y)$ increases with y. Note, however, that at ages 65+ the share of *background mortality* in total mortality becomes smaller and smaller.

The increasingly higher share of senescent mortality at older ages results in a steeper bell-shaped LAR curve. However, estimating $c(y)$ is essential to capture the latter (Horiuchi and Wilmoth 1997), and the smaller the estimates of $c(y)$, the weaker is the associated LAR's curvature (see estimates for Russia and Ukraine). Steady improvements in life expectancy result in bell-shaped patterns with stronger curvature that aid estimating LAR patterns by a ΓGM model with high accuracy (see results for France and Japan).

The estimation accuracy for the age of mortality deceleration depends on LAR's curvature: detecting x^* for flatter the LAR patterns is more problematic. If x^* is well approximated, the obtained results suggest a stronger correlation between this measure and both life expectancy at birth e_0 (see also Ribeiro and Missov 2014) and the median age at death Md. The high correlations in these pairs can be attributed to the fact that all three measures (e_0, Md and x^*) are sensitive to mortality at younger ages as opposed to M, the modal age at death. This study identifies the following relationship between statistically significant piece-wise changes in e_0 and x^*: every change in the slope of e_0's linear increase is reflected in steeper LAR patterns and higher associated x^*.

ΓGM LAR patterns for cohorts provide important insight, too. First, they confirm that mortality deceleration does not pertain (as previously hypothesized Gavrilov and Gavrilova 2011) to period mortality only. Second, ΓGM LAR patterns seem to be flatter for more recent cohorts. This can be explained, on the one hand, by the lower estimates of the Makeham term for the latest cohorts and, on the other hand, by the steady mortality improvements at ages 65+ on a period basis (especially after 1950) that postpones deaths in each cohort to later ages. Finally, the evolution of cohort life expectancy seems to be related to the cohort age of mortality deceleration. Although the changes in the two mortality measures might not be the same in direction and magnitude, the statistically significant segments of curvilinear increase in e_0 correspond to segments of different functional behavior for x^*. Finding a formal relationship that links changes in x^* to changes in life expectancy requires future study.

7.5 Conclusion

Empirical and model-based period LARs are consistent with point *b)* of the heterogeneity hypothesis by Horiuchi and Wilmoth (1998) as mortality deceleration shifts to older ages while the level of total adult mortality declines. The age of mortality deceleration x^*, as well as life expectancy e_0 and the median age at death *Md* seem to be strongly influenced by changes in the overall pattern of mortality, while M, the modal age at death, is highly affected by old-age mortality. Following the preliminary findings in Ribeiro and Missov (2014), this study finds a relationship between changes in the rate of life-expectancy increase with time and the corresponding LAR patterns: each breakpoint in the curvilinear evolution of e_0 results in steeper LAR patterns. A similar pattern is observed for cohort LAR patterns, too. However, the evolution of the age of mortality deceleration for cohorts is more erratic, and, in the case of Sweden, x^* is even decreasing. Nevertheless mortality deceleration is clearly seen for both periods and cohorts, and the link between its onset and different longevity measures is still to be formalized.

Appendix

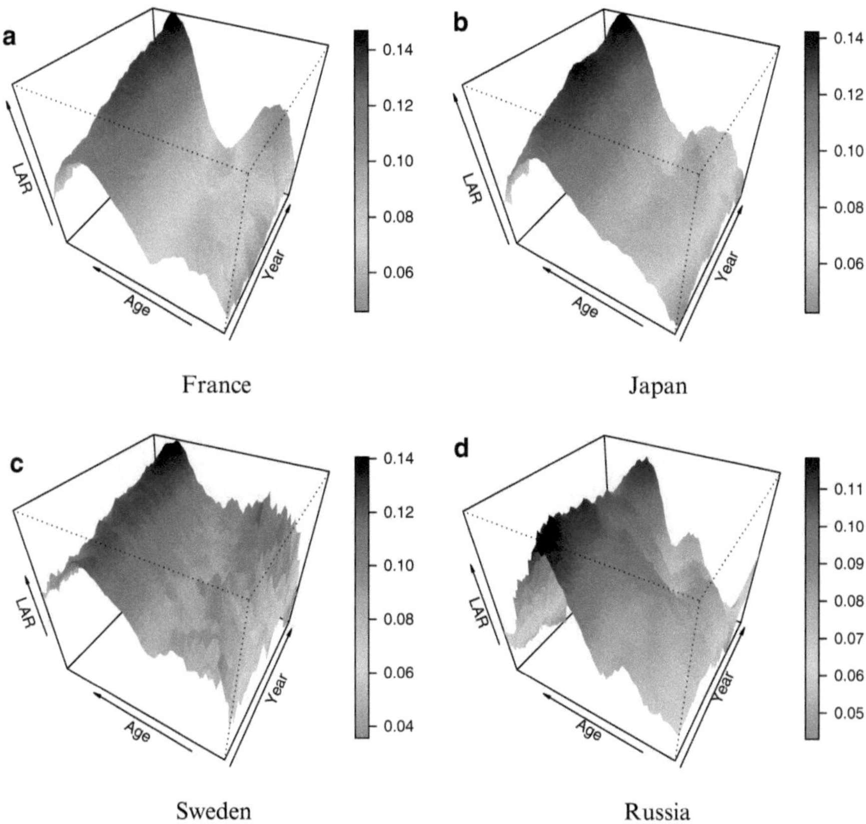

Fig. 7.13 Empirical LAR for females aged 30–94 in Japan (**a**) and France (**b**) 1970–2012, Sweden (**c**) 1970–2011 and Russia (**d**) 1970–2010 (Source: HMD 2015, own calculation)

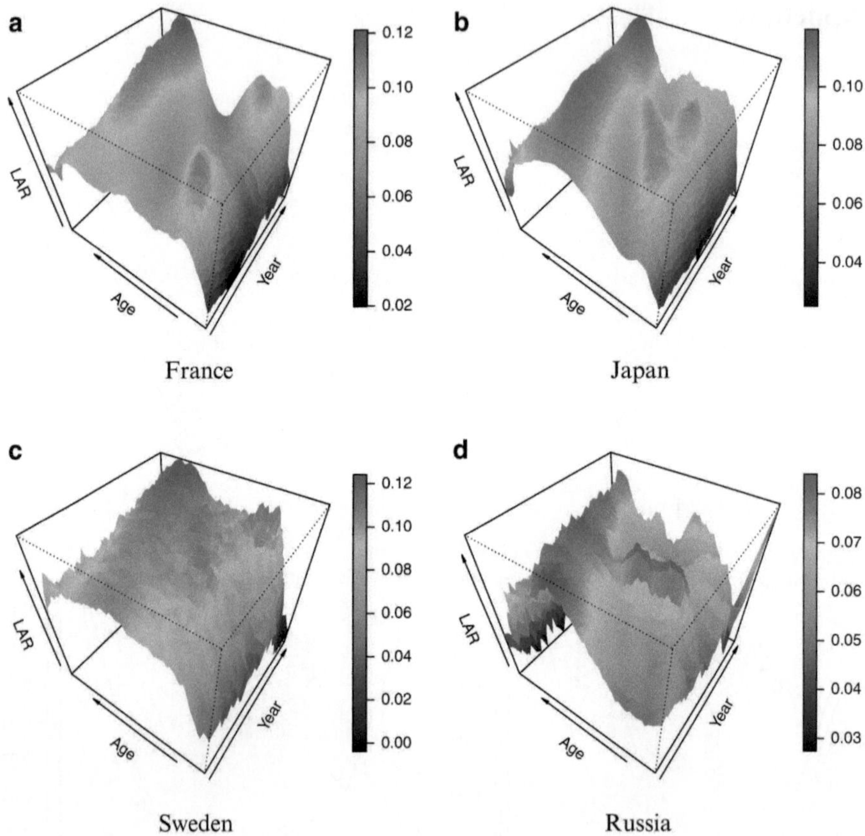

Fig. 7.14 Empirical LAR for males aged 30–94 in Japan (**a**) and France (**b**) 1970–2012, Sweden (**c**) 1970–2011 and Russia (**d**) 1970–2010 (Source: HMD 2015, own calculation)

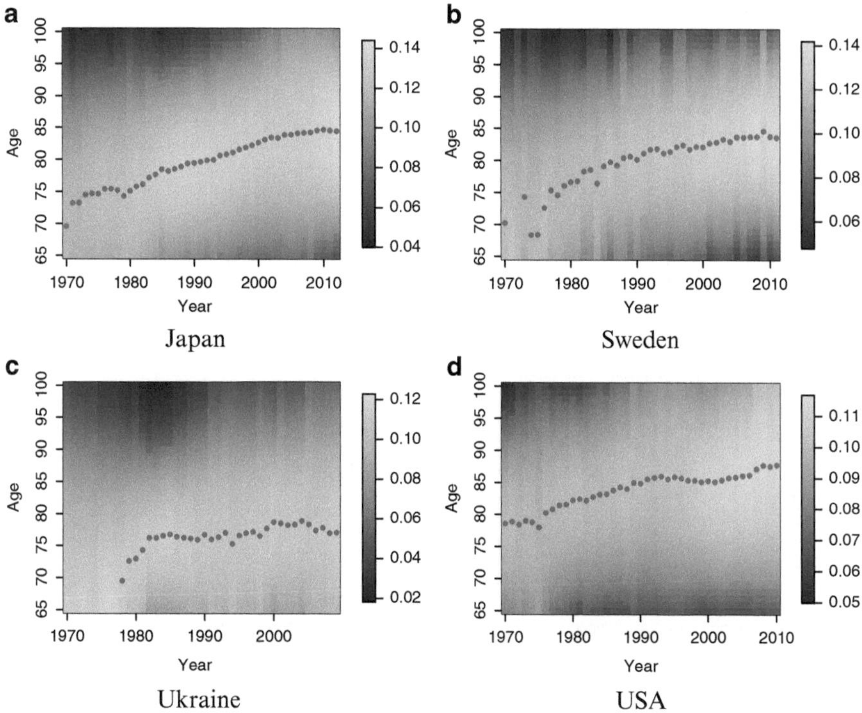

Fig. 7.15 Model-based LAR and the age of mortality deceleration (x^*) for females in Japan (**a**), Sweden (**b**), Ukraine (**c**) and USA (**d**) (Source: HMD 2015, own calculation). Image surface denotes LAR over age and year, and *black circles* x^*

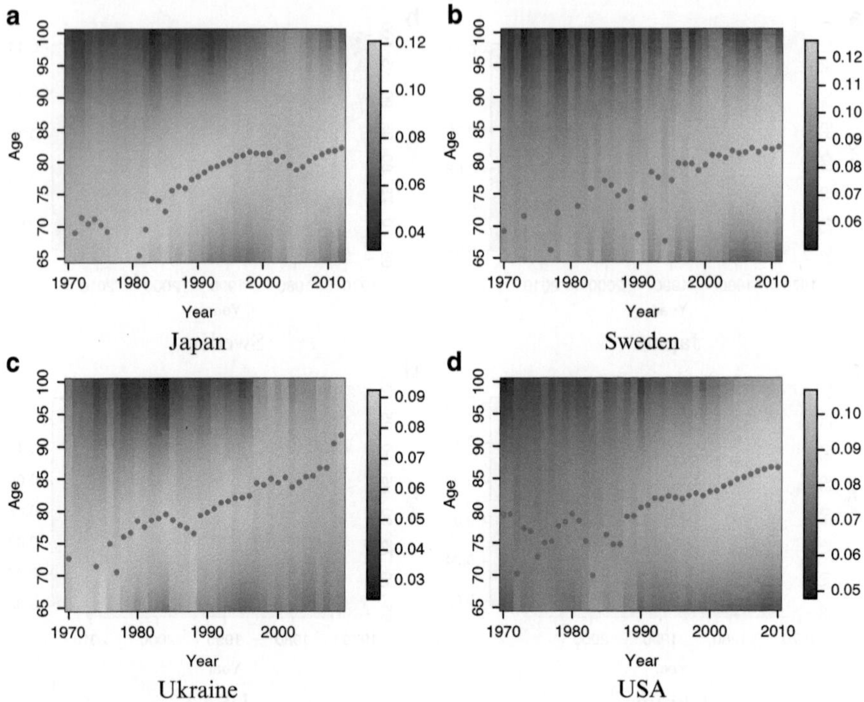

Fig. 7.16 Model-based LAR and the age of mortality deceleration (x^*) for males in Japan (**a**), Sweden (**b**), Ukraine (**c**) and USA (**d**) (Source: HMD 2015, own calculation). Image surface denotes LAR over age and year, and *black circles* x^*

References

Brillinger, D. R. (1986). The natural variability of vital rates and associated statistics. *Biometrics, 42*, 693–734.

Canudas-Romo, V. (2008). The modal age at death and the shifting mortality hypothesis. *Demographic Research, 19*(30), 659–686.

Carey, J. R., & Liedo, P. (1995). Sex-specific life table aging rates in large medfly cohorts. *Experimental Gerontology, 30*, 315–325.

Cheung, S. L. K., & Robine, J. M. (2007). Increase in common longevity and the compression of mortality: The case of Japan. *Population Studies, 19*(30), 85–97.

Cheung, S. L. K., Robine, J. -M., Tu, E. J. C., & Caselli, G. (2005). Three dimensions of the survival curve: Horizontalization, verticalization, and longevity extension. *Demography, 42*(2), 243–258.

Gampe, J. (2010). Human mortality beyond age 110. In H. Maier, J. G. B. Jeune, J. M. Robine, & J. W. Vaupel (Eds.), *Supercentenarians* (pp. 219–230). Heidelberg: Springer.

Gavrilov, L. A., & Gavrilova, N. S. (2011). Mortality measurement at advanced ages: A study of the social security administration death master file. *North American Actuarial Journal, 15*(3), 442–447.

Golubev, A. (2004). Does Makeham make sense? *Biogerontology, 5*, 159–167.

Gompertz, B. (1825). On the nature of the function expressive of the law of human mortality, and on a new mode of determining the value of life contingencies. *Philosophical Transactions of the Royal Society of London, 115*, 513–585.

HMD (2015). *The human mortality database*. Accessed January 23, 2015. http://www.mortality.org/

Horiuchi, S., & Coale, A. J. (1990). Age patterns of mortality for older women: An analysis using the age-specific rate of mortality change with age. *Mathematical Population Studies, 2*(4), 245–267.

Horiuchi, S., & Wilmoth, J. R. (1997). Age patterns of the life table aging rate for major causes of death in Japan, 1951–1990. *Journal of Gerontology: Biological Sciences, 52A*(1), B67–B77.

Horiuchi, S., & Wilmoth, J. R. (1998). Deceleration in the age pattern of mortality at older ages. *Demography, 35*(4), 391–412.

Horiuchi, S., Cheung, S. L. K., & Robine, J. M. (2012). Cause-of-death decomposition of old-age mortality compression. In *2012 Annual Meeting of the Population Association of America (PAA)*, San Francisco.

Horiuchi, S., Ouellette, N., Cheung, S. L. K., & Robine, J. M. (2013). Modal age at death: Lifespan indicator in the era of longevity extension. *Vienna Yearbook of Population Research, 11*, 37–69.

Kannisto, V. (2001). Mode and dispersion of length of life. *Population: An English Selection, 13*, 159–171.

Makeham, W. M. (1860). On the law of mortality. *Journal of the Institute of Actuaries, 13*, 283–287.

Missov, T. I., & Finkelstein, M. (2011). Admissible mixing distributions for a general class of mixture survival models with known asymptotics. *Theoretical Population Biology, 80*(1), 64–70.

Missov, T. I., & Ribeiro, F. (2016, forthcoming). Do individuals age at the same rate? Findings from cause-of-death data. *Demography*.

Missov, T. I., & Vaupel, J. W. (2015). Mortality implications of mortality plateaus. *SIAM Review, 57*(1), 61–70.

Missov, T. I., Lenart, A., Nemeth, L., Canudas-Romo, V., & Vaupel, J. W. (2015). The Gompertz force of mortality in terms of the modal age at death. *Demographic Research, 32*(36), 1031–1048.

Muggeo, V. (2003). Estimating regression models with unknown break-points. *Statistics in Medicine, 22*, 3055–3071.

Nemeth, L., & Missov, T. I. (2014). How wrong could parameter estimates be? Statistical consequences of fitting the wrong model to human mortality data. In *2014 European Population Conference*, Budapest.

Oeppen, J., & Vaupel, J. W. (2002). Broken limits to life expectancy. *Science, 296*, 1029–1031.

Pollard, J. H. (1998a). *An old tool–modern applications* (Research paper series no. 001/98). School of Economic and Financial Studies, Macquarie University, Sydney.

Pollard, J. H. (1998b). *Keeping abreast of mortality change* (Research paper series no. 002/98). School of Economic and Financial Studies, Macquarie University, Sydney.

Preston, S., Heuveline, P., & Guillot, M. (2001). *Demography: Measuring and modeling population processes*. Malden: Wiley-Blackwell.

Pyrozhkov, S., Foygt, N., & Jdanov, D. (2011). About mortality data for Ukraine. In *Human mortality database background and documentation*. http://www.mortality.org/hmd/UKR/InputDB/UKRcom.pdf

Ribeiro, F., & Missov, T. I. (2014). Mortality inferences from estimated life-table aging rates in a gamma-Gompertz-Makeham framework. In *2014 Annual Meeting of the Population Association of America (PAA)*, Boston.

Thatcher, A. R., Kannisto, V., & Vaupel, J. W. (1998). *The force of mortality at ages 80 to 120* (Monographs on population aging, Vol. 5). Odense: Odense University Press.

Vaupel, J. W., & Missov, T. I. (2014). Unobserved heterogeneity: A review of formal relationships. *Demographic Research, 31*(22), 659–686.

Vaupel, J. W., & Yashin, A. I. (1985). Heterogeneity's ruses: Some surprising effects of selection on population dynamics. *The American Statistician, 39*, 176–185.

Vaupel, J. W., & Zhang, Z. (2010). Attrition in heterogeneous cohorts. *Demographic Research, 23*(26), 737–748.

Wilmoth, J. R., & Horiuchi, S. (1999). Rectangularization revisited: Variability of age at death within human populations. *Demography, 4*(36), 475–495.

Vaupel, J. W., Manton, K., & Stallard, E. (1979). The impact of heterogeneity in individual frailty on the dynamics of mortality. *Demography, 16*, 855–860.

Chapter 8
Demographic Consequences of Barker Frailty

Alberto Palloni and Hiram Beltrán-Sánchez

8.1 Introduction

In this paper we develop a formal model to represent the effects of early life conditions with delayed health impacts on patterns of old age mortality. The model can capture three types of mechanisms: those involving exposure to early malnutrition and deprivation (Barker 1998; Gluckman and Hanson 2006; Langley-Evans 2004), those linking early infectious diseases and adult chronic conditions (Elo and Preston 1992; Fong 2000) and, finally, those working through sustained inflammation to promote adult chronic illnesses (Finch 2007; Finch and Crimmins 2004; Crimmins and Finch 2006; Danesh et al. 2000; McDade et al. 2010). In our model the excess mortality implications of these processes are identical and we use the term 'Barker frailty' to refer to each one of them, all of them, or specific combinations including some of them.[1] The formal model is an extension of the standard frailty model in demographic analysis (Vaupel et al. 1979; Vaupel and Yashin 1987; Vaupel and Missov 2014; Manton et al. 1986; Hougaard 1986; Steinsaltz and Wachter 2006;

[1]Our model is a particular case of a more general representation that includes all the nuances associated with each of these three mechanisms.

A. Palloni (✉)
Center for Demography of Health & Aging, University of Wisconsin-Madison, 4434 Social Sciences Building, 1180 Observatory Drive, Madison, WI 53726, USA
e-mail: palloni@ssc.wisc.edu

H. Beltrán-Sánchez
Community Health Sciences and California Center for Population Research (CCPR), University of California, Los Angeles, CA 90095, USA
e-mail: beltrans@ucla.edu

R. Schoen (ed.), *Dynamic Demographic Analysis*,
The Springer Series on Demographic Methods and Population Analysis 39,
DOI 10.1007/978-3-319-26603-9_8

Aalen 1988). We show that adult mortality patterns in populations with Barker frailty are equivalent to adult mortality patterns in populations with a class of time-varying and/or age dependent frailty. We demonstrate formally and via simulations that populations with Barker frailty could, in theory at least, experience unchanging or increasing adult mortality even when background mortality has been declining for long periods of time. Most commonly, however, these populations will experience slower rates of mortality decline than the background mortality regime. We argue that Barker frailty should be pervasive in low-to-middle income populations, e.g. those that experienced a mortality decline fueled largely by post-1950 medical innovations that reduced the load and lethality of infectious and parasitic diseases.

The plan of the paper is as follows: in Sect. 8.2 we define the theoretical underpinnings of the notion of Barker frailty; in Sect. 8.3 we propose alternative models to formalize Barker frailty and Barker effects; in Sect. 8.4 we develop generalized models better suited for capturing dynamics associated with changes in the size and/or heterogeneity of populations expressing Barker frailty; in Sect. 8.5 we introduce simulations to assess the impact of Barker frailty on adult mortality rates and review selected results. The last section summarizes findings, discusses implications, proposes extensions of the model, and concludes.

8.2 Nature of Barker Frailty

There is solid empirical evidence and persuasive theoretical argumentation supporting the idea that early life conditions–*in utero*, around birth and during early childhood–exert an important influence on adult health and mortality (Barker 1998; Anson 2002; Gluckman and Hanson 2006; McDade and Kuzawa 2004; Langley-Evans 2004; Beltrán-Sánchez et al. 2012; Barouki et al. 2012). The mechanisms that trigger these delayed effects involve redirection of processes of organ-specific cell growth and functional differentiation, various types of epigenetic changes (histone covalent modifications, non-coding RNA expression, methylation) manifesting developmental plasticity, and more general conditions including exposure to acute poverty, deprivation, and stress (Forsdahl 1977, 1978). The best known, albeit not the most comprehensive, version of the hypothesis of delayed health effects, was articulated by Barker (1998). The cornerstone idea of this variant of the theory ('fetal programming') is that nutritional deprivation *in utero*, and soon after birth, disrupts processes of organ formation (cell division, cell growth and functional specialization) and exposes survivors to excess risk of developing a number of chronic conditions during late adulthood, including type-2 diabetes *mellitus* (T2D), hypertension and other circulatory disorders, kidney and heart disease, and some diseases of the respiratory system.

An older, related, but better established strand of the literature identifies linkages between exposure and contraction of specific early life infections and the development of adult chronic conditions. Examples of this mechanism include the relation between *helicobacter pylori* and colon cancer, HPV and uterine cancer, *Hepatitis B* and liver cirrhosis and cancer, and rheumatic heart fever and mitral valve stenosis. In all these cases exposure and contraction of well-defined infections induces organ damage manifested in adult chronic illnesses (Fong 2000; Elo and Preston 1992).

Finally, a third mechanism linking early conditions and adult health via delayed response involves sustained inflammation that results from recurrent episodes of infectious diseases, persistent and durable but low-level infections and, more generally, continuous exposure to multiple infectious and parasitic diseases (Finch 2007; Finch and Crimmins 2004; Crimmins and Finch 2006; Danesh et al. 2000). The theory suggests that when the inflammatory processes promoted by these exposures are long lasting, such as those associated with *periodontitis* or *chlamydia pneumoniae* (Fong 2000) they are likely to generate physiological damage that increases susceptibility to chronic illnesses (Finch 2007).

A common theme of all three mechanisms identified above is the presence of organ damage or abnormal immune/metabolic responses triggered by adverse early experiences, long latency periods, and delayed manifestations in late adulthood. The expression of the damage inflicted early in life is observable only if the following conditions are met: (a) members of a birth cohort who experience the offending early conditions survive beyond some critical age Y_1 after which manifestation of the original damage begins to unfold; (b) the delayed effect must be significant in the sense that it should implicate a broad range of illnesses and conditions with high fatality rates; (c) the beneficial mortality-related effects of medical technological advances adopted and diffused between the time of onset of adverse experiences and the time at which the birth cohort attains the critical age Y_1 are less than the excess mortality risks implied by chronic conditions related to Barker frailty.

In a population that meets all the foregoing conditions, time trends of adult mortality may experience singular features. In a secular mortality decline, the rate of mortality reduction over time will slow-down (at all ages), even if the background rate of mortality decline remains invariant. This should always occur when the mortality regime is influenced by standard frailty. When, in addition, Barker frailty and effects are present, the rate of mortality decline at ages above Y_1 can decrease, converge to zero or even increase. In an unchanging mortality regime the impact of Barker frailty on adult mortality rates crucially depends on the fraction of individuals who experience early damage and survive past age Y_1. Two invariant mortality regimes with identical mortality rates above age Y_1 but different survival probabilities to age Y_1 could result from very different Barker frailty dynamics, none of which is identifiable from conventionally available data. By contrast, mortality regimes affected by a secular decline offer an opportunity to identify the imprints of Barker frailty. In fact, when reductions in mortality are partially or totally sustained by massive improvements in infant and child survival there will be potentially large but lagged increases of the fraction of individuals surviving to ages past Y_1. This opens the gates for expression of Barker frailty. Declining mortality regimes such as those that emerged after 1950 in low-to-middle income countries are enablers of Barker frailty and, as a result, mortality trajectories may undergo stages in which adult mortality rates decline more slowly than background mortality rates, remain steady, or even increase.

In a mortality regime with declining mortality the selection pressure of standard frailty on members of a cohort weakens over time. When this is combined with Barker frailty the levels of adult mortality will be subject to two sources of upward pressure: the first is rooted in the increase of (standard) mean frailty of individuals who attain adult ages, a natural product of mortality decline. The second originates in excess mortality risks shared by at least a fraction of 'new' survivors who were exposed to adverse early life conditions and who attain adult ages in the lower mortality regime. Thus, Barker frailty will magnify the decelerating force that naturally arises when only standard frailty prevails.

Although we do not explore them in this paper, the presence of Barker frailty leads to predictions for aggregate outcomes (rate of mortality change, slope of adult mortality) that could be tested with empirically observed patterns to identify some of the mechanisms described above.

8.3 Formal Models for Barker Frailty and Barker Effects

To fix concepts and set up notation we begin with a brief review of the standard frailty model in continuous and discrete form. We then introduce a general form of the model, continuous and discrete versions, and variants using the Gamma distribution.

8.3.1 Standard Frailty: Continuous and Discrete Models

The standard frailty model assumes the existence of an age-invariant trait ε assigned to individuals at birth according to some known probability distribution. The role of this trait is to shift individual mortality levels by the same amount at all ages. Although the interpretation of the trait is the subject of some controversy, in most cases it is assumed to refer to some fixed, perhaps genetic, endowment and not to an acquired one (Vaupel et al. 1979; Vaupel and Missov 2014; Vaupel and Yashin 1987; Kannisto 1994; Aalen 1988; Hougaard 1986; Manton et al. 1986).

Let $\mu_i(y)$ represent the force of mortality at age y for individual i, $\mu_s(y)$ be a baseline mortality rate, and $\bar{\mu}(y)$ the average mortality rate at age y . The standard frailty model is fully defined by a pair of related equations:

$$\mu_i(y) = \varepsilon_i \mu_s(y) \tag{8.1}$$
$$\bar{\mu}(y) = E_y(\varepsilon)\mu_s(y)$$

where ε_i is random frailty with density $f(\epsilon)$. The expected value of the frailty trait at age y under density $f(\varepsilon)$, $E_y(\varepsilon)$, depends on the conditional density of the trait at age y and this, in turn, is a function of probabilities of surviving to age y. The key insight is that the population composition by trait ε changes as mortality differentially selects out individuals with higher values of ε. In particular, the slope of average mortality rates should increase less rapidly than the slope of the baseline mortality rates.

A very special case of (8.1) is when ε attains only one of two values, say $\varepsilon_1 > \varepsilon_2$ with probabilities f and $(1-f)$ and the average force of mortality is

$$\bar{\mu}(y) = \left[\varepsilon_1 \left(\frac{1}{1 + h\,exp(-(\varepsilon_2 - \varepsilon_1)\Lambda_s(0,y))}\right) + \varepsilon_2 \left(\frac{h\,exp(-(\varepsilon_2 - \varepsilon_1)\Lambda_s(0,y))}{1 + h\,exp(-(\varepsilon_2 - \varepsilon_1)\Lambda_s(0,y))}\right)\right] \mu_s(y) \tag{8.2}$$

where $h = (1-f)/f$ and $\Lambda_s(0,y) = \int_0^y \mu_s(x)dx$. Because $\varepsilon_1 > \varepsilon_2$, the average mortality rate will always be closer to mortality rates among those with frailty ε_2 (Vaupel and Yashin 1987).

8.3.2 Barker Frailty: General Continuous Model

A model for Barker frailty includes and defines two key properties. First, individuals who could express Barker frailty experience excess mortality (Barker effects) at adult ages (and perhaps in early childhood) but not necessarily in the rest of the life course. Second, the fraction of individuals at adult ages who could potentially express Barker frailty must increase whenever mortality declines. This can take place via two different but not always dependent mechanisms. The first operates by simply reducing selective pressure: individuals who would have died in a higher mortality regime can survive to adult ages in a more beneficial regime and some of them may be carriers of Barker frailty. The second mechanism increases the fraction of births that are carriers of Barker frailty and, therefore, augments the pool of individuals who can potentially express Barker frailty, irrespective of mortality changes. This mechanism operates when there are improvements associated with mortality decline, such as reduced maternal exposure to parasitic and infectious diseases and better prenatal care, that translate into lower fetal and perinatal mortality rates. These mechanisms can prevail alone or in combination and the models below offer room to incorporate either of them.

8.3.2.1 Barker Frailty and Barker Effects with Unchanging Mortality

The simplest model assumes the existence of a trait, δ, acquired as early as during conception and gestation that continues to be shaped during early childhood. Thus, in theory at least, the trait itself can be changing at least for some time before full adulthood. Once the trait is shaped and its value fixed, it will mark individual carriers throughout life. As in the standard frailty model, the effects of the trait on mortality rates will be multiplicative but could vary with age.

We introduce the function $R(\delta, y)$ to reflect the impact of Barker frailty on adult mortality, e.g. the Barker effect (for $y > Y_1$). A very general form is $R(\delta, y) = 1 + I_i(\alpha_1(y - Y_1) + \alpha_2(\delta - \delta_0))$ with $\alpha_1 > 0, \alpha_2 > 0$, and I_i a random indicator function attaining value 1 if $\delta > \delta_0$ and 0 otherwise, where δ_0 is a threshold Barker frailty value for expression of Barker effects and for $y > Y_1$. The force of mortality of individual i is

$$\mu_i(y) = \mu_s(y)\{\delta_i(1 - I_i) + \delta_i[1 + I_i(\alpha_1(y - Y_1) + \alpha_2(\delta_i - \delta_0))]\}, \quad \forall\, y > Y_1 \quad (8.3)$$

and the average mortality rate at age $y > Y_1$ is

$$\bar{\mu}(y) = \frac{\mu_s(y)[E_y(\delta|\delta > \delta_0)(1 + (\alpha_1(y - Y_1)) + \alpha_2 E_y(\delta(\delta - \delta_0)|\delta > \delta_0) + E_y(\delta|\delta \leq \delta_0)D_y]}{1 + D_y} \quad (8.4)$$

where D_y is a function of the baseline survival from 0 to y and the expected value of the probability of surviving to age y among those with $\delta > \delta_0$.

Expression (8.4) contains the contributions of two subpopulations to excess mortality relative to the baseline: the first is associated with the subpopulation that meets the condition $\delta > \delta_0$, namely, $(E_y(\delta|\delta > \delta_0)(1 + (\alpha_1(y - Y_1)) + \alpha_2 E_y(\delta(\delta - \delta_0)|\delta > \delta_0))$, and the second is associated with the subpopulation that does expresses only standard frailty, namely, $E_y(\delta|\delta \leq \delta_0)$. The first part includes the influence of standard frailty, the effect of age ($\alpha_1' s$), and of excess triggered by the level of individual frailty ($\alpha_2' s$). As is the case of standard frailty, the expected value of the force of mortality is an implicit function of the integrated hazards up to age y that contribute to the moments of the conditional distribution of δ and embedded in D_1. Unlike the case of standard frailty, Eq. (8.4) involves the first as well as second moments of the conditional distribution for all ages above Y_1.

The expression in (8.3) captures a number of desirable features identified above. First, Barker frailty is a random trait acquired early in life and influencing mortality at all ages. Second, there is a critical age Y_1 above which Barker frailty expresses itself as excess mortality risks (Barker effects). Third, the excess mortality risk can potentially increase after attaining the critical age Y_1 and could do so as a function of the individual level of Barker frailty. The latter feature implies that survivors at older ages who express Barker frailty will do so in direct proportion to their level of vulnerability.

Note that, at the outset, the model above (8.3) imposes the restriction that the critical age, Y_1, is a fixed quantity. This need not be the case. It is conceivable that the critical age itself is a function of individual experiences or traits as well as of features of the background epidemiological regime. To reflect this we could extend the model so that Y_1 is also a randomly distributed trait with a systematic component associated with background mortality. Because this extension complicates the model and obscures its main features we pursue the generalization elsewhere. In a comparative statics exercise (Sect. 8.5.4), our simulations show that variation in the threshold age Y_1 has considerable impact on the slope of average mortality rates and on average levels of Barker frailty.

8.3.2.2 Barker Frailty and Barker Effects with Secular Mortality Decline

Introducing generalized Barker frailty as in (8.3) when the mortality regime is changing complicates the algebra. To simplify developments we will assume $\alpha_1 = \alpha_2 = 0$ and $\alpha_0 = R > 1$. Although less rich, this formulation enables us to identify the main properties of Barker frailty and minimizes clutter without great loss of generality.[2] In this simpler formulation the average mortality level at age y is given by

$$\bar{\mu}(y) = R(y)E_y(\delta)\mu_s(y) \tag{8.5}$$

where, for simplicity, $R(y) = R$ for $y \geq Y_1$ and $R(y) = 1$ otherwise.[3] The survival function to age $y \geq Y_1$ for individual i that is implicit in (8.5) is given by

$$\begin{aligned} S_i(y) &= exp\left(-\delta_i\left(\int_0^{Y_1} \mu_s(x)dx + R\int_{Y_1}^{y} \mu_s(x)dx\right)\right) \\ &= exp\left(-\delta_i \Lambda_{sB}(y)\right) \end{aligned} \tag{8.6}$$

where $\Lambda_{sB}(y)=\int_0^{Y_1} \mu_s(x)dx+R\int_{Y_1}^{y} \mu_s(x)dx$ is the integrated force of standard mortality from age 0 to age y with Barker effects. In this variant of the model, the *cumulated* Barker effects reflected on the integrated force of mortality to any age $y \geq Y_1$ are larger when R increases and when the critical age Y_1 decreases.

Assume now that the mortality regime undergoes a secular change with onset at $t = 0$ and that each birth cohort is exposed to a mortality level corresponding to the year when they were born. Thus, members of a birth cohort born t years after the onset of the secular decline experience throughout their lives mortality rates from the life table for year t. To define each birth cohort's life table we assume there is a standard mortality pattern characterized by mortality rates $\{\mu_s(y)\}$ and that the force of mortality for year t ('background' mortality) is $\mu(y,t) = k(t) * \mu_s(y)$,where $k(t)$

[2] As we show later, the simpler functional form adopted here represents lower bounds of Barker effects in the sense that time trends in both adult mortality levels and age patterns are least affected by their presence.

[3] In a recent paper Vaupel and Missov (2014) proposes an equivalent age dependent effect of standard frailty but with no association to Barker's conjecture. Coincidentally, we are using the same symbol, R, to express extra mortality in the special case when $R(\delta,y) = R(y) = R$ is constant.

is a monotonically decreasing function of time (linear or exponential).[4] The force of mortality at age y for a member of a cohort born in year t is given by

$$\mu_i(y,t) = \begin{cases} \delta_i k(t)\mu_s(y), & \forall y \leq Y_1 \\ R\delta_i k(t)\mu_s(y), & \forall y > Y_1,\ R \geq 1. \end{cases}$$

The expression for the average mortality at any age $y \geq Y_1$ in a cohort born t years after the onset of the mortality decline is:

$$\bar{\mu}(y,t) = RE_{yt}(\delta)k(t)\mu_s(y)$$

and

$$E_{yt}(\delta) = \frac{\int_0^\infty \delta f(\delta)\ exp(-k(t)\delta\Lambda_{sB}(y))d\delta}{\int_0^\infty f(\delta)\ exp(-k(t)\delta\Lambda_{sB}(y))\ d\delta}$$

where $f(\delta)$ is the density of Barker's frailty. The time dependency of Barker effects—$R * E_{yt}(\delta)$—is a result of changes in mortality that increase both survival probabilities and the conditional mean of Barker frailty.[5]

How does the average force of mortality at age y change over time? Note that when there is neither standard nor Barker frailty the change over time of average mortality rates depends only on changes in the function that controls the secular mortality decline or background mortality. In contrast, when there is Barker frailty the derivative of the force of mortality at age $y > Y_1$ at time t depends on changes in the expected value of a time dependent function:

$$\frac{\partial \ln(\bar{\mu}(y,t))}{\partial t} = \frac{\partial \ln(k(t))}{\partial t} + \frac{\partial \ln(E_{yt}(\delta))}{\partial t} \tag{8.7}$$

with $\frac{\partial \ln(k(t))}{\partial t} < 0$. Since mortality is declining the time derivative of $E_{yt}(\delta)$ is positive because as time passes the fraction of a cohort that survives to age Y_1 will have higher Barker frailty (values of δ). This occurs by virtue of the mortality decline itself that allows more children and young adults with higher values of Barker frailty to survive to adult ages (see below). Thus, according to (8.7) it is possible that changes in $\bar{\mu}(x,t)$ over time can be negative, 0, or positive for some values of y and t. At the very least, though, the presence of Barker frailty will offer resistance to the rate of mortality improvements. We explore this below.

[4]This simplified functional form for mortality decline avoids cumbersome algebra but leads to no loss of precision or generality.

[5]Below we explore the case when time dependency of Barker effects is linked not to mortality decline but to changes in the distribution $f(\delta)$.

8.3.2.3 The Nature of Changes in $E_{yt}(\delta)$ and $\bar{\mu}(y,t)$

(a) *Time dependent changes at ages* $y > Y_1$

When mortality declines rapidly, the derivative of $k(t)$ will be large and negative. However, because a faster mortality decline also increases more rapidly the fraction of individuals with higher levels of Barker frailty who survive to adult age Y_1, the change in $E_{yt}(\delta)$ will be larger and, consequently, the absolute value of the derivative of the second term in (8.7) will also be larger. [6]

One can show that

$$\frac{\partial \ln(E_{yt}(\delta))}{\partial t} = -\frac{\partial \ln(k(t))}{\partial t}[\bar{\Lambda}(y,t)(CV_{yt}(\delta))^2] \tag{8.8}$$

where $\bar{\Lambda}(y,t) = \Lambda_{sB}(y)k(t)E_{yt}(\delta)$ is the average integrated hazard at age y for the cohort born at time t and $CV_{yt}(\delta)$ is the coefficient of variation of the conditional density δ at age y and time t. Thus, larger increases in average Barker frailty take place when the distribution of δ has higher variance and at higher levels of mortality, that is, in the earlier stages of the secular mortality decline. From (8.8) and (8.7) we get

$$\frac{\partial \ln(\bar{\mu}(y,t))}{\partial t} = \frac{\partial \ln(k(t))}{\partial t}[1 - \bar{\Lambda}(y,t)(CV_{yt}(\delta))^2]. \tag{8.9}$$

Since $\bar{\Lambda}(y,t)(CV_{yt}(\delta))^2$ can potentially attain values higher than 1, adult mortality decline could slow down, stop altogether or even reverse. Expression (8.9) is general and also applies to standard frailty. The main difference is that in the case we study here the value of $\bar{\Lambda}(y,t)$ will always be larger for ages $y > Y_1$. Since the relative magnitude of $(CV_{yt}(\delta))^2$ depends in both cases on the original and conditional, age-specific distributions of δ, the differences between effects of standard and Barker frailty cannot be determined *a priori*. If excess risks are large or Y_1 is low, the influence of the integrated hazard will dominate and the effects of Barker frailty on the average rate of mortality decline will be substantially larger than under standard frailty alone. Finally, since the variance of the conditional distribution of frailty must converge to 0 at very old ages, the rate of mortality decline will converge to the rate of background mortality decline.

---

[6]Expression (8.7) also holds with standard frailty. The difference between it and the standard frailty case is in the quantities that come into play: in the case of Barker frailty the value of $\partial \ln(E_y(\delta,t))/\partial t$ depends on R (not just on δ) via the dependence of the integrated survival function on R (see Eq. (8.6)).

(b) *Age dependent changes at ages $y > Y_1$*

We now explore the influence of Barker effects on the age-related slope of average mortality rates. The first age derivative of $E_{yt}(\delta)$ is given by[7]:

$$\frac{\partial \ln(E_{yt}(\delta))}{\partial y} = -\bar{\mu}(y,t)(CV_{yt}(\delta))^2 = -RE_{yt}(\delta)k(t)\mu_s(y)(CV_{yt}(\delta))^2.$$

As expected, this derivative will always be negative, that is, the mean level of Barker frailty will decrease with age and the rate of decrease will be faster in this regime than in one with standard frailty at all ages where $R > 1$. Note that R also influences the magnitude of $CV_{yt}(\delta)$. The slope of the mean mortality rate at age y and time t is:

$$\frac{\partial \ln(\bar{\mu}(y,t))}{\partial y} = \frac{\partial \ln(\mu_s(y))}{\partial y} - \bar{\mu}(y,t)(CV_{yt}(\delta))^2. \tag{8.10}$$

Thus, as a result of frailty the slope of the average mortality pattern will differ from the slope of the background mortality pattern and will do all the more so at ages where $R > 1$. By the same token, since $[CV_{yt}(\delta)]$ decreases over time and by age, departures from the slope of background mortality will be smaller at older ages and at more advanced stages of the secular mortality decline.[8]

8.3.2.4 Implications of Barker Frailty

We now summarize the most important implications of the expressions derived before. Survivors to age Y_1 can be thought of as a newly 'born' cohort exposed to a mortality regime at ages $y \geq Y_1$ with standard frailty dependent on random frailty equal to $R\delta$ acquired 'at birth', e.g. when reaching the Y_1th birthday, with density

[7]This expression is also derived by Vaupel and Missov (2014) in the case of constant mortality.

[8]Note that, by construction, $\frac{\partial \ln(\mu_s(y))}{\partial y} = \beta_s(y)$ is invariant over time. The implication of this expression seems to have gone unnoticed in the literature (but see Vaupel and Missov (2014) for an analogous expression and recent discussion). Even in the absence of Barker effects and with an age-invariant $\beta_s(y)$ at adult ages (as in a Gompertz baseline adult mortality pattern), the age-derivative of the average mortality pattern cannot be constant (across ages or across time when there is a mortality decline). The regime of frailty assumed here will always induce an age dependent slope smaller than the standard slope. This has important consequences for the study of old age mortality in that the standard interpretation of an empirical slope estimated after fitting, for example, a Gompertz function to a cohort' s adult mortality rates is probably always incorrect. As suggested by (8.10), such estimate contains an age and time dependent downward bias. To avoid this bias one needs to estimate a Gompertz model controlling *both* for age and for the value of the (age and time varying) negative term in the expression. To our knowledge this has never been done in empirical studies. Elsewhere, we show that Barker effects and mortality decline *will always induce a negative correlation between the levels of child mortality experienced by a cohort and the cohort's adult mortality slope* (Palloni and Beltrán-Sánchez 2015).

$f_{Y_1}(\delta)$, e.g. the conditional density of δ among survivors to age Y_1. At ages older than Y_1 the cohort experiences mortality with standard frailty and extra mortality $R\delta$ and all the algebra of standard frailty applies. Selective survival to age Y_1, as well as mortality decline, operate as factors that reshape the distribution at "birth" (age Y_1) of δ. This interpretation isolates the mechanisms through which Barker effects are manifested. First, to the extent that the secular mortality decline increases the probability of survival to age Y_1, a higher fraction of the initial birth cohort will be exposed to expression of Barker frailty. Second, the force of mortality at ages above Y_1 is shifted upwards directly (through R) and indirectly, via the increased expected values of δ that result from mortality improvements before age Y_1. Third, the conditional survival and expected value of frailty after age Y_1 decreases more rapidly than under standard frailty.[9]

As in the case of standard frailty, the slope of average mortality rates at older ages will deviate away from the slope of the standard mortality pattern and will do so more at ages closer to Y_1. From (8.10) one can show that deviations from the slope of background mortality set in and then vanish earlier when the mortality decline is faster. An intriguing aspect of this dynamics is that when R is large, members of birth cohorts who attain ages $y > Y_1$ will be more severely selected out than when R is lower and the selection pressure will be even harsher among those with high values of δ. Standard frailty arguments imply that the opportunity for expressing Barker frailty shrinks rapidly with age since those who are more likely to have a large impact on mortality rates (higher values of δ) will be weeded out sooner after age Y_1. Thus, the resistance that Barker frailty opposes to adult mortality decline will only apply to some ages y in the neighborhood of Y_1 and its durability will be short-lived. A faster mortality decline decreases the windows of time and age within which Barker frailty can visibly slow-down mortality decline. This is an example of negative feedback whereby stronger Barker effects generate mortality conditions that undermine their continued operation and result in more transient and evanescent adult mortality manifestations. The implied non-linear dynamics of this negative feedback remains to be explored. These are surely consequential for empirical investigation since they will define conditions under which the presence of Barker frailty may not be well identified.

8.3.2.5 Special Case I: Gamma Distributed Barker Frailty

Suppose that $f(\delta)$ is *Gamma*(r, λ) with mean r/λ and variance r/λ^2. The expected value $E_{yt}(\delta)$ is

$$E_{yt}(\delta) = \frac{r}{\lambda + k(t)\Lambda_{sB}(y)}.$$

[9] An alternative way of interpreting Barker effects defined above is that they are tantamount to a shift of the standard mortality rates at older ages ($y > Y_1$), e.g. from $\mu_s(y)$ to $R\mu_s(y)$.

The time derivative of the expected value $E_y(\delta, t)$ is always positive

$$\frac{\partial \ln(E_{yt}(\delta))}{\partial t} = -\frac{\partial \ln(k(t))}{\partial t}\frac{k(t)\Lambda_{sB}(y)}{\lambda + k(t)\Lambda_{sB}(y)} \tag{8.11}$$

and the age derivative is

$$\frac{\partial \ln(E_{yt}(\delta))}{\partial y} = -\frac{k(t)\mu_s(y)}{\lambda + k(t)\Lambda_{sB}(y)}. \tag{8.12}$$

Furthermore, the rate of change over time of average mortality at age y and time t is

$$\frac{\partial \ln(\bar{\mu}(y,t))}{\partial t} = \frac{\partial \ln(k(t))}{\partial t}[1 - k(t)\Lambda_{sB}(y)E_{yt}^{exp}(\delta)] \tag{8.13}$$

where $E_{yt}^{exp}(\delta)$ is the conditional expectation of δ under an exponential density, e.g., $r = 1$. The minimum value of expression (8.13) is close to 0, that is, background mortality decline will not be reversed unless the variance of $f(\delta)$ is very large. Note that $\Lambda_{sB}(y)E_{yt}^{exp}(\delta)k(t)$ is the average integrated hazard up to age y in a mortality regime with Barker frailty distributed as $Gamma(1, \lambda)$.

Barker and standard frailty with secular mortality decline generate departures from the standard slope as average mortality will increase more slowly with age. The rate of change of mortality rates is given by:

$$\frac{\partial \ln(\bar{\mu}(y,t))}{\partial y} = \beta_s(y) - k(t)\mu_s(y)E_{yt}^{exp}(\delta) \tag{8.14}$$

where $\beta_s(y)$ is the slope of the mortality curve at age y in the baseline mortality pattern.

When frailty is gamma distributed and for ages over Y_1, $E_{yt}^{exp}(\delta)$ is larger under standard frailty and, therefore, the increased attenuation of the mortality slope will be less with Barker frailty than with standard frailty: excess adult mortality brought into the mix by those who express Barker effects will offset the downward bias imparted by standard frailty on the slope of average mortality. In the limit, when $y \rightarrow \infty$ and $E_{yt}^{exp} \rightarrow 0$, the slope of baseline mortality is restored under both standard and Barker frailty. One can show that when $f(\delta)$ is $Gamma(r, \lambda)$ the second age-derivative of the average force of mortality at age y and time t can be positive or negative. This implies that the slope of average mortality rates attains *minima* and *maxima* above age Y_1. But since the slope of the average age pattern of mortality at older ages must converge to the slope of the standard age pattern of mortality, the magnitude of the impact of Barker effects on the slope of average mortality will oscillate, weaken, and then vanish altogether at very old ages.

8.3.2.6 Special Case II: Discrete Barker Frailty

Suppose there are two groups, one that expresses Barker frailty, (*B*), with an excess mortality risk at ages over Y_1 equal to λ_B ($\lambda_B > 1$) and the other does not (*NB*), e.g. $\lambda_{NB} = 1$. Suppose also that the fraction of births who express Barker frailty is a constant g. The average mortality rate at age y and time t is given by

$$\bar{\mu}(y,t) = \mu_s(y)k(t)(P_B(y,t)(\lambda_B - 1) + 1) \tag{8.15}$$

where λ_B is applied to all ages $y \geq Y_1$, and $P_B(y,t)$ is the fraction of the population who expresses Barker at age y and time t. The expressions for $P_B(y,t)$ and its derivatives with respect to t are

$$P_B(y,t) = \frac{1}{1 + h\, exp(-k(t)(1-\lambda_B)\phi_s(y))} \tag{8.16}$$

$$\frac{\partial P_B(y,t)}{\partial t} = -\frac{\partial \ln(k(t))}{\partial t}[h\, k(t)\phi_s(y)(\lambda_B - 1)P_B(y,t)(1 - P_B(y,t))] \tag{8.17}$$

$$\frac{\partial \ln(P_B(y,t))}{\partial t} = \frac{\partial \ln(k(t))}{\partial t}[k(t)\phi_s(y)(1-\lambda_B)(1-P_B(y,t))]$$

where $\phi_s(y) = \int_{Y_1}^{y} \mu_s(x)dx$ and $h = (1-g)/g$.

The mean value of frailty at age y and time t is:

$$E_{yt}(\lambda_B) = P_B(y,t)(\lambda_B - 1).$$

Taking logs in (8.15) and then derivatives with respect to time we get

$$\frac{\partial \ln(\bar{\mu}(y,t))}{\partial t} = \frac{\partial \ln(k(t))}{\partial t} - \left\{\frac{(\lambda_B - 1)(\partial P_B(y,t)/\partial t)}{P_B(\lambda_B - 1) + 1}\right\}. \tag{8.18}$$

The term in curly brackets is always positive and expression (8.18) will always be less that the average rate of background mortality decline. Furthermore, the quantity is dependent on changes in $P_B(y,t)$, and these changes proceed faster at the outset of mortality decline and are gradually spent at advanced stages of the secular decline.

Discrete Barker frailty will also reduce the slope of average adult mortality. The age derivative of the average force of mortality is exactly analogous to (8.18) but with the roles of $k(t)$ and $\mu_s(y)$ interchanged:

$$\frac{\partial \ln(\bar{\mu}(y,t))}{\partial y} = \frac{\partial \ln(\mu_s(y))}{\partial y} + \left\{\frac{(\lambda_B - 1)(\partial P_B(y,t)/\partial y)}{P_B(\lambda_B - 1) + 1}\right\}. \tag{8.19}$$

Since $(\partial P_B(y,t)/\partial y)$ is always negative, the age-specific slope of average adult mortality will be reduced (relative to the standard) due to Barker effects. The differences will be higher at ages closer to Y_1 and at more advanced stages of the secular decline, when $P_B(y,t)$ attains a maximum. In the limit, as $t \to \infty$ or $y \to \infty$, the slopes of background and average mortality will be identical.

A model with discrete frailty is appealing since it captures well the idea, implicit in Barker theory, that individuals vulnerable to the impact of adverse early conditions on adult health and mortality are those who experience organ damage *above a given threshold*. The disadvantage of the discrete model is that it requires specification of the threshold, a quantity that is, for all purposes, difficult to either theorize about or empirically estimate.

8.4 Generalized Distributions

An important shortcoming of the continuous and discrete models is that they assume that the allocation of Barker frailty at birth is carried out according to a fixed rule. In fact, in the continuous case the distribution of δ is time-invariant and all births are allocated excess risks (in proportion to δ) by the same probability distribution. In the discrete case the parameter g is the same for all birth cohorts. These are simplifications that fail to translate with high fidelity the nature of Barker frailty defined at the outset.

Two extensions are possible. In the continuous case we can introduce time dependencies in $f(\delta)$ by letting its mean or variance increase hand-in-hand with mortality decline. In the discrete case we can let g be an increasing function of time. Furthermore, one could also include dependencies between the rate of change of the distribution of frailty and/or the rate of mortality decline. These extensions are better suited to capture scenarios where the size and/or heterogeneity of birth cohorts at risk of expressing Barker frailty increases as a result of new epidemiological regimes with better maternal health, fetal, perinatal and early child survival but not triggered by improved nutritional status.

8.4.1 Gamma Barker Frailty with Time Dependence

The average mortality rate at age y and time t is

$$\bar{\mu}(y,t) = RE_{yt}(\delta(t))\, k(t)\mu_s(y) \tag{8.20}$$

where $E_{yt}(\delta(t))$ refers to the expectation of a random variable $\delta(t)$ that follows a $Gamma(r(t), \lambda)$, $Gamma(r, \lambda(t))$, or $Gamma(r(t), \lambda(t))$.

8.4.1.1 Time Dependence of the "Shape" Parameter

When the density of random Barker frailty, $f(\delta(t))$, is $Gamma(r(t), \lambda)$ and $r(t)$ is a increasing function of time the dynamic of mortality decline changes by a factor equal to the rate of increase of the mean. In fact,

$$\frac{\partial \ln(\bar{\mu}(y,t))}{\partial t} = \frac{\partial \ln(k(t))}{\partial t}\left[1 - k(t)\Lambda_{sB}(y)E_{yt}^{exp}(\delta(t))\right] + \frac{\partial \ln(r(t))}{\partial t} \quad (8.21)$$

The drag force on the background rate of mortality is boosted by the rate of change of *r(t).* Admittedly, $r(t)$ cannot grow indefinitely and should eventually attain a maximum midway through the process. If so, the additional tug exerted by a growing mean frailty will disappear many years after $r(t)$ attains a maximum, when cohorts born under an increasing regime of $r(t)$ complete their passage to ages older than Y_1.

The age specific slope of average mortality at age y and time t is:

$$\frac{\partial \ln(\bar{\mu}(y,t))}{\partial y} = \beta_s(y) - \mu_s(y)k(t)E_{yt}^{exp}(\delta(t)). \quad (8.22)$$

Departures from the background or baseline mortality age-specific slopes will be more marked at ages closer to Y_1 (values of $E_{yt}^{exp}(\delta(t))$ for a given t are larger when y remains close to the threshold age Y_1) and among more recent birth cohorts (values of $E_{yt}^{exp}(\delta(t))$ for any age y are larger for higher values of t).

8.4.1.2 Time Dependence of the "Rate" Parameter of Frailty

Assume now that the parameter λ is time dependent so that its value decreases over time (and the variance of $f(\delta)$ increases). The rate of decline of average mortality is:

$$\frac{\partial \ln(\bar{\mu}(y,t))}{\partial t} = \frac{\partial \ln(k(t))}{\partial t}\left[1 - k(t)\Lambda_{sB}(y)E_{yt}^{exp}(\delta(t))\right] - \frac{\partial \lambda(t)}{\partial t}E_{yt}^{exp}(\delta(t)). \quad (8.23)$$

The implication of this expression is that a growing variance of at-birth distribution of frailty leads to rates of decline of adult average mortality that decrease by a larger amount when frailty is fixed: the extra drag force on the rate of decline is directly proportional to the age-time specific mean of frailty in an exponential distribution with rate $1/(\lambda(t)+k(t)\Lambda_{sB}(y))$. Comparing the last two expressions suggests that the additional downward pressure on the rate of mortality decline imposed by changes in the mean of the frailty distribution is age invariant. By contrast the additional downward pressure associated with changes in the variance of the distribution of frailty is age dependent.

The age specific slope of average mortality at age y and time t is:

$$\frac{\partial \ln(\bar{\mu}(y,t))}{\partial y} = \beta_s(y) - \mu_s(y)k(t)E_{yt}^{exp}(\delta(t)) \tag{8.24}$$

and departures from the background slope are now dependent on the changing conditional frailty distribution.

8.4.2 Discrete Barker Frailty with Changing Size of Vulnerable Population

Suppose there are two groups, one that expresses Barker frailty, (B), with an excess mortality risk λ_B ($\lambda_B > 1$) and the other does not (NB), e.g. $\lambda_{NB} = 1$, and that the fraction of births who express it is $g(t)$, an increasing function of t. The average morality rate at age y and time t is given by (8.15) above but $P_B(y,t)$ is now a function of the time dependent fraction of individuals who are vulnerable to Barker effects, ($h(t) = (1 - g(t))/g(t)$). The expressions for $P_B(y,t)$ and its derivative with respect to t are

$$P_B(y,t) = \frac{1}{1 + h(t)\ exp(-k(t)(1-\lambda_B)\phi_s(y))} \tag{8.25}$$

and

$$\frac{\partial P_B(y,t)}{\partial t} = -\frac{\partial \ln(k(t))}{\partial t}[h(t)k(t)\phi_s(y)(\lambda_B - 1)P_B(y,t)(1 - P_B(y,t)] \\ -(\lambda_B - 1)\frac{\partial \ln(h(t))}{\partial t}. \tag{8.26}$$

Since the rate of change of $h(t)$ is negative, expression (8.26) will be larger than expression (8.17). This shows that the presence of an increasing fraction of births that could express Barker frailty ($g(t)$) acts as an additional brake on the rate of background mortality decline at older ages. The slope of average adult mortality is identical to expression (8.18) except that $P_B(y,t)$ and its derivative with respect to y are now dependent on $h(t)$: because an increase of the fraction of births who are vulnerable to Barker frailty leads to a positive change in $P_B(y,t)$, there will be a growing tug and deceleration of the rate of adult mortality decline.

8.5 Simulation of Mortality Regimes

To provide a sense of the magnitude of Barker effects and some insight into the relations described above, we simulate a series of cohorts undergoing a secular mortality decline using the simple formulation shown in Sect. 8.3.2.2, that is, $\bar{\mu}(y) = R\, E_{yt}(\delta)\, k(t)\, \mu_s(y)$.

8.5.1 Simulation Steps

We assume the critical age to be 50 ($Y_1 = 50$) after which manifestation of early damage begins to take place in the form of excess mortality risk ($R > 1$) and a Gamma distributed barker frailty (see Sect. 8.3.2.5). We simulate 50 cohorts starting with a baseline life table with a life expectancy at birth of 40 years at time 0 (from Model Life Tables (Coale and Demeny 1983)). We then define a yearly mortality decline, $k(t)$, so that the sequence of age-specific mortality rates after 100 years corresponds to a life table with life expectancy at birth of 80 years (i.e., cohort specific life expectancy at birth doubles over a century). The simulation proceeds in three steps as follows:

1. For each of the 50 cohorts we create 300 copies and each of these copies has a random frailty value, $(1 + \delta)$, where δ is drawn from a gamma distribution $Gamma(r, \lambda)$ with $r = 1$ and $1/\lambda$ (the standard deviation of the distribution) varying from 1.5 to 4 in 0.5 increments. That is, we create 8 different variants of frailty regimes and each of these is represented by 300 copies.[10]
2. We define six different regimes of Barker effects with excess mortality R ranging from 1.5 to 4 in increments of 0.5. Excess mortality applies to all ages $y \geq Y_1 = 50$. The combination of different gamma distributions for Barker frailty and levels for Barker effects yields 8×6 = 48 different sets of 50 cohorts each and each of these contains 300 copies.
3. Each of the $i = 1, \ldots, 300$ copies of cohorts contained in the 48 variants is survived forward with mortality rates $\mu_i(y)$ and survival probabilities $S_i(y)$ that reflect the regime of Barker frailty and Barker effects defined for that copy. At each age $y \leq 100$ we compute the conditional distribution of δ_i, its mean and variance, and the mean mortality rate ($\bar{\mu}(y)$) across all 300 copies of each cohort.

[10]By design, the random terms for frailty, δ, $\delta = \iota + 1$ where $\iota \sim Gamma(1, \lambda)$. Thus, the frailty term we use has a minimum value of 1 and its mean is equal to 1 plus the conditional mean of the gamma random term.

8.5.2 Illustration 1: Secular Mortality Decline, R = 1.5, and Time-Invariant Gamma Frailty with Mean and Standard Deviation 1.5

We first illustrate a scenario with Barker effect of $R = 1.5$ with a time-invariant Gamma frailty distribution with mean and standard deviation 1.5 that applies to all 50 cohorts.[11] We show two sets of results from the simulation, changes over time (Fig. 8.1) and changes over age (Fig. 8.2) of the (log of) average mortality rate, $(\bar{\mu}(y))$, and the (log of) expected value of δ, $(E_{yt}(\delta))$. For simplicity, we show two ages when looking at changes over time (ages 55 and 80) and two cohorts (cohorts 1 or 5 and 50) when examining changes over age. We show average mortality rates in the regime with Barker frailty (solid line), with standard frailty (dashed line), and background mortality in the absence of both Barker and standard frailty (dots).

Figure 8.1 contains two results. First, panel (a) shows that the average force of mortality at ages 55 and 80 is always higher with Barker frailty. While virtually all the excess mortality at age 55 in a Barker frailty regime is explained by the magnitude of Barker effect ($R = 1.5$), the excess at age 80 is a result of Barker effects as well as of the tightening of selective pressure due to Barker frailty applied to ages over 50. Since these two forces operate in opposite direction, the differences between the average rates at age 80 in a regime with Barker frailty and a regime with standard frailty or background mortality are slightly lower than those at age 55. Furthermore, panel (a) also shows that, as expected from expression (8.13), the decline of the average force of mortality $(\bar{\mu}(x, t))$ is flatter for both ages in either the Barker or standard frailty regimes but more so in the latter than in the former.

The second result in panel (b) is that there are virtually no differences between a standard and Barker frailty regimes regarding the behavior of mean frailty at age 55 because before age 50 both regimes are identical. At age 80, however, Barker frailty has had some room to operate and the levels and trajectories are different: the absolute values of mean frailty are always higher in a standard frailty regime since there are no penalties associated with it past age 50 as there are in a regime with Barker frailty.

Panel (a) of Fig. 8.2 displays effects of frailty regimes on the age slope of adult mortality. As expected from expression (8.14), these panels, corresponding to mortality experiences of the oldest (1) and youngest (50) cohorts, show that the slope of average mortality is flatter in both frailty regimes.

[11]The simulated scenario is very easy to implement but it has an odd implication. Note that the mortality experience of the birth cohort born 50 years after the onset of secular mortality decline experiences a baseline life table with life expectancy at birth of roughly 60 years. Thus, the period life table corresponding to the year of their birth has a life expectancy at birth lower than 60 years. The sequence of baseline (period) life tables implied by the simulated birth cohorts includes a range of life expectancies at birth from 40 to less than 60 years. This range is only a small fraction of the observed improvements in period life expectancy of low to middle income countries after 1950.

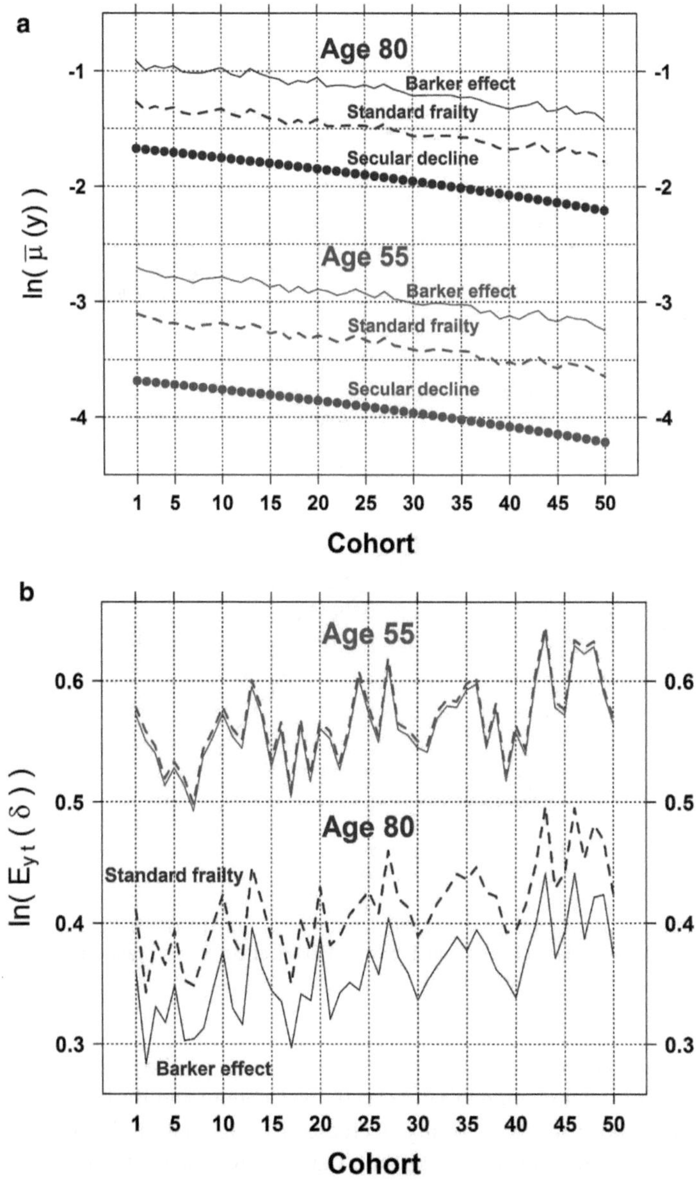

Fig. 8.1 Plots of (**a**) average mortality rates in log-scale, $\ln(\bar{\mu}(y,t))$, and (**b**) average δ in log-scale, $\ln(E_{yt}(\delta))$, over time for ages 55 and 80 with Barker effect $R = 1.5$ and time-invariant frailty $Gamma(r = 1, 1/\lambda = 1.5)$, respectively

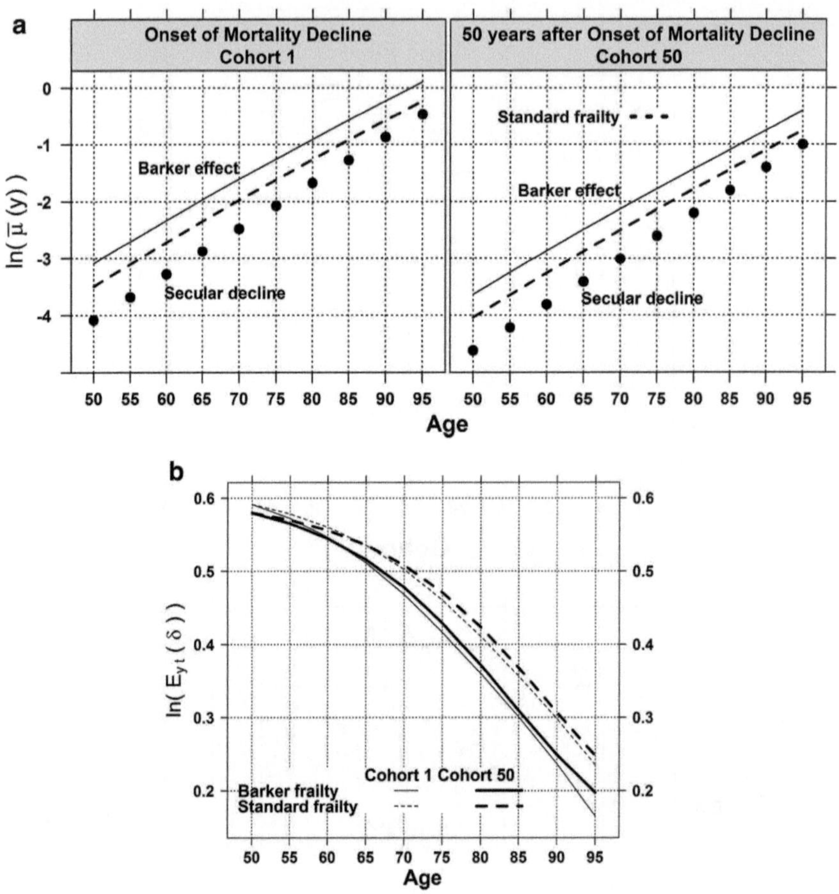

Fig. 8.2 Plots of (**a**) average mortality rates in log-scale, $\ln(\bar{\mu}(y,t))$, and (**b**) average δ in log-scale, $\ln(E_{yt}(\delta))$, by age for cohorts 1 and 50 with Barker effect $R = 1.5$ and time-invariant frailty $Gamma(r = 1, 1/\lambda = 1.5)$, respectively

Finally, Panel (b) of the same figure displays the age profile of the average level of frailty in the oldest and youngest cohorts in the standard and Barker frailty regimes. As expected, mean levels of frailty decrease with age but at slightly slower pace when mortality is lower (youngest cohort) and in a regime of standard frailty.

8.5.3 Illustration 2: Secular Mortality Decline, R = 1.5, and Time-Varying Gamma Frailty

This case corresponds to the description in Sect. 8.4.1. We simulate cohorts assuming a fixed Barker effect equivalent to $R = 1.5$ with a *Gamma* distribution with fixed scale parameter, ($r = 1$), and increasing standard deviation ranging from 1.5 to about 4 over 50 cohorts. As before, we display results showing effects on the rate of average mortality decline and on the expected value of δ (Fig. 8.3), and on the age dependency of mortality rates (Fig. 8.4). In order to highlight the significance of a changing frailty dispersion, these figures also show results from a time-invariant frailty regime with a *Gamma* distribution with scale parameter $r = 1$ and standard deviation 1.5.

Figure 8.3 displays logs of average mortality (panel (a)) and logs of average frailty (panel (b)) across cohorts at ages 55 and 80 when the parameter λ increases over time (the variance of the gamma distribution increases). Panel (a) shows that a time-varying distribution of Barker frailty exerts a powerful drag on background mortality decline as successive birth cohorts could experience increased mortality rates at ages older than 50. This implies that increased dispersion of the initial distribution of Barker frailty can derail the secular mortality decline. As expected from previous discussion, these changes are non-lineal. Panel (b) of Fig. 8.3 reveals that average frailty increases as background mortality declines and attains a maximum among cohorts in the later stages of the mortality decline. The magnitude of average frailty is higher for younger cohorts, e.g. when Barker effects had time to operate.

Figure 8.4 panel (a) shows the effects of a growing variance of frailty for the youngest and one of the oldest cohorts (cohorts 50 and 20, respectively) on the slopes of adult mortality. This figure confirms that mortality increases more slowly with age than was the case when the distribution of Barker frailty was time-invariant. Panel (b) of the same figure displays age profiles of the logs of average level of frailty with time-variant and time-invariant Barker frailty regimes. As expected, mean levels of frailty are higher when there is time-variant Barker frailty and they decrease with age everywhere but at slightly faster pace in time-variant Barker frailty regimes.

8.5.4 Illustration 3: The Impact of Y_1 with Secular Mortality Decline, $R = 4.0$, and Time-Invariant Gamma Frailty with Mean and Standard Deviation 1.5

As pointed out at the outset, the basic model developed here contains a massive simplification, namely, we assume that the threshold age Y_1 is constant. Although the assumption is useful because it leads to simpler analytic expressions and to more transparent implications, it can be removed without altering most of the conclusions described before. To illustrate the role of changing Y_1 we resort to a comparison of scenarios with two different values of Y_1. This comparative statics is suggestive

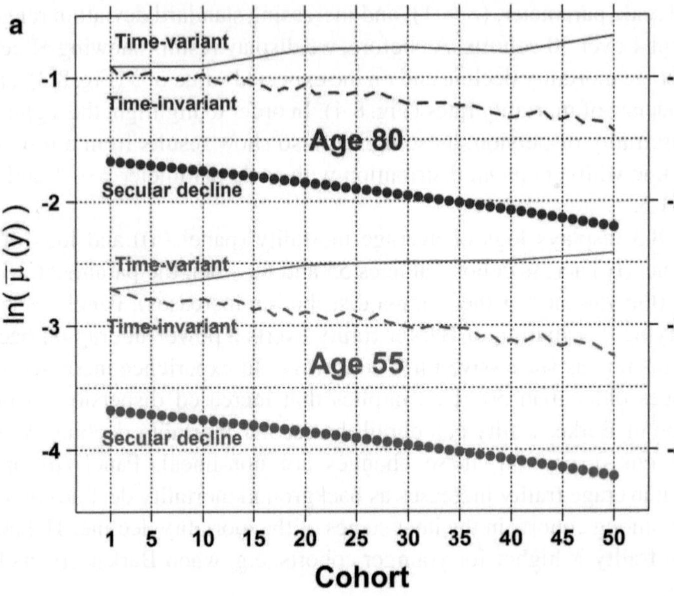

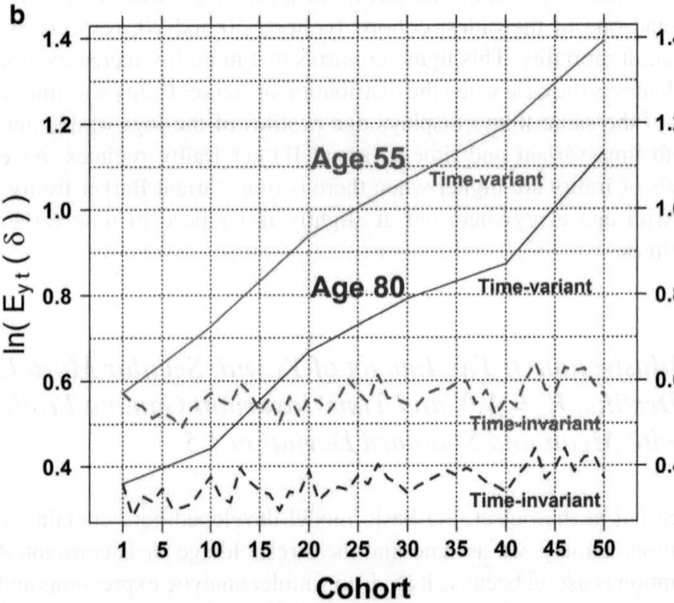

Fig. 8.3 Plots of (**a**) average mortality rates in log-scale, $\ln(\bar{\mu}(y,t))$, and (**b**) average δ in log-scale, $\ln(E_{yt}(\delta))$, over time for ages 55 and 80 with Barker effect $R = 1.5$ and time-variant frailty *Gamma*($r = 1, 1.5 \leq 1/\lambda \leq 4.0$) between cohorts 1 and 50 (see text for further details)

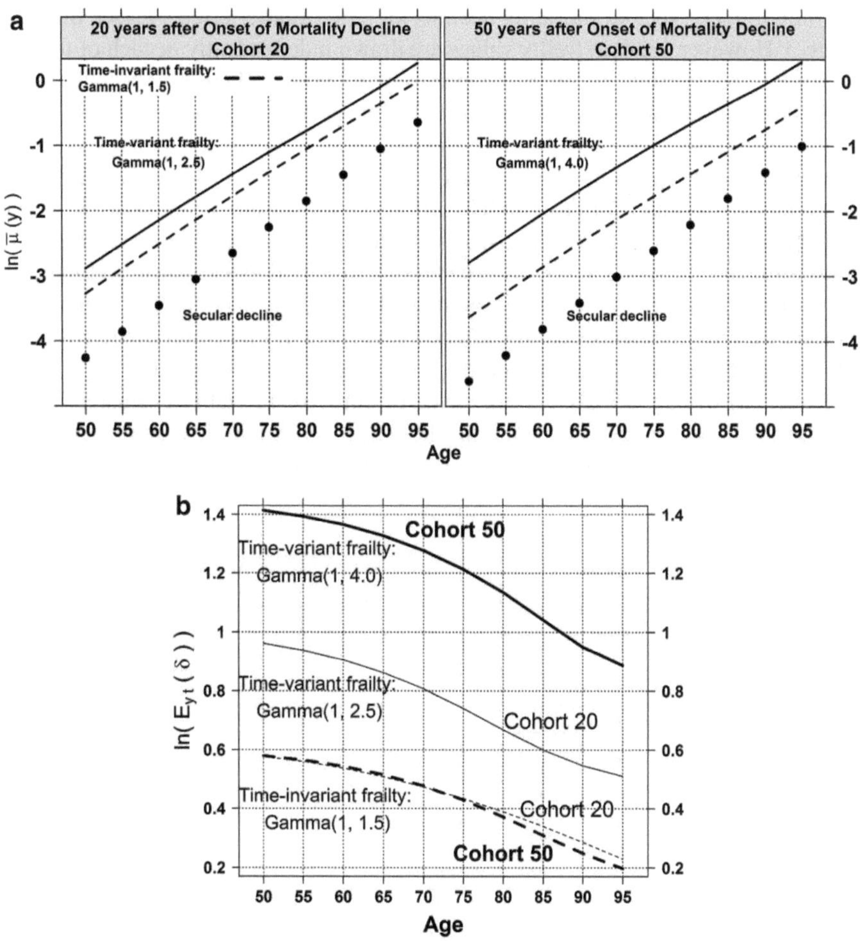

Fig. 8.4 Plots of (**a**) average mortality rates in log-scale, $\ln(\bar{\mu}(y,t))$, and (**b**) average δ in log-scale, $\ln(E_{yt}(\delta))$, by age for selected cohorts with Barker effect $R = 1.5$ and time-variant frailty $Gamma(r = 1, 1.5 \leq 1/\lambda \leq 4.0)$ between cohorts 20 and 50 (see text for further details)

and does not replace a more elaborate model where Y_{1i} is a random variable with a systematic and a random individual component for individuals $i = 1, \ldots, N$.[12]

In these simulations we use two different threshold or ages of onset for Barker effects, early ($Y_1 = 40$) and late ($Y_1 = 50$) and continue to use the same background

[12]The idea that Y_1 should be random is consistent with theories of fetal origins according to which there are several chronic illnesses that could play a role and each of them has different time of onset after which Barker effects begin to be felt. Another interpretation is that Barker effects exert an impact on the rate of senescence itself and the slope of the mortality curve begins to accelerate by a random amount after a fixed age Y_1.

mortality, $\mu_s(x)$, secular mortality decline, $k(t)$, and Barker effects R (see Sect. 8.5 above). However, random frailty values are drawn independently in each of the two scenarios. As a result, when R is small, say $R = 1.5$, there can be only small differences in average mortality rates and average δ at ages close to Y_1 between the two scenarios. This is so because changes in $\bar{\mu}(y,t)$ between the two scenarios are driven by changes in $E_{yt}(\delta)$ (see Eq. (8.7)) and these differ only marginally between ages 40 and 50. To amplify differences we use a value of $R = 4$ and we obtain illustrations that help visualize better the dynamics of Barker effects. In these simulations we use a time-invariant frailty distribution with mean and standard deviation 1.5 that applies to all 50 cohorts with both early and late Barker onset. As before, we present results showing effects on average mortality, on expected values of δ (Fig. 8.5), and on the age dependency of mortality rates (Fig. 8.6).

Panel (a) of Fig. 8.5 displays the average force of mortality evaluated at two ages (55 and 80) for scenarios with threshold ages 40 and 50. It shows that average mortality rate is always higher when Barker effects start at a later age. The mortality

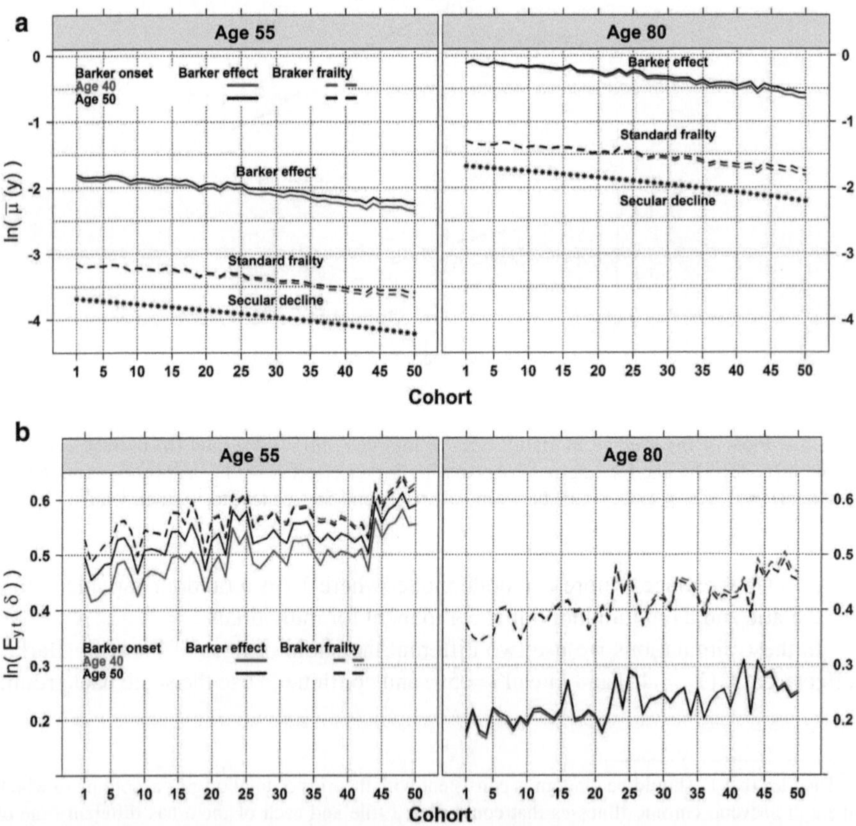

Fig. 8.5 Plots of (**a**) average mortality rates in log-scale, $\ln(\bar{\mu}(y,t))$, and (**b**) average δ in log-scale, $\ln(E_{yt}(\delta))$, over time for early ($Y_1 = 40$) and late ($Y_1 = 50$) Barker onset for ages 55 and 80 with Barker effect $R = 4$ and time-invariant frailty $Gamma(r = 1, 1/\lambda = 1.5)$, respectively

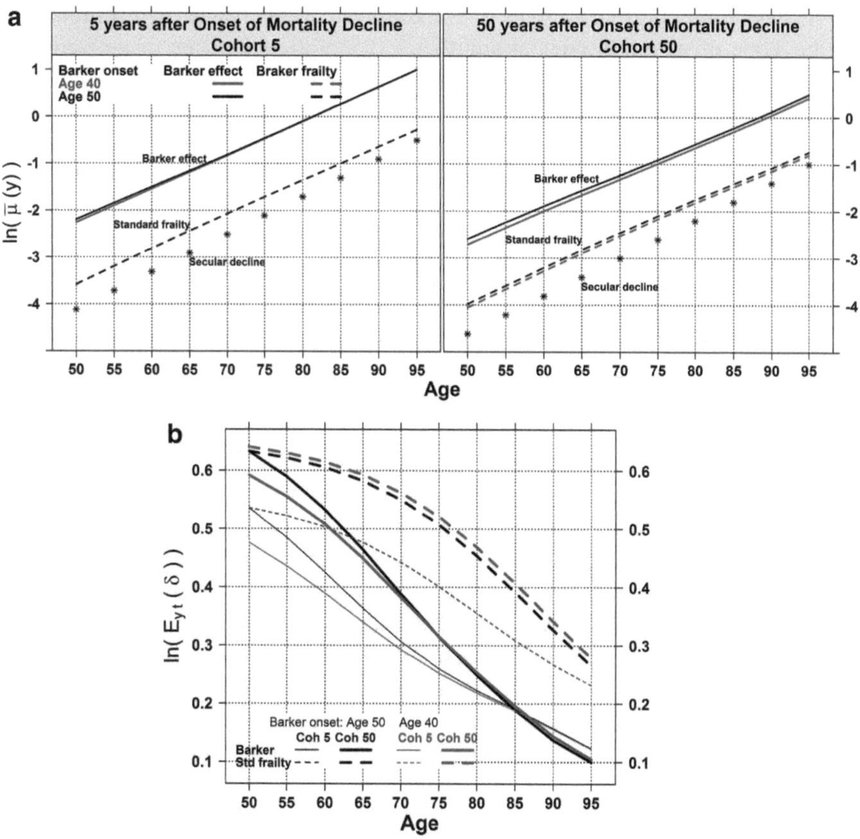

Fig. 8.6 Plots of (**a**) average mortality rates in log-scale, $\ln(\bar{\mu}(y, t))$, and (**b**) average δ in log-scale, $\ln(E_{yt}(\delta))$, by age for early ($Y_1 = 40$) and late ($Y_1 = 50$) Barker onset for cohorts 5 and 50 with Barker effect $R = 4$ and time-invariant frailty $Gamma(r = 1, 1/\lambda = 1.5)$, respectively

difference between scenarios increases over time with a larger gap at age 55 than at age 80. The result at age 55 is mainly driven by the magnitude of Barker effects ($R = 4$) that elevates overall mortality rates at ages closer to Y_1. As background mortality improves over time, a higher fraction of survivors is exposed to Barker frailty and Barker effects are stronger at age 55 when the onset is at 50 than at 40. At older ages, however, there is an additional force due to the tightening of selective pressure associated with Barker frailty which reduces the difference in mortality rates between the two scenarios.

As shown in panel (b) of the same Figure there are virtually no differences in the behavior of mean frailty at age 80 between early vs. late Barker age at onset. As Barker frailty has had some room to operate up through age 80, it leads to similar frailty values between scenarios. At age 55, however, the levels are different: the absolute values of mean frailty are always higher in late vs. early Barker onset. This is so because there is higher fraction of survivors exposed to Barker frailty at age 55 when the onset is at age 50 than at age 40.

The last figure, Fig. 8.6, permits to assess the effects of changes in threshold ages on the slope of mortality rates at adult ages (Panel a) and on the mean frailty by age (Panel b). The panels, plotting mortality experiences of cohorts 5 and 50, show that the slope of average adult mortality rates is somewhat sharper when the onset of Barker effects occurs earlier in life. Panel (b) of the same figure displays the age profile of the average level of frailty in the same cohorts in the standard and Barker frailty regimes. Mean levels of frailty always decrease with age but at a faster pace when the onset of Barker effects is later. The behavior of the two scenarios converges at older ages.

8.6 Concluding Remarks

We propose a formal treatment of the demographic consequences of Barker effects on adult mortality rates. We show, mathematically and via simulations, that in a mortality regime with declining mortality, Barker frailty will compound the decelerating force that naturally arises when only standard frailty prevails. This is so because as mortality declines, the selection pressure of standard frailty always weakens over time while Barker frailty imposes an additional force in the form of excess mortality among those who were exposed to adverse early life conditions and attain adult ages. The formulation reconciles standard frailty with Barker frailty as survivors to a certain age (e.g., Y_1) become a cohort that is 'newly born' at age Y_1 that will experience mortality rates with standard frailty and mortality multiplier $R\delta$ (rather than δ as in standard frailty) (Vaupel et al. 1979; Vaupel and Yashin 1987; Vaupel and Missov 2014; Aalen 1988; Steinsaltz and Wachter 2006). Furthermore, Barker frailty exhibits a distinct dynamic: while standard frailty always leads to deceleration of mortality rates at older ages, Barker frailty unleashes forces that work in the opposite direction and promote increases in the rate of aging at ages above Y_1. When background mortality declines, average adult mortality may go through stages in which mortality will decline more slowly than background mortality rates, remain steady, or even increase. We use standard frailty arguments and suggest, but did not demonstrate, that Barker frailty generates non-linear dynamics in the sense that the manifestation of excess adult mortality implied by Barker frailty undermines its continued operation in a mortality regime.[13]

The impact of Barker effects on adult life expectancies is more likely to be observed in countries that experienced mortality declines that were at least partially sustained by massive improvements in infant and child survival. Preliminary findings from Latin American countries, for example, suggest that foregone gains in life expectancy at age 60 associated with Barker effects may be as high as 20 % of the projected values over a period of 30–50 years (Palloni and Souza 2013). The

[13]The study of the precise dynamics of Barker effects necessitates additional investigation.

changing composition of cohorts by early exposures represents a powerful force that could drag down or halt short-run progress in life expectancy at older ages. The methods developed here facilitate the study of these type of effects with simple, parsimonious models. The task that remains is to translate relations embedded in the model(s) into predicted outcomes and to design empirical tests to falsify such predictions.

The models described here could be generalized in several directions. Among the most important is to attempt to establish a tight connection between variants of developmental origins theories and the models. Thus, the existence of variable threshold ages and their properties should be deduced from predictions derived from the theories. Similarly, the magnitude of excess mortality associated with Barker frailty should be specified in accordance to the types of chronic illnesses that are known or suspected to be influenced by adverse effects of early conditions. Furthermore, and as suggested by researchers working on senescence (Finch 2007), adverse early conditions may affect the rate of senescence itself with the implication that our models should impose random effects on the slope of adult mortality, not just on its level. Finally, there is a burgeoning literature (Kuzawa and Eisenberg 2014) showing that expression of poor early conditions may implicate germ cells in which case the risk of adult manifestation of early conditions is passed on from one generation to the next. The models proposed here completely ignore this aspect but there is no inherent reason why they could not be extended to incorporate such relations through application of generalized stable population models.

Appendix: Main Definitions

$$\Lambda_{sB}(y) = \int_0^{Y_2} \mu_s(x)dx + R\int_{Y_1}^{y} \mu(x)dx$$

$$\bar{\Lambda}(y,t) = \Lambda_{sB}(y)k(t)E_{yt}(\delta)$$

$$E_{yt}^{exp}(\delta) = \frac{1}{\lambda + k(t)\Lambda_{sB}(y)}$$

$$E_{yt}^{exp}(\delta(t)) = \frac{1}{\lambda(t) + k(t)\Lambda_{sB}(y)}$$

$$\left[CV_{yt}(\delta)\right] Gamma(r,\lambda) = \frac{\sqrt{r}}{r}$$

$$\left[CV_{yt}(\delta)\right]^2 Gamma(r,\lambda) = \frac{1}{r}$$

Table A.1 Summary of formal relations for the rate of change of average mortality rates at age $y > Y_1$

General form	$Gamma(r, \lambda)$	$Gamma(r, \lambda(t))$	$Gamma(r(t), \lambda)$
$\bar{\mu}(y,t) = RE_{yt}(\delta)k(t)\mu_s(y)$	$\bar{\mu}(y,t) = RE_{yt}(\delta)k(t)\mu_s(y)$	$\bar{\mu}(y,t) = RE_{yt}(\delta(t))k(t)\mu_s(y)$	$\bar{\mu}(y,t) = RE_{yt}(\delta(t))k(t)\mu_s(y)$
Rate of change with respect to time (t): $\frac{\partial \ln(\bar{\mu}(y,t))}{\partial t} =$			
$\frac{\partial \ln(k(t))}{\partial t}\left[1 - \bar{\Lambda}(y,t)(CV_{yt}(\delta))^2\right]$	$\frac{\partial \ln(k(t))}{\partial t}\left[1 - \frac{k(t)\Lambda_{sB}(y)}{\lambda + k(t)\Lambda_{sB}(y)}\right] =$	$\frac{\partial \ln(k(t))}{\partial t}\left[1 - \frac{k(t)\Lambda_{sB}(y)}{\lambda(t) + k(t)\Lambda_{sB}(y)}\right] - \frac{\partial\lambda(t)/\partial t}{\lambda(t) + k(t)\Lambda_{sB}(y)} =$	$\frac{\partial \ln(k(t))}{\partial t}\left[1 - \frac{k(t)\Lambda_{sB}(y)}{\lambda(t) + k(t)\Lambda_{sB}(y)}\right] + \frac{\partial \ln(r(t))}{\partial t} =$
	$\frac{\partial \ln(k(t))}{\partial t}\left[1 - k(t)\Lambda_{sB}(y)E_{yt}^{exp}(\delta)\right]$	$\frac{\partial \ln(k(t))}{\partial t}\left[1 - k(t)\Lambda_{sB}(y)E_{yt}^{exp}(\delta(t))\right] - \frac{\partial\lambda(t)}{\partial t}E_{yt}^{exp}(\delta(t))$	$\frac{\partial \ln(k(t))}{\partial t}\left[1 - k(t)\Lambda_{sB}(y)E_{yt}^{exp}(\delta(t))\right] + \frac{\partial \ln(r(t))}{\partial t}$
Rate of change with respect to age (y): $\frac{\partial \ln(\bar{\mu}(y,t))}{\partial y} =$			
$\frac{\ln(\mu_s(y))}{\partial y} - \bar{\mu}(y,t)(CV_{yt}(\delta))^2 =$	$\beta_s(y) - \frac{\mu_s(y)k(t)}{\lambda + k(t)\Lambda_{sB}(y)} =$	$\beta_s(y) - \frac{\mu_s(y)k(t)}{\lambda(t) + k(t)\Lambda_{sB}(y)} =$	$\beta_s(y) - \frac{\mu_s(y)k(t)}{\lambda + k(t)\Lambda_{sB}(y)} =$
$\beta_s(y) - \bar{\mu}(y,t)(CV_{yt}(\delta))^2$	$\beta_s(y) - \mu_s(y)k(t)E_{yt}^{exp}(\delta)$	$\beta_s(y) - \mu_s(y)k(t)E_{yt}^{exp}(\delta(t))$	$\beta_s(y) - \mu_s(y)k(t)E_{yt}^{exp}(\delta(t))$

References

Aalen, O. O. (1988). Heterogeneity in survival analysis. *Statistics in Medicine, 7*, 1121–1137.

Anson, J. (2002). Of entropies and inequalities: Summary measures of the age distribution of mortality. In G. Wunsch & M. Mouchart (Eds.), *Life tables: Data, method and models* (pp. 95–116). Dordrecht: Kluwer.

Barker, D. J. P. (1998). *Mothers, babies, and health in later life* (2nd ed.). Edinburgh/New York: Churchill Livingstone.

Barouki, R., Gluckman, P. D., Grandjean, P., Hanson, M., & Heindel, J. J. (2012). Developmental origins of non-communicable disease: implications for research and public health. *Environmental Health, 11*, 10–1186.

Beltrán-Sánchez, H., Crimmins, E. M., & Finch, C. E. (2012). Early cohort mortality predicts the rate of aging in the cohort: A historical analysis. *Journal of Developmental Origins of Health and Disease, 3*, 380–386.

Coale, A. J., & Demeny, P. (1983). *Regional model life tables and stable populations* (2nd ed.). Princeton: Princeton University Press.

Crimmins, E. M., & Finch, C. E. (2006). Infection, inflammation, height, and longevity. *Proceedings of the National Academy of Sciences of the United States of America, 103*, 498–503.

Danesh, J., Whincup, P., Walker, M., Lennon, L., Thomson, A., Appleby, P., Gallimore, J. R., & Pepys, M. B. (2000). Low grade inflammation and coronary heart disease: Prospective study and updated meta-analyses. *BMJ, 321*, 199–204.

Elo, I. T., & Preston, S. H. (1992). Effects of early-life conditions on adult mortality: A review. *Population Index, 58*, 186–212.

Finch, C. (2007). *The biology of human longevity: Inflammation, nutrition, and aging in the evolution of life spans* (1st ed.). Burlington: Academic.

Finch, C. E., & Crimmins, E. M. (2004). Inflammatory exposure and historical changes in human life-spans. *Science, 305*, 1736–1739.

Fong, I. W. (2000). Emerging relations between infectious diseases and coronary artery disease and atherosclerosis. *Canadian Medical Association Journal, 163*, 49–56.

Forsdahl, A. (1977). Are poor living conditions in childhood and adolescence an important risk factor for arteriosclerotic heart disease? *British Journal of Preventive & Social Medicine, 31*, 91–95.

Forsdahl, A. (1978). Living conditions in childhood and subsequent development of risk factors for arteriosclerotic heart disease. The cardiovascular survey in Finnmark 1974–75. *Journal of Epidemiology and Community Health, 32*, 34–37.

Gluckman, P. D., & Hanson, M. A. (2006). *Developmental origins of health and disease*. Cambridge/New York: Cambridge University Press.

Hougaard, P. (1986). Survival models for heterogeneous populations derived from stable distributions. *Biometrika, 73*, 387–396.

Kannisto, V. (1994). *Development of oldest-old mortality, 1950–1990: Evidence from 28 developed countries* (Monographs on population aging). Odense: Odense University Press.

Kuzawa, C. W., & Eisenberg, D. T. A. (2014). The long reach of history: Intergenerational and transgenerational pathways to plasticity in human longevity. In M. Weinstein & M. Lane (Eds.), *Sociality, hierarchy, health: Comparative biodemography* (pp. 65–94). Washington, DC: National Research Council Press. Book section 4.

Langley-Evans, S. C. (2004). *Fetal nutrition and adult disease: Programming of chronic disease through fetal exposure to undernutrition* (Frontiers in nutritional science). Wallingford/Oxfordshire/Cambridge: CABI Publication.

Manton, K. G., Stallard, E., & Vaupel, J. W. (1986). Alternative models for the heterogeneity of mortality risks among the aged. *Journal of the American Statistical Association, 81*, 635–44.

McDade, T. W., & Kuzawa, C. W. (2004). Fetal programming of immune function: The early origins of immunity in Filipino adolescents. In S. C. Langley-Evans (Ed.), *Fetal nutrition and adult disease: Programming of chronic disease through fetal exposure to undernutrition* (Frontiers in nutritional science, book section 13, pp. 311–332). Wallingford/Oxfordshire/Cambridge: CABI Publication.

McDade, T. W., Rutherford, J., Adair, L., & Kuzawa, C. W. (2010). Early origins of inflammation: Microbial exposures in infancy predict lower levels of C-reactive protein in adulthood. *Proceedings of the Royal Society B: Biological Sciences, 277*, 1129–1137.

Palloni, A., & Beltrán-Sánchez, H. (2015). Mortality regimes with Barker frailty and the warped dynamics of old age mortality. *work-in-progress*.

Palloni, A., & Souza, L. (2013). The fragility of the future and the tug of the past: Longevity in Latin America and the Caribbean. *Demographic Research, 29*, 543–578.

Steinsaltz, D. R., & Wachter, K. W. (2006). Understanding mortality rate deceleration and heterogeneity. *Mathematical Population Studies, 13*, 19–37.

Vaupel, J. W., & Missov, T. (2014). Unobserved population heterogeneity: A review of formal relationships. *Demographic Research, 31*, 659–686.

Vaupel, J. W., & Yashin, A. I. (1987). Repeated resuscitation – how lifesaving alters life-tables. *Demography, 24*, 123–135.

Vaupel, J. W., Manton, K. G., & Stallard, E. (1979). Impact of heterogeneity in individual frailty on the dynamics of mortality. *Demography, 16*, 439–454.

Chapter 9
Mortality Crossovers from Dynamic Subpopulation Reordering

Elizabeth Wrigley-Field and Felix Elwert

Mortality crossovers—reversals of mortality differentials—between advantaged and disadvantaged populations are common across the world (e.g., Wilmoth and Dennis 2007; Zeng and Vaupel 2003). For example, in the United States, black mortality exceeds white mortality for most of the life course before it crosses below white mortality in old age (Berkman et al. 1989; Fenelon 2013; Kestenbaum 1992; Lynch et al. 2003; Masters 2012; Sautter et al. 2012; Vaupel et al. 1979). Demographers have proposed several non-exclusive explanations for mortality crossovers, including measurement error (Preston et al. 2003). The dominant explanation for mortality crossovers in recent work, however, is mortality selection (Fenelon 2013; Lynch et al. 2003; Masters 2012; Vaupel et al. 1979; Vaupel and Yashin 1985).

Most discussions of mortality selection and mortality crossovers between populations appeal to starkly parsimonious theoretical models. These models assume proportional hazards and a single dimension of fixed heterogeneity that divides each population into "frail" and "robust" subpopulations (e.g., Lynch et al. 2003; Vaupel et al. 1979; Vaupel and Yashin 1985). The frail subpopulations have higher mortality than the robust subpopulations within each population. Because,

E. Wrigley-Field (✉)
Robert Wood Johnson Foundation Health and Society Scholars Program, INCITE, Columbia University, 606 West 122nd Street, New York, NY 10027, USA

Department of Sociology and Minnesota Population Center, University of Minnesota, Twin Cities, Minneapolis, MI, USA
e-mail: ewf@umn.edu

F. Elwert
Department of Sociology, University of Wisconsin-Madison, Sewell Social Science Building, 1180 Observatory Drive, Madison, WI 53706, USA

Social Science Research Center Berlin, Berlin, Germany
e-mail: felwert@ssc.wisc.edu

R. Schoen (ed.), *Dynamic Demographic Analysis*,
The Springer Series on Demographic Methods and Population Analysis 39,
DOI 10.1007/978-3-319-26603-9_9

conditional on frailty, the disadvantaged population has higher mortality than the advantaged population, the disadvantaged population loses its frail members more quickly than the advantaged population. Eventually, the disadvantaged population may be so heavily selected for robustness that its aggregate mortality drops below the aggregate mortality of the advantaged population, which retains comparatively more frail members. The result is a crossover in aggregate mortalities between the populations.

The contribution of conventional mortality selection models with a single dimension of fixed heterogeneity, including Vaupel and Yashin's (1985) classic model in "Heterogeneity's Ruses" as well as gamma-Gompertz models (e.g., Gampe 2010; Vaupel et al. 1979), is that they highlight the central logic of mortality selection as a cause of mortality crossovers: a crossover can occur because mortality differentially changes population composition over age. The weakness of these unidimensional heterogeneity models is that they are not especially realistic: real populations are heterogeneous with respect to multiple, crosscutting dimensions of heterogeneity.

In this chapter, we analyze mortality crossovers in a model with multidimensional heterogeneity (Wrigley-Field and Elwert 2015). Specifically, we analyze a model in which two populations (e.g., blacks and whites) are crosscut by two binary dimensions of fixed heterogeneity (e.g., frail vs. robust and chronically ill vs. not chronically ill). We employ the black-white mortality crossover in the United States as our running example, but our results equally apply to other crossovers.[1]

Our central result is the identification of a new mechanism by which mortality selection can induce a mortality crossover. The multidimensional heterogeneity model can generate mortality crossovers in more ways than the unidimensional heterogeneity model because the multidimensional model exhibits an additional source of dynamics over age. Here, we preview our argument by contrasting the central differences between the conventional unidimensional heterogeneity model and our multidimensional heterogeneity model.

The basic difference between the unidimensional model and the multidimensional model is that the subpopulations are internally homogenous in the unidimensional model but internally heterogeneous in the multidimensional model. That is, in the unidimensional model, all frail blacks are assumed to have the same mortality, but in the multidimensional model, some frail blacks are chronically ill and have high mortality, whereas other frail blacks are not chronically ill and have lower mortality.

Because the subpopulations are homogeneous in the unidimensional model but heterogeneous in the multidimensional model, the multidimensional model adds a new source of dynamics to racial differences in aggregate mortalities. In the unidimensional model, dynamics in aggregate mortality differentials arise exclusively from changes over age in the subpopulation share within each population. In the multidimensional model, by contrast, dynamics in aggregate mortality differentials

[1]For important prior work on mortality selection with multidimensional heterogeneity (albeit not analyzing mortality crossovers), see e.g., Manton et al. (1994, 1995) and Woodbury and Manton (1983).

can arise from two sources. First, as in the unidimensional model, the share of subpopulations within the black and white populations changes over age (*dynamic subpopulation share*).[2] Second, unlike in the unidimensional model, the rank order of subpopulation mortalities can change over age (*dynamic subpopulation mortality order*). For example, in the multidimensional model, frail blacks may have higher mortality than frail whites at birth, but lower mortality than frail whites later in life. The same is not possible in the unidimensional model. In the unidimensional model, subpopulation mortalities cannot cross (*static subpopulation mortality order*).

Because the multidimensional model has a dynamic order of subpopulation mortalities, while the unidimensional model has a static order of subpopulation mortalities, the multidimensional model can produce an aggregate black-white mortality crossover in more ways than the unidimensional model. In the unidimensional model, the black-white crossover can only result from changes in subpopulation share: as frail blacks become a relatively small portion of black survivors compared to frail whites, blacks in the aggregate can have lower mortality than whites in the aggregate. In the multidimensional model, the black-white crossover can similarly result from changes in the share of subpopulations within each race. But in the multidimensional model, the crossover can also result from a change in the rank order of the subpopulations over age. In particular, the subpopulations defined by one dimension of heterogeneity (e.g., frail blacks regardless of chronic illness vs. frail whites regardless of chronic illness) can change rank order such that, in some age interval, both white subpopulations have higher mortality than both black subpopulations: frail whites and robust whites can both have higher mortality than frail blacks and robust blacks. We refer to this event as *racial segregation of subpopulation mortalities*. When subpopulations are racially segregated in the direction of greater white mortality, the share of the subpopulations within each race is irrelevant to the crossover: aggregate white mortality will exceed aggregate black mortality no matter how many blacks and whites are frail. Aggregate mortality crossovers due to racial segregation of subpopulation mortalities have no analogue in the unidimensional model.

The multidimensional model is also more flexible than the unidimensional model in one additional sense: the two-dimensional model allows black and white aggregate mortality to cross and uncross twice.

In summary, this chapter is the first to point out that when subpopulations are heterogeneous, mortality crossovers between populations can arise through the reordering of subpopulation mortalities, regardless of subpopulation share of the population. This observation is important because real populations are assuredly heterogeneous along multiple dimensions of heterogeneity, which means that

---

[2]Throughout, when we refer to the makeup of a larger aggregation in terms of its lower-level components, we use *composition* from the perspective of the aggregation and *share* from the perspective of the components. Thus, the *population composition of subpopulations* and the *subpopulation shares of the population* both refer to the proportions of the population that are in each subpopulation.

the ordering of subpopulations may well be dynamic. Thus, the possibility that crossovers can arise from the dynamic reordering of subpopulation mortalities should inform the way that demographers explain crossovers when they analyze heterogeneous subpopulations.

In the remainder of this chapter, we demonstrate these contrasts between a conventional unidimensional heterogeneity model and our multidimensional heterogeneity model analytically and by illustration. Section 9.1 reviews the unidimensional heterogeneity model and introduces the distinctions between *static* and *dynamic mortality orders*, and between *racially segregated* and *racially mixed mortality orders*. A simulated example illustrates that an aggregate crossover in the unidimensional model with static subpopulation mortalities ordering requires a racially mixed subpopulation mortality order. Section 9.2 analyses the multidimensional heterogeneity model and compares it to the unidimensional model. We decompose aggregate mortality differentials in the multidimensional model to demonstrate that, due to dynamic reordering of subpopulation mortalities, the black-white crossover can occur both under a racially mixed and under a racially segregated subpopulation mortality order. We also note that multidimensional heterogeneity models permit multiple crossovers. Section 9.3 concludes.

A Note on Terminology Throughout, we refer to populations, subpopulations, and groups. *Populations* represent the highest level of aggregation, e.g. blacks vs. whites. *Aggregate mortality* refers to population-level mortality. *Subpopulations* divide populations along a single dimension of heterogeneity, e.g., frail blacks vs. robust blacks. *Groups* divide populations along two crosscutting dimensions of heterogeneity, e.g., frail blacks who are chronically ill vs. frail blacks who are not chronically ill.

9.1 Subpopulation Ordering and Mortality Crossovers with Unidimensional Heterogeneity

9.1.1 Unidimensional Heterogeneity Model

Standard models of mortality selection posit two populations (e.g., blacks and whites) and a single dimension of heterogeneity that divides each population into frailer or more robust subpopulations with proportional hazards. Equation 9.1 gives a specific unidimensional-heterogeneity model with four subpopulations defined by race and frailty:

$$\begin{aligned} \mu(a)_{w,r} &= \alpha e^{\beta a} && \text{White, robust} \\ \mu(a)_{w,f} &= f\alpha e^{\beta a} && \text{White, frail} \\ \mu(a)_{b,r} &= b\alpha e^{\beta a} && \text{Black, robust} \\ \mu(a)_{b,f} &= bf\alpha e^{\beta a} && \text{Black, frail} \end{aligned} \tag{9.1}$$

In this model, the mortality, $\mu(a)$, of each subpopulation at age a follows a Gompertz law with shared intercept α and log-slope β that is scaled up for blacks and the frail by their respective mortality multipliers $b > 1$ and $f > 1$.

Aggregate mortality, $\overline{\mu}_k(a)$ for race $k = b, w$, is the average of the mortalities of the frail and robust subpopulations within each race, weighted by the proportions of the race that are in the frail or robust subpopulations, respectively:

$$\overline{\mu}_k(a) = \pi_k(a) \cdot \mu_{k,f}(a) + (1 - \pi_k(a)) \cdot \mu_{k,r}(a) \tag{9.2}$$

where $\pi_k(a)$ is the proportion frail at age a in race k, $0 \leq \pi_k(a) \leq 1$.

Since blacks are the disadvantaged population, $b > 1$, an aggregate mortality crossover occurs when aggregate black mortality falls below white aggregate mortality, $\overline{\mu}_b(a) < \overline{\mu}_w(a)$. Note that this model does not guarantee an aggregate mortality crossover.

We highlight three essential facts about this unidimensional heterogeneity model, and others like it. First, the frail and robust subpopulations have proportional (i.e., log-parallel) hazards, and therefore do not cross over age. This represents what we call a *static subpopulation mortality order*: every subpopulation (e.g., frail blacks) that has higher mortality than some other subpopulation (e.g., robust whites) at one age will have higher mortality at all ages. A static mortality order contrasts with a *dynamic subpopulation mortality order*, in which the hazards of two subpopulations cross over age, much as aggregate mortality can. A dynamic subpopulation mortality order is precluded in the unidimensional heterogeneity model.

Second, depending on the specific parameter values, the ranking of subpopulation mortalities at any given age may be racially mixed or racially segregated. In a *racially mixed subpopulation mortality order*, at least one black subpopulation has higher mortality than at least one white subpopulation, and at least one black subpopulation has lower mortality than at least one white subpopulation. By contrast, in a *racially segregated subpopulation mortality order*, all black subpopulations have higher mortality than all white subpopulations, or vice-versa. In the unidimensional heterogeneity model of Eq. 9.1, the subpopulation mortality order is racially mixed at all ages if the mortality multiplier on frailty exceeds that on being black, $f > b > 1$, and the subpopulation mortality order is racially segregated at all ages (in the direction of higher black mortality) if $b > f > 1$.

Third, when the subpopulation mortality order is static, a racially mixed mortality order is a necessary (but not a sufficient) condition for an aggregate mortality crossover. The aggregate crossover occurs because the black population becomes progressively more robust, while the white population does not change composition to the same degree. This changing composition can produce an aggregate crossover only if the mortality of robust blacks is lower than the mortality of frail whites. If there is no subpopulation of blacks with lower mortality than any subpopulation of whites, then it follows from Eq. 9.2 that no amount of compositional change can push aggregate black mortality below aggregate white mortality.

These three facts are not peculiarities of our specific model, but are shared by all proportional hazards models with a single dimension of heterogeneity, i.e., to our knowledge, by all models considered in the demographic literature on mortality crossovers, including gamma-Gompertz models (e.g., Gampe 2010, Horiuchi and Wilmoth 1998, Missov and Finkelstein 2011, Steinsaltz and Wachter 2006, Vaupel et al. 1979), and the famous crossover model in "Heterogeneity's Ruses" (Vaupel and Yashin 1985: 179–180).[3] In all of these unidimensional heterogeneity models, the subpopulation mortality order is static, the rank-order of subpopulation mortalities may be mixed or segregated depending on specific parameter values, and a mixed ordering of subpopulation mortalities is a necessary precondition for an aggregate mortality crossover. When the order of subpopulation mortalities is racially mixed, changes in subpopulation share due to mortality selection can create aggregate mortality crossovers.

9.1.2 Examples: Unidimensional Heterogeneity, Mortality Orders, and the Aggregate Crossover

We give two numerical examples of the model in Eq. 9.1 to illustrate the role of mixed and segregated subpopulation mortality orders in enabling and preventing, respectively, aggregate mortality crossovers in proportional hazards models with unidimensional heterogeneity. For clarity, and without loss of generality, all numerical illustrations in this chapter assume that within-race heterogeneity is equally distributed across races at baseline, $\pi_b(0) = \pi_w(0)$.

Figure 9.1a shows a static and racially mixed ordering of subpopulation mortalities, in which frail whites have higher mortality than robust blacks at all ages, $\mu_{b,f}(a) > \mu_{w,f}(a) > \mu_{b,r}(a) > \mu_{w,r}(a)$.[4] The order of subpopulation mortalities is racially mixed because the frailty multiplier exceeds the race multiplier, $f > b$. This racially mixed subpopulation ordering enables a black-white crossover in aggregate mortality. As frail blacks are selected out of the population more quickly than frail whites, $\pi_b(a) < \pi_w(a)$ for all $a > 0$, by proportionally higher mortality, $\mu_{b,f}(a) = b \cdot \mu_{w,f}(a)$, $b > 1$, the composition of the aggregate black population shifts to robust individuals more quickly than that of the aggregate white population. And since robust blacks have lower mortality than frail whites, $\mu_{b,r}(a) < \mu_{w,f}(a)$, aggregate black mortality eventually crosses below aggregate white mortality for age interval above some critical age a_c, $\overline{\mu}_b(a^*) < \overline{\mu}_w(a^*)$, $a^* > a_c$. For the parameter values chosen in Fig. 9.1a, aggregate black and white mortalities cross at age 84.9.

[3]Gamma-Gompertz models replace the fixed binary frailty term of our model with a fixed gamma-distributed continuous frailty term. Vaupel and Yashin (1985) replace the frailty multiplier of Eq. 9.1 with an interacted multiplier: robust people have the same mortality regardless of whether they are in the advantaged (e.g., white) or disadvantaged (e.g., black) population, and the penalty for frailty is large among the disadvantaged and small among the advantaged.

[4]The parameter values for all examples are presented and discussed in the Appendix.

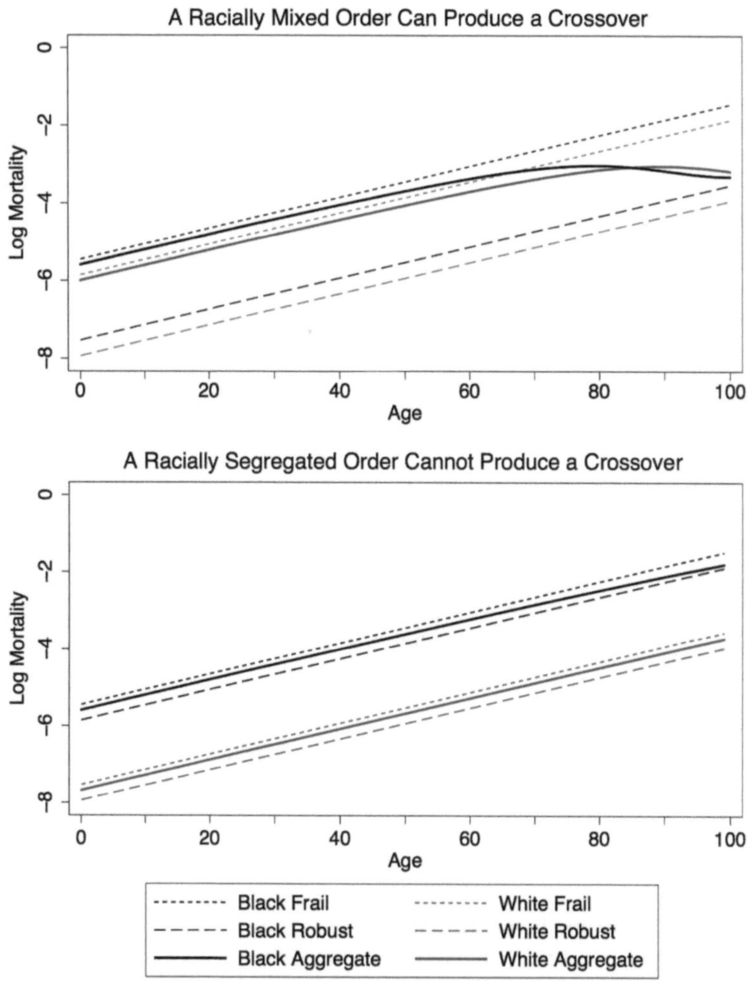

Fig. 9.1 With unidimensional heterogeneity, a racially mixed subpopulation mortality order can produce a population mortality crossover (*top*), while a racially segregated subpopulation mortality order cannot produce a crossover (*bottom*)

By contrast, Fig. 9.1b shows a static and racially segregated subpopulation ordering in which all blacks have higher mortality than all whites at all ages, $\mu_{b,f}(a) > \mu_{b,r}(a) > \mu_{w,f}(a) > \mu_{w,r}(a)$. The order of subpopulation mortalities is racially segregated because the race multiplier exceeds the frailty multiplier, $b > f$. As in the racially mixed ordering of Fig. 9.1a, the frailty share of the black population depletes more quickly than the frailty share of the white population. But since all blacks have higher mortality than all whites, the racially segregated ordering prevents a crossover in aggregate mortalities.

9.2 Subpopulation Ordering and Mortality Crossovers with Multidimensional Heterogeneity

In proportional hazards models with a single dimension of heterogeneity, the subpopulations defined by race and frailty are internally homogeneous and the racial ordering of subpopulation-specific mortalities is static. By contrast, in proportional-hazards models with multidimensional heterogeneity, the subpopulations within race are internally heterogeneous and the ordering of subpopulation mortalities can become dynamic. As a consequence, aggregate black and white mortalities can cross either because of racial differences in the subpopulation share, or because the order of subpopulation mortalities becomes racially segregated in a way that necessitates a crossover.

In the following, we posit a mortality model with multidimensional heterogeneity and analyze its crossover conditions, first from the perspective of internally homogenous groups and then from the perspective of internally heterogeneous subpopulations. The subpopulation analysis is particularly informative because it corresponds to the perspective taken in some recent work on mortality crossovers (e.g., Dupre et al. 2006; Sautter et al. 2012; Yao and Robert 2011), where analysts stratify populations along a single dimension of observed heterogeneity while implying that the resulting subpopulations remain internally heterogeneous.

9.2.1 *Multidimensional Heterogeneity Model*

Consider a model in which the aggregate populations are internally divided by two crosscutting dimensions of fixed binary heterogeneity. For concreteness, consider black and white populations that are internally divided by being frail or robust, and also by being chronically ill or not chronically ill. Thus, each racial population contains four internally homogenous and disjunct groups (e.g., frail blacks who are chronically ill vs. frail blacks who are not chronically ill, etc.) and four internally heterogeneous and overlapping subpopulations (e.g., frail blacks regardless of chronic illness vs. chronically ill blacks regardless of frailty, etc.).

Analogous to the subpopulation mortalities in the unidimensional heterogeneity model in Eq. 9.1, group-level mortalities at age a in the two-dimensional heterogeneity model are defined by a shared Gompertz function with parameters α and β that is scaled up by mortality multipliers for being black, $b > 1$, being chronically ill, $c > 1$, and being frail, $f > 1$, as shown in Eq. 9.3:

$$
\begin{aligned}
\mu(a)_{w,n,r} &= \alpha e^{\beta a} && \text{White, not chronically ill, robust}\\
\mu(a)_{w,n,f} &= f\alpha e^{\beta a} && \text{White, not chronically ill, frail}\\
\mu(a)_{w,c,r} &= c\alpha e^{\beta a} && \text{White, chronically ill, robust}\\
\mu(a)_{w,c,f} &= cf\alpha e^{\beta a} && \text{White, chronically ill, frail}\\
\mu(a)_{b,n,r} &= b\alpha e^{\beta a} && \text{Black, not chronically ill, robust}\\
\mu(a)_{b,n,f} &= bf\alpha e^{\beta a} && \text{Black, not chronically ill, frail}\\
\mu(a)_{b,c,r} &= bc\alpha e^{\beta a} && \text{Black, chronically ill, robust}\\
\mu(a)_{b,c,f} &= bcf\alpha e^{\beta a} && \text{Black, chronically ill, frail}
\end{aligned}
\tag{9.3}
$$

This multidimensional heterogeneity model has proportional hazards at the group level.

Analogous to Eq. 9.2, Eq. 9.4a gives aggregate population-level mortalities in the multidimensional model, $\overline{\mu}_k$, as weighted averages of the four group-specific mortalities within each population,

$$
\begin{aligned}
\overline{\mu}_k(a) = {} & \mu_{k,n,r}(a) \cdot \rho_{k,n,r}(a) + \mu_{k,n,f}(a) \cdot \rho_{k,n,f}(a)\\
& + \mu_{k,c,r} \cdot \rho_{k,c,r}(a) + \mu_{k,c,f} \cdot \rho_{k,c,f}(a)
\end{aligned}
\tag{9.4a}
$$

where $\rho_{k,j,i}(a)$ is the proportion of each race $k = b, w$ that is in the group defined by chronic illness $j = n, c$ and frailty $i = r,f$. Equivalently, one can express aggregate mortality as the weighted average of two subpopulation-specific mortalities within each race,

$$
\overline{\mu}_k(a) = \pi_k(a) \cdot \overline{\mu}_{k,f}(a) + (1 - \pi_k(a)) \cdot \overline{\mu}_{k,r}(a), \tag{9.4b}
$$

where, as before, $\pi_k(a)$ is the proportion of race k that is frail, $0 \leq \pi_k(a) \leq 1$. Subpopulation mortalities, in turn, are a weighted average of the group-specific mortalities within each subpopulation,

$$
\overline{\mu}_{k,i}(a) = \tau_{k,i}(a) \cdot \mu_{k,c,i}(a) + (1 - \tau_{k,i}(a)) \cdot \mu_{k,n,i}(a), \tag{9.5}
$$

where $\tau_{k,i}(a)$ is the proportion of each subpopulation of race k and frailty i that is chronically ill, $0 \leq \tau_{k,i}(a) \leq 1$.

Equation 9.4b highlights that subpopulation-level mortalities in the multidimensional model are not proportional over age; disparities in subpopulation mortalities change as a function of shifts in the subpopulation composition of groups due to mortality selection. Consequently, the ordering of subpopulation mortalities in the multidimensional model can be dynamic, because subpopulation mortalities can cross over age.

9.2.2 Crossover Conditions in the Multidimensional Model

From a group-level perspective, the multidimensional model behaves much like the unidimensional model. In the multidimensional model, the eight groups—like the four subpopulations of the unidimensional model—are internally homogenous, and their mortality ordering is static over age. In the unidimensional model, the necessary condition for the crossover was a racially mixed order of subpopulation mortalities. Analogously, in the multidimensional model, the necessary condition for an aggregate mortality crossover is a racially mixed order of group-level mortalities; unless some black group has lower mortality than some white group, no amount of compositional shifts within the races can push aggregate black mortality below white aggregate mortality.

From a subpopulation-level perspective, however, the multidimensional model behaves quite differently from the unidimensional model. In the unidimensional model, the ordering of subgroup mortalities is static. In the multidimensional model, by contrast, the ordering of subpopulation mortalities can be dynamic, as any two subpopulation mortalities may cross over age. Subpopulation mortality order in the multidimensional model is dynamic because all subpopulations are internally heterogeneous: within each race, each subpopulation comprises two groups that are subject to different selective pressures. For example, frail blacks and robust blacks can have their own crossover, driven by the greater selection against the chronically ill in the frail subpopulation.

A consequence of the dynamic subpopulation mortality order is that the multidimensional model permits aggregate crossovers via two different scenarios. In the multidimensional model, any crossover is driven by changes in the share of homogeneous groups. However, changes in the share of these homogeneous groups can result in two different kinds of changes at the subpopulation level. Changes in the share of groups in the population can result in racial differences in subpopulation share, which can produce an aggregate crossover much as racial differences in subpopulation share produce an aggregate crossover in the unidimensional model. But changes in the share of groups can also result in changes in the ordering of subpopulation mortalities. When the subpopulation mortalities become racially segregated along one dimension of heterogeneity, then the share of each race that is in each of those subpopulations is irrelevant to whether a crossover occurs.[5] This is important because empirical work often engages with heterogeneous subpopulations, and

[5]It follows from Eq. 9.4b that racial segregation of the subpopulations along one dimension of heterogeneity determines the direction of aggregate mortality differentials: if black and white populations are exhaustively partitioned into non-overlapping subsets (e.g., subpopulations defined along one dimension of heterogeneity) such that all white subsets have higher mortality than all black subsets (or vice-versa), then aggregate white mortality must exceed aggregate black mortality (or vice-versa). (If subpopulations are racially segregated along one dimension of heterogeneity, they may or may not also be segregated along the other dimension of heterogeneity.)

heterogeneous subpopulations can produce a crossover via more mechanisms than the homogeneous subpopulations in the unidimensional mortality selection models.

We now show how a crossover can occur while subpopulations are racially segregated by expressing the aggregate crossover as a function of subpopulations defined along one dimension of heterogeneity, e.g., frailty. An aggregate black-white crossover occurs when $\overline{\mu}_b(a) < \overline{\mu}_w(a)$. Expanding aggregate mortalities in this crossover condition as a function of subpopulation-specific mortalities and rearranging terms gives Eq. 9.6:

$$\left[\overline{\mu}_{b,r}(a) - \overline{\mu}_{w,r}(a)\right] + b \cdot \pi_b(a)\left[\overline{\mu}_{b,f}(a) - \overline{\mu}_{b,r}(a)\right] + \pi_w(a)\left[\overline{\mu}_{w,r}(a) - \overline{\mu}_{w,f}(a)\right] < 0 \tag{9.6}$$

The sign of the left-hand side of Eq. 9.6 depends on the absolute magnitudes and the signs of the three terms, and the signs of the three terms depend on whether particular lower-level crossovers have occurred.

- The first term is negative when robust blacks have lower mortality than robust whites—that is, when the robust subpopulation has had a black-white crossover, $\overline{\mu}_{b,r}(a) < \overline{\mu}_{w,r}(a)$.
- The second term is negative when frail blacks have lower mortality than robust blacks—that is, when blacks have had a frail-robust crossover, $\overline{\mu}_{b,f}(a) < \overline{\mu}_{b,r}(a)$.
- The third term is negative when frail whites have higher mortality than robust whites—that is, when whites have *not* had a frail-robust crossover, $\overline{\mu}_{w,r}(a) < \overline{\mu}_{w,f}(a)$.

The fact that three different subpopulation crossovers contribute to the aggregate black-white crossover gives the multidimensional model a measure of flexibility. Whether or not aggregate mortalities are crossed depends on the sign and magnitude of the three terms in Eq. 9.6, as summarized in Fig. 9.2.

Black and white aggregate mortalities are crossed (i.e., white mortality is higher) when the sum of the three left-hand side terms in Eq. 9.6 is negative. Black and white aggregate mortalities are therefore necessarily crossed when all three terms are negative, i.e., when the robust have a black-white crossover, blacks have a frail-robust crossover, and whites do not have a frail-robust crossover. These conditions guarantee an aggregate black-white crossover because they imply a racially segregated ordering of subpopulation mortalities, in which the mortalities of both white subpopulations exceed the mortalities of both black subpopulations, $\overline{\mu}_{w,f}(a) > \overline{\mu}_{w,r}(a) > \overline{\mu}_{b,r}(a) > \overline{\mu}_{b,f}(a)$.

Conversely, black and white aggregate mortalities cannot be crossed when all three terms are positive, i.e., when there is no black-white crossover among the robust, there is no frail-robust crossover among blacks, and there is a frail-robust crossover among whites. These conditions prevent an aggregate crossover because they imply a racially segregated subpopulation ordering in which the mortalities of both black subpopulations exceed the mortalities of both white subpopulations, $\overline{\mu}_{b,f}(a) > \overline{\mu}_{b,r}(a) > \overline{\mu}_{w,r}(a) > \overline{\mu}_{w,f}(a)$.

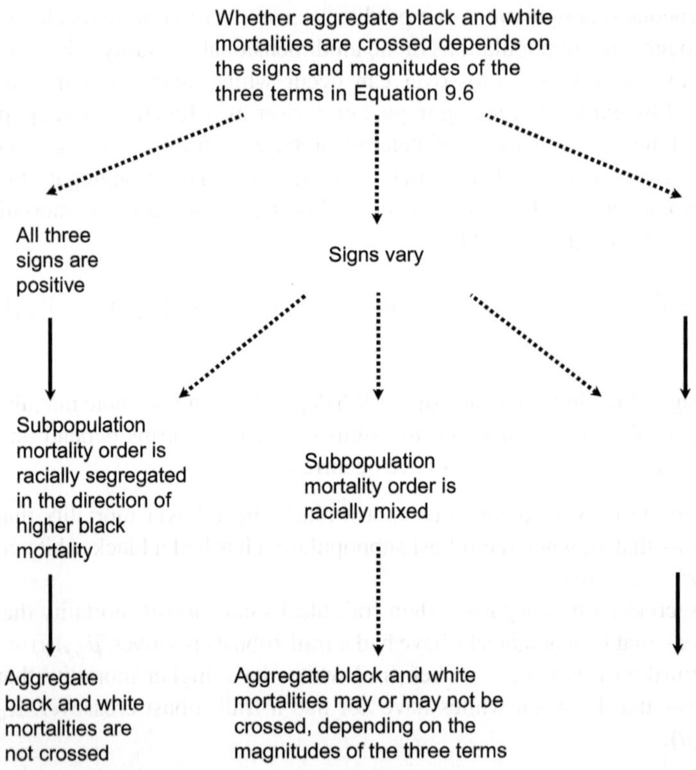

Fig. 9.2 Whether aggregate black and white mortalities are crossed depends on three subpopulation-level crossovers, which can produce a racially segregated or a racially mixed subpopulation mortality order

When the signs of the three terms in Eq. 9.6 differ, the mortality order of the subpopulations is not fully constrained: it may be either racially segregated or racially mixed. For example, if the robust have had a black-white crossover and both blacks and whites have had a frail-robust crossover (so that the first two terms on the left-hand side of Eq. 9.6 are negative and the third term is positive), then the subpopulation mortality order may be either $\overline{\mu}_{w,r}(a) > \overline{\mu}_{w,f}(a) > \overline{\mu}_{b,r}(a) > \overline{\mu}_{b,f}(a)$ —a racially segregated order—or $\overline{\mu}_{w,r}(a) > \overline{\mu}_{b,r}(a) > \overline{\mu}_{w,f}(a) > \overline{\mu}_{b,f}(a)$ —a racially mixed order. If the subpopulation mortality order is racially segregated, then a crossover necessarily occurs (if the white subpopulations have higher mortality) or necessarily does not occur (if the black subpopulations have higher mortality) at those ages. If, on the other hand, the subpopulation mortality order is racially mixed, then a crossover is possible but not assured, depending on the magnitudes of the three terms.

9.2.3 Example: Multidimensional Heterogeneity, Mortality Orders, and the Aggregate Crossover

This section uses a simulated example cohort to illustrate the dynamics of subpopulation ordering in the multidimensional model, highlighting that a crossover can arise while the subpopulations are racially segregated and while the subpopulations are racially mixed. The example cohort is composed of black and white populations, which are divided along two crosscutting binary dimensions, chronic illness and frailty, yielding eight internally homogenous groups of individuals. We focus on the mortalities of the aggregate black and white populations and the mortalities of the internally heterogeneous subpopulations defined by frailty, as in Eqs. 9.4b, 9.5, and 9.6.

Figure 9.3 shows logged aggregate mortalities and subpopulation mortalities for blacks and whites from birth to age 100.[6] Aggregate mortalities cross from ages 80–99. Figure 9.4 zooms in on old age when aggregate mortalities cross and uncross and shows the dynamics of subpopulation ordering allowed under Eq. 9.6. Figure 9.4a shows logged aggregate mortalities, and Fig. 9.4b shows logged subpopulation mortalities, from ages 65–100. The two panels of Fig. 9.4

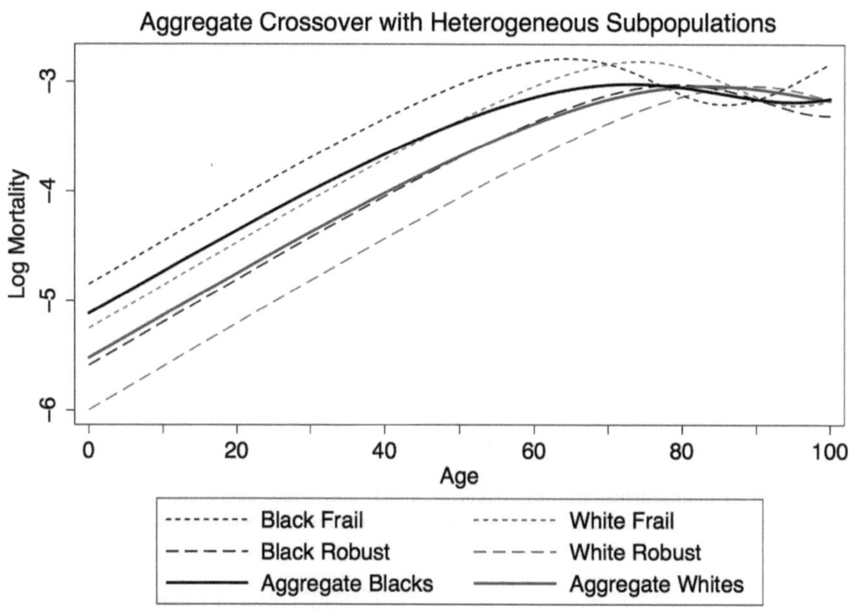

Fig. 9.3 With multidimensional heterogeneity, aggregate black and white mortalities can cross and uncross as subpopulations cross and uncross

[6]Nothing new happens after age 100.

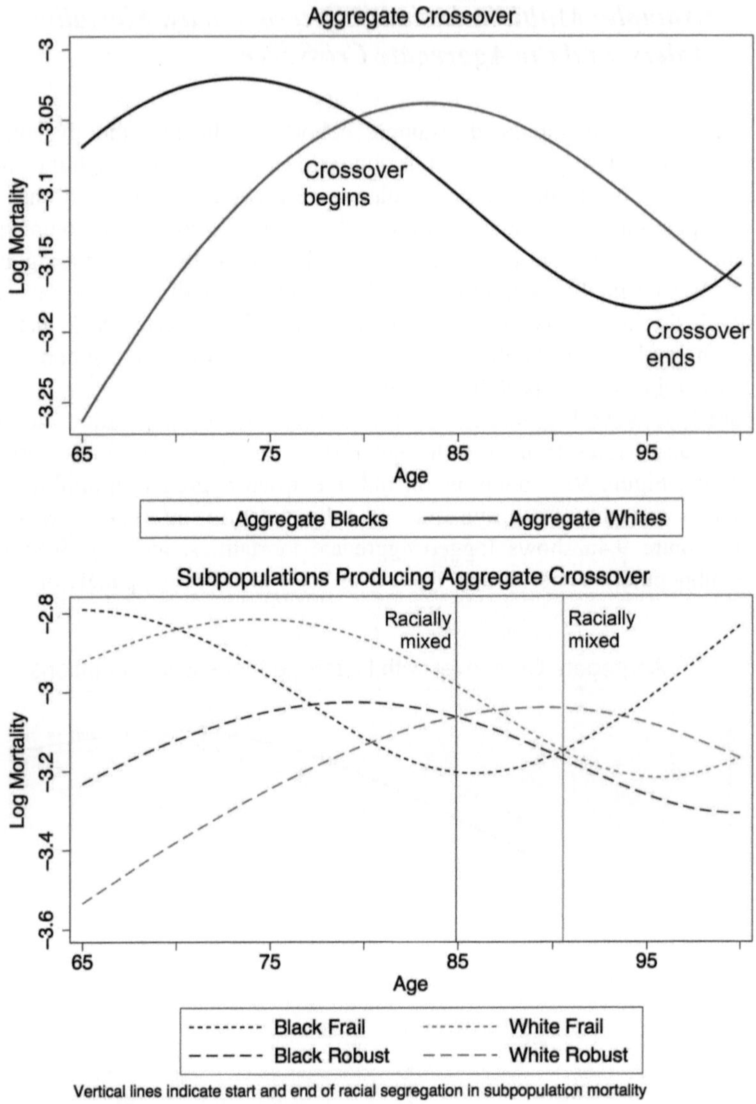

Fig. 9.4 With multidimensional heterogeneity, black and white aggregate mortalities can be crossed while subpopulation mortalities are racially segregated or while they are racially mixed

highlight the important changes in the decomposition terms of Eq. 9.6, the presence or absence of an aggregate black-white crossover, and the racial ordering of subpopulation mortalities. Here, we describe in detail the changes in this cohort over age and then summarize the implications of these changes.

As Fig. 9.3 shows, for most of the age range, up to age 70, frail blacks have the highest mortality, followed respectively by frail whites, robust blacks, and robust whites. The order of subpopulation mortalities is mixed. The first two terms of Eq. 9.6 are positive, while the third is negative—there has been no black-white crossover among the robust, no frail-robust crossover among blacks, and no frail-robust crossover among whites. As discussed in the previous section, the fact that a frail-robust crossover among whites has *not* occurred makes an aggregate crossover more likely, but this is outweighed by the magnitude and positive sign of the other two terms, which keep aggregate black mortality above aggregate white mortality at these ages.

The first subpopulation-level crossover occurs at age 70: a black-white crossover among the frail. This is the subpopulation-level crossover that does not appear in Eq. 9.6. Hence, from the perspective of Eq. 9.6, it is not directly relevant to the aggregate crossover. The subpopulation mortality order remains racially mixed, and no aggregate crossover occurs.

The next subpopulation-level crossover begins around age 77, when frail blacks and robust blacks switch places. This causes the second term on the left-hand side of Eq. 9.6 to switch its sign to negative. This term now brings aggregate black and white mortalities closer to a crossover, as does the continued lack of a frail-robust crossover among whites. But the sum of these two negative terms is still outweighed by the lack of a black-white crossover among the robust (which leaves the first term in Eq. 9.6 positive). The subpopulation mortality order remains racially mixed, and no aggregate crossover occurs.

The aggregate crossover finally occurs around age 80, when the second and third terms of Eq. 9.6 become sufficiently large to outweigh the first term. Nonetheless, the order of subpopulation mortalities remains racially mixed until age 85, when two things occur. First, robust black mortality falls below robust white mortality, such that all three terms in Eq. 9.6 are now negative (i.e., all three terms now work in the direction of a crossover). Second, as a result, the subpopulations now have a racially segregated mortality order in the direction of higher white mortality: both white subpopulations have higher mortality than both black subpopulations.

The last remaining subpopulation-level crossover occurs at age 87, when robust white mortality falls below frail white mortality. This is the subpopulation crossover that mitigates the tendency toward an aggregate crossover, per the third term of Eq. 9.6. But this crossover alone cannot uncross the aggregate crossover, since it only switches the order of the two white subpopulations, both of whose mortality currently exceeds that of both black subpopulations. The mortality of the subpopulations remains racially segregated.

The interval of racial segregation continues beyond the end of the crossover between the black frail and black robust subpopulations at age 90 and ends only with the end of the black-white crossover in the frail subpopulation at age 91. After age 91, the aggregate crossover continues, with a racially mixed mortality order, until age 99, when aggregate mortalities uncross. The end of the aggregate crossover is not precipitated by any subpopulation-level crossover beginning or ending, but by changes in the magnitudes of the mortality disparities, and their weights, in Eq. 9.6.

This example illustrates several important points about the black-white crossover in the context of heterogeneous subpopulations. First, subpopulation mortality order is dynamic in this multidimensional model of heterogeneous subpopulations, because mortality selection can produce crossovers at the subpopulation level as well as at the aggregate population level. This contrasts with the unidimensional heterogeneity model, in which subpopulations are homogeneous and their mortality order is static.

In the multidimensional model, as the cohort ages, not only does the order of subpopulation mortality change, but it also can switch between racially mixed and racially segregated orders. In our example, the subpopulation mortality order is racially segregated from ages 85 to 91. Within this interval, three different subpopulation mortality orders occur, all racially segregated in the direction of higher white mortality. Subpopulation-level mortality ordering is racially mixed before age 85 and after age 91, although the specific racially mixed subpopulation orders vary.

Second, aggregate mortalities can be crossed either while the subpopulation mortality order is racially segregated (in the direction of higher white mortality) or while it is racially mixed. In this example cohort, the aggregate crossover begins before the interval of racial segregation, and ends after it.

Third, whether subpopulation mortalities have a racially segregated or a racially mixed order is partially determined by the subpopulation crossovers in Eq. 9.6. When the interval of racial segregation begins, all three decomposition terms have the sign that contributes to the crossover, and the order of subpopulation mortality is fully constrained to be the racially segregated order $\overline{\mu}_{w,f}(a) > \overline{\mu}_{w,r}(a) > \overline{\mu}_{b,r}(a) > \overline{\mu}_{b,f}(a)$. Before the interval of racial segregation ends, two of the terms have switched signs, changing the subpopulation mortality order but not the fact that it is racially segregated. When the decomposition terms all have the same sign, the subpopulation order must be racially segregated. But when the terms have different signs, either a racially segregated or a racially mixed subpopulation mortality order is possible. Thus, in this example, the interval of racially segregated subpopulation mortality occurs both when the decomposition terms necessitate it (ages 85–87) and when the decomposition terms allow it but do not necessitate it (ages 87–90).

Fourth, when the subpopulations have a mixed mortality order, the magnitudes of the weighted mortality differences in Eq. 9.6 determine whether an aggregate crossover occurs. All the terms representing the mortality differences and the proportion frail (but not the black mortality multiplier) are dynamic in this model. Thus, the crossover interval begins (at age 80) and ends (at age 99) without any immediate change in the mortality order in either case. What initially pushes the aggregate black-white mortality difference below zero and then above zero is not a change in the subpopulation mortality order, but changes in the magnitudes of the mortality differences (and in the weights provided by the proportion of survivors of each race that are frail).

In the Appendix, we provide additional detail about the parameters that generate the patterns demonstrated in Figs. 9.3, 9.4, and 9.5.

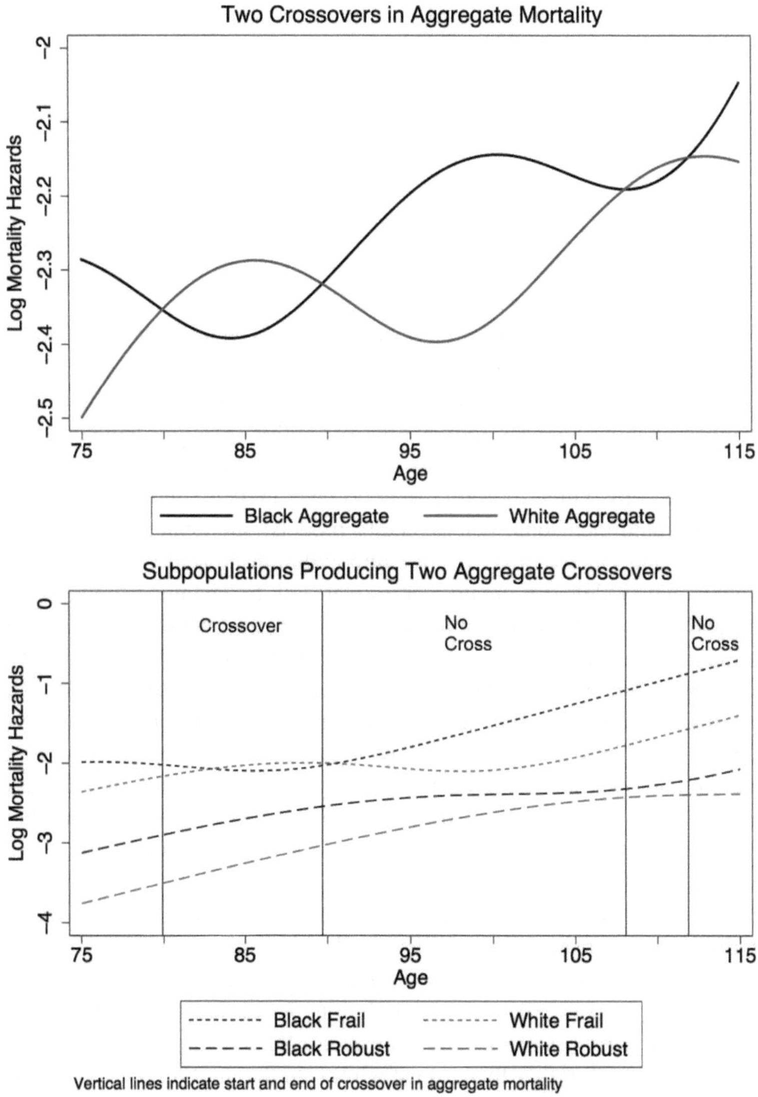

Fig. 9.5 With multidimensional heterogeneity, black and white mortalities can cross and uncross twice

9.2.4 *Multiple Crossovers*

As an aside, we note that the multidimensional heterogeneity model, in contrast to the unidimensional heterogeneity model, enables multiple crossovers. Aggregate mortality can switch from higher black to higher white mortality, then back to higher black mortality, then back to higher white mortality, before entering a final period of higher black mortality.

Figure 9.5 shows an example of multiple aggregate mortality crossovers. We note, however, that the order of subpopulations mortalities in this example is always racially mixed. Thus, the two crossovers occur within a relatively stable ordering of subpopulation mortality, rather than arising from a dynamic reordering of subpopulation mortalities over age.

9.3 Conclusion

This chapter analyzed mortality crossovers in a mortality selection model with multiple, crosscutting dimensions of heterogeneity (Wrigley-Field and Elwert 2015). We discussed the black-white mortality crossover in the United States for concreteness, but our analysis applies equally to other selection-induced crossovers in other places and different domains. Our central result is the identification of a new mechanism by which mortality selection can induce a mortality crossover. In conventional unidimensional heterogeneity models, subpopulations are internally homogenous and cannot cross (static subpopulation mortality order). Therefore, unidimensional models can create a crossover only by changing the relative share of frail and robust subpopulations within the black and white populations. In our multidimensional heterogeneity model, by contrast, subpopulations are internally heterogeneous, and subpopulation mortalities can cross because mortality selection acts on this heterogeneity (dynamic subpopulation mortality order). Therefore, the multidimensional model can create an aggregate crossover via crossovers between the subpopulations. When aggregate mortality crosses because of subpopulation crossovers, the relative shares of frail and robust subpopulations in the black and white populations may be irrelevant—the ordering of subpopulation mortalities alone can compel a crossover.

Subpopulation crossovers, i.e., the dynamic ordering of subpopulation mortalities in the multidimensional model, also challenge another key facet of the conventional unidimensional model. In the unidimensional model, aggregate crossovers require a racially mixed order of subpopulation mortalities: an aggregate crossover can only occur if robust blacks have lower mortality than frail whites, and frail blacks have higher mortality than robust whites. In the multidimensional model, by contrast, aggregate crossovers can also occur when the order of subpopulation mortalities (defined along a specific dimension of heterogeneity) is racially segregated. Specifically, an aggregate crossover can occur when subpopulation-mortality order has changed such that both the frail and the robust subpopulations among whites have higher mortality than the frail and the robust subpopulations among blacks.

These results highlight that empirical analyses of mortality selection, including the crossover, should engage with the dynamics of mortality differentials between heterogeneous collectives at all levels, not only the most aggregated level of interest. In suggesting this conclusion, our analysis dovetails with scattered but important prior work on multidimensional mortality selection models (e.g., Manton et al. 1994, 1995; Woodbury and Manton 1983).

In addition to offering new insights into the creation of aggregate mortality crossovers, multidimensional heterogeneity models deserve more attention because they are more realistic than unidimensional models. Real human populations are crosscut by numerous dimensions of difference (Manton et al. 1994). For example, blacks in the United States do not just differ by how frail they are at birth, as unidimensional heterogeneity models would have it, but also by whether they happen to suffer from chronic illness, live in urban or rural areas, have acquired educational degrees, and so forth.

Although multidimensional models are more realistic than conventional unidimensional heterogeneity model in general, our specific multidimensional model with just two dimensions of heterogeneity obviously remains an abstraction. Realistic heterogeneity within black and white populations surely consists of many dimensions of stratification, some continuous and others categorical. Nevertheless, our results point to the benefits of an explicitly multidimensional analysis of the crossover.

Against our multidimensional representation, one might object that the mortality differentials contained in any multidimensional model could be projected into a single dimension of heterogeneity, thus restoring the simplicity of a unidimensional heterogeneity model. For example, our multidimensional model of Eq. 9.3 with two binary dimensions of heterogeneity could be expressed as a unidimensional model with four levels of frailty, and a multidimensional model with three binary dimensions of heterogeneity could be represented by a unidimensional model with eight levels of frailty, and so forth, culminating, perhaps, in a gamma-Gompertz model with a single, continuous, dimension of heterogeneity.

But such a projection of multiple dimensions of heterogeneity into a single, omnibus dimension of heterogeneity would miss a critical feature of most empirical analyses of mortality crossovers: the subpopulations considered in empirical work (e.g., subpopulations defined by illness, poverty, or religiosity) are always internally heterogeneous. By contrast, subpopulations defined by the single dimension of heterogeneity in unidimensional models are by definition internally homogenous. Multidimensional heterogeneity models better represent empirical work on mortality crossovers because the subpopulations defined by a single dimension of heterogeneity in a multidimensional model are internally heterogeneous.

In this chapter, we showed how multidimensional heterogeneity models can generate aggregate crossovers via a dynamic reordering of subpopulation mortalities, regardless of the share of the population that each subpopulation represents (and regardless of whether those subpopulation shares differ substantially by race). But more remains to be said about mortality crossovers in models with multidimensional heterogeneity. For example, in other work (Wrigley-Field and Elwert 2015), we show how the dynamics of multidimensional models can undermine inferences about the age at crossover.

More broadly, our analysis suggests a new line of inquiry for mortality selection theory. Mortality selection between subpopulations is premised on a relatively static subpopulation mortality order. Compositional changes arise because mortality differentials (e.g., between the frail and the robust) compound over age; but if subpopulation mortalities were frequently dynamically reordered, then the frail might not accumulate so much excess mortality that they would be disproportionately selected out of populations. The disproportionate selection of certain subpopulations out of the black population therefore requires that those subpopulations have high mortality (relative to other subpopulations) for a long stretch of ages. This requirement of a relatively static order of subpopulation mortalities for enabling selection between subpopulations contrasts with the fact, emphasized here, that the subpopulation mortality order can be dynamic when the subpopulations are heterogeneous. The tension between selection *between* and selection *within* heterogeneous subpopulations remains an issue for a truly multidimensional mortality selection theory to grapple with. This tension merits further attention so that demographic theories of mortality selection can reflect the empirical reality of populations riven by differences and inequalities at many levels.

Appendix: Notes on the Simulation Parameters

This chapter identified a new mechanism by which mortality selection can generate aggregate mortality crossovers. In models with multidimensional heterogeneity, mortality selection can generate a dynamic subpopulation mortality ordering, and changes in the subpopulation mortality ordering can, in turn, generate one or more aggregate crossovers. Our numerical examples in Figs. 9.3, 9.4, and 9.5 focused on demonstrating these theoretical possibilities.

Calibrating these possibilities against data is challenging because the parameters in mortality selection models, whether unidimensional or multidimensional, refer to latent constructs (e.g., frailty). Nevertheless, it may be informative to consider how common the patterns shown in Figs. 9.3, 9.4, and 9.5 are within a larger universe of simulated cohorts. Table 9.A1 presents the parameter values for the four simulated cohorts used as illustrations in the chapter.

Table A.1 Parameter values for simulated cohorts

Model	Figs.	Parameters
Unidimensional heterogeneity, Racially mixed subpopulation order	9.1a	$\alpha = .00035818$, $\beta = 0.4$, $b = 1.5$, $f = 8$, $\pi_k(0) = 0.85$
Unidimensional heterogeneity, Racially segregated subpopulation order	9.1b	$\alpha = .00035818$, $\beta = 0.4$, $b = 8$, $f = 1.5$, $\pi_k(0) = 0.6$
Multidimensional heterogeneity, Racially mixed and racially segregated intervals	9.3 and 9.4	$\alpha = .00035818$, $\beta = 0.4$, $b = 1.5$, $f = 2$, $c = 8$, $\tau_{k,r}(0) = .085$, $\tau_{k,f}(0) = 0.9$, $\pi_k(0) = 0.55$
Multidimensional heterogeneity, Multiple Crossovers	9.5	$\alpha = .00011111$, $\beta = .055$, $b = 2$, $f = 4$, $c = 4$, $\tau_{k,r}(0) = 0.85$, $\tau_{k,f}(0) = 0.95$, $\pi_k(0) = 0.95$

We simulated cohorts with mortality multipliers on being black at values of $b = 1.2,\ 1.5, 2$, and mortality multipliers on being frail at values of $f = 2, 4, 6,\ 8$,[7] with values of the initial proportion frail in each race, $\pi_k(0)$, ranging from .55 to .95 in units of .05. The mortality multiplier on being chronically ill, c, and the initial proportion chronically ill within each subpopulation defined by race and frailty, $\tau_{k,i}(0)$, were simulated over the same ranges as those for frailty. (The shared mortality intercept α and slope β are irrelevant to whether a crossover occurs, although they influence the age at which it occurs in cohorts that have a crossover.)

In the unidimensional model, crossovers occur in 39 of the 108 simulated cohorts (36 %). Each of these cohorts has a frailty multiplier of at least 4. In the multidimensional model, a crossover occurs in 18,889 out of 34,992 simulated cohorts (54 %). In these cohorts, the mortality multipliers on being black, frail, or chronically ill can take any of the simulated values, but a crossover never occurs unless either the multiplier on being frail or the multiplier on being chronically ill exceeds 2. In other words, it is easier to produce an aggregate mortality crossover in the multidimensional model than in the unidimensional model.

In the simulation universe we explored, racial segregation of the subpopulations is quite rare. An interval in which subpopulation mortality (defined along the frail-robust dimension) is racially segregated in the direction of higher white mortality occurs in 90 of the simulated multidimensional cohorts (0.48 % of those with a crossover). These cohorts uniformly have a small mortality multiplier on being frail ($f = 2$) and a large mortality multiplier on being chronically ill ($c = 6$ or $c = 8$). They also have high initial values of the proportion chronically ill (.8–.9 for the robust and .85–.95 for the frail).

Cohorts experiencing two distinct crossover intervals are more common; this occurs in 3,374 multidimensional simulated cohorts (18 % of those with at least one crossover). These cohorts each have mortality multipliers on being frail and being chronically ill of at least 4, and initial proportion chronically ill among the frail of at least .65, but otherwise occur across the range of parameter values examined in these simulations.

The relative rarity of these crossover outcomes in the simulation universe we explored partly reflects that generating aggregate crossovers from a small number of subpopulations requires fairly extreme parameter values. For example, the unidimensional heterogeneity model in "Heterogeneity's Ruses" (Vaupel and Yashin 1985) worked with an extreme black-frailty interaction: whereas frail whites had only 1.25 times the mortality of robust whites, frail blacks had mortality 5 times larger than the mortality of robust blacks. This black-frailty interaction helps to produce a crossover in two ways: it exaggerates the extent to which mortality selection against frailty occurs more extremely for blacks than whites, and it also means that the same percentage decline in frailty share reduces white mortality by

[7] We simulated cohorts with small values of the black mortality multiplier and large values of the frailty mortality multiplier (and chronically ill mortality multiplier) in order to find cohorts that might have an aggregate black-white crossover.

only a small amount, but black mortality by a large amount. Our model eschews a black-frailty interaction: the frailty multiplier is the same for blacks and for whites. Consequently, the mortality multipliers required to generate a crossover are larger.

References

Berkman, L., Singer, B., & Manton, K. (1989). Black-white differences in health status and mortality among the elderly. *Demography, 26*(4), 661–678. doi:10.2307/2061264.

Dupre, M. E., Franzese, A. T., & Parrado, E. A. (2006). Religious attendance and mortality: Implications for the Black-White mortality crossover. *Demography, 43*(1), 141–164. doi:10.1353/dem.2006.0004.

Fenelon, A. (2013). An examination of black/white differences in the rate of age-related mortality increase. *Demographic Research, 29*(17), 441–472. doi:10.4054/DemRes.2013.29.17.

Gampe, J. (2010). Human mortality beyond age 110. In H. Maier, J. Gampe, B. Jeune, J.-M. Robine, & J. Vaupel (Eds.), *Supercentenarians* (Demographic Research Monographs 7, pp. 219–230). Berlin: Springer. doi:10.1007/978-3-642-11520-2.

Horiuchi, S., & Wilmoth, J. R. (1998). Deceleration in the age pattern of mortality at older ages. *Demography, 35*, 391–412.

Kestenbaum, B. (1992). A description of the extreme aged population based on improved medicare enrollment data. *Demography, 29*(4), 565–580. doi:10.2307/2061852.

Lynch, S. M., Brown, J. S., & Harmsen, K. G. (2003). Black-white differences in mortality compression and deceleration and the mortality crossover reconsidered. *Research on Aging, 25*(5), 456–483. doi:10.1177/0164027503254675.

Manton, K. G., Stallard, E., Woodbury, M. A., & Dowd, J. E. (1994). Time-varying covariates in models of human mortality and aging: Multidimensional generalizations of the Gompertz. *Journals of Gerontology, 49*(4), B169–B190. doi:10.1093/geronj/49.4.b169.

Manton, K. G., Woodbury, M. A., & Stallard, E. (1995). Sex differences in human mortality and aging at late ages: The effect of mortality selection and state dynamics. *Gerontologist, 35*(5), 597–608. doi:10.1093/geront/35.5.597.

Masters, R. K. (2012). Uncrossing the U.S. Black-white mortality crossover: The role of cohort forces in life course mortality risk. *Demography, 49*(3), 773–796. doi:10.1007/s13524-012-0107-y.

Missov, T. I., & Finkelstein, M. S. (2011). Admissible mixing distributions for a general class of mixture survival models with known asymptotics. *Theoretical Population Biology, 80*, 64–70.

Preston, S. H., Elo, I. T., Hill, M. E., & Rosenwaike, I. (2003). *The demography of African Americans, 1930–1990*. Norwell: Kluwer/Springer. doi:10.1007/978-94-017-0325-3.

Sautter, J. M., Thomas, P. A., Dupre, M. E., & George, L. K. (2012). Socioeconomic status and the black-white mortality crossover. *American Journal of Public Health, 102*(8), 1566–1571. doi:10.2105/ajph.2011.300518.

Steinsaltz, D. R., & Wachter, K. W. (2006). Understanding mortality rate deceleration and heterogeneity. *Mathematical Population Studies, 13*, 19–37.

Vaupel, J. W., & Yashin, A. I. (1985). Heterogeneity's ruses: Some surprising effects of selection on population dynamics. *American Statistician, 39*(3), 176–185. doi:10.1080/00031305.1985.10479424.

Vaupel, J. W., Manton, K. G., & Stallard, E. (1979). Impact of heterogeneity in individual frailty on the dynamics of mortality. *Demography, 16*(3), 439–454. doi:10.2307/2061224.

Wilmoth, J. R., & Dennis, M. (2007). Social differences in older adult mortality in the United States: Questions, data, methods, and results. In J. M. Robine, E. I. Crimmins, S. Horiuchi, & Y. Zeng (Eds.), *Human longevity, individual life duration, and the growth of the oldest-old population* (pp. 297–332). Dordrecht: Springer. doi:10.1007/978-1-4020-4848-7_14.

Woodbury, M. A., & Manton, K. G. (1983). A mathematical model of the physiological dynamics of aging and correlated mortality selection. 1: Theoretical development and critiques. *Journals of Gerontology, 38*, 398–405. doi:10.1093/geronj/38.4.398.
Wrigley-Field, E. & Elwert, F. (2015). *Multidimensional mortality selection and the black-white mortality crossover.* Paper presented at the Population Association of America: April, San Diego.
Yao, L., & Robert, S. A. (2011). Examining the racial crossover in mortality between African American and White older adults: A multilevel survival analysis of race, individual socioeconomic status, and neighborhood socioeconomic context. *Journal of Aging Research, 1–8.*
Zeng, Y., & Vaupel, J. W. (2003). Oldest-old mortality in China. *Demographic Research, 8*(7), 215–244. doi:10.4054/demres.2003.8.7.

Part IV
Extending Stationary and Stable Population Analysis

Chapter 10
The Continuing Retreat of Marriage: Figures from Marital Status Life Tables for United States Females, 2000–2005 and 2005–2010

Robert Schoen

10.1 Introduction

Family formation and dissolution, because of their close connection to both mortality and fertility, have long been subjects of demographic analysis. Census data typically provides information on the number of persons by age, sex, and marital status. In the United States, marriages and divorces, traditionally considered vital events, have been collected by the states and assembled and published by the National Center for Health Statistics (NCHS). From the vital statistics numerators and the census population denominators, occurrence/exposure rates of marriage, divorce, and remarriage could be calculated, and their implications displayed in a life table context.

Unfortunately, after 1995, NCHS largely abandoned the collection and publication of marriage and divorce data. Trends in American marriage and divorce since then have been difficult to follow, since without vital statistics numerator data, age-sex-marital status-specific rates of marriage and divorce could not be calculated. That is particularly regrettable because, as Raley (2000, p. 36) wrote, "Recent trends in marriage, cohabitation, and sexual relationships demonstrate that what we know about intimate sexual unions can quickly become outdated."

It is fair to say that the United States and most Western countries have experienced profound changes in marriage and family patterns over the past 50 years. Cohabitation has become widespread, marriage later and less likely, out-of-wedlock childbearing common, and divorce risks extremely high. The causes are multiple and complex. Cherlin (2004) saw a "deinstitutionalization" of marriage, where its practical benefits and normative supports declined though its symbolic

---

R. Schoen (✉)
Population Research Institute, Pennsylvania State University, University Park, PA 16802, USA
e-mail: rschoen309@att.net

R. Schoen (ed.), *Dynamic Demographic Analysis*,
The Springer Series on Demographic Methods and Population Analysis 39,
DOI 10.1007/978-3-319-26603-9_10

value endured. Ruggles (2015) stressed macroeconomic factors, particularly the shift from male breadwinner to dual earner families and the recent stagnation in male wages. Given the shift to the service sector and widespread labor force participation by women, Schoen (2010) focused on social dynamics, especially the gender competition for power and control that destabilized intimate unions. It is generally acknowledged that, in the words of Axinn and Thornton (2000), there has been a "transformation in the meaning of marriage."

Based on the vital rates observed in the United States in 1995, 88 % of women would ever marry with an average age at first marriage of 26.6 years, and 42.5 % of those marriages would end in divorce (Schoen and Standish 2001). There is some reason to believe that marriage is continuing to retreat. The National Survey of Family Growth found that the probability of first marriage for young women, by age, was less in 2002 that in 1995, and had declined even more by 2006–2010 (Copen et al. 2012, p. 6). Crude marriage rates calculated by NCHS show a steady decline of some 24 % from 1995 to 2010, a pattern found in nearly every state (NCHS 1996). Regarding divorce, Goldstein (1999) saw a leveling of divorce rates from 1980 to 1995. In contrast, recent work by Kennedy and Ruggles (2014), using new data from the American Community Survey (ACS), found a substantial rise in rates of marital dissolution from 1980 to 2010, with the age-standardized divorce rate rising some 25–40 %.

The purpose of this chapter is two-fold. Substantively, we consider recent levels and trends in marriage and divorce behavior in the light of new marital status life tables for 2000–2005 and 2005–2010. Methodologically, we make use of a new approach, Rate Estimation from Adjacent Populations (REAP), to overcome the absence of vital statistics marriage data. To begin, we describe the marital status life table model and the new REAP methodology.

10.2 Marital Status Life Tables from Population Data

Marital status life tables follow a hypothetical closed cohort from birth to the death of the cohort's last member, and can reflect how cohort members move between the states of Never Married (s), Presently Married (m), Widowed (w), and Divorced (v) (see Fig. 10.1). A marital status life table is usually calculated from observed age-sex-specific rates, and can provide measures of the number of persons in each marital status and the number of deaths, marriages, widowhoods, divorces, and remarriages experienced by members of the life table cohort. Trends in such measures as the likelihood and average age at first marriage, and the probability that a marriage ends in divorce can be examined. A number of such tables have been calculated for the United States (e.g. Schoen and Standish 2001) using vital statistics data on deaths, marriages, and divorces. In the absence of vital statistics on marriage and divorce, we now make greater use of the relationships between population stocks and flows.

Fig. 10.1 Diagram of a four living state marital status life table model

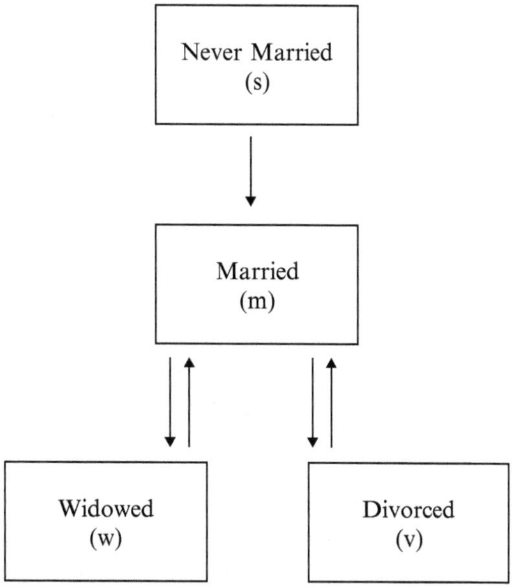

10.2.1 Finding Rates from Population Data Using REAP

Population data that give the number of persons by age, sex, and marital status at the beginning and end of a 5-year interval constrain the rates of interstate movement during that interval. Given four living states, four flow equations can be written to describe the movements between states, equating the number of persons in a state at the end of the interval to the starting number of persons, plus the entries from the other states, and minus the exits to the other states.

A general approach for multistate transfer rate estimation from adjacent populations was developed in Schoen (2015). That Quadratic Estimation of Rates of Transfer (QERT) approach imposed the constraint that the product of selected pairs of rates was constant over time. In the present marital status analysis, however, that constant product assumption is not needed.

The 4-state marital status model can be described by the four flow equations

$$
\begin{aligned}
P_s(x+5,5,t+5) &= P_s(x,5,t) - PP_s(x,5)\,[m_{sm}(x,5) + m_{sd}(x,5)] \\
P_m(x+5,5,t+5) &= P_m(x,5,t) - PP_m(x,5)\,[m_{mw}(x,5) + m_{mv}(x,5) + m_{md}(x,5)] \\
&\quad + PP_s(x,5)\,m_{sm}(x,5) + PP_w(x,5)\,m_{wm}(x,5) + PP_v(x,5)\,m_{vm}(x,5) \\
P_w(x+5,5,t+5) &= P_w(x,5,t) - PP_w(x,5)\,[m_{wm}(x,5) + m_{wd}(x,5)] \\
&\quad + PP_m(x,5)\,m_{mw}(x,5) \\
P_v(x+5,5,t+5) &= P_v(x,5,t) - PP_v(x,5)\,[m_{vm}(x,5) + m_{vd}(x,5)] \\
&\quad + PP_m(x,5)\,m_{mv}(x,5)
\end{aligned}
\tag{10.1}
$$

where $P_j(x,5,t)$ is the population in state j (j = s,m,w,v) between the ages of x and x + 5 at time t, $PP_j(x,5)$ is the number of person-years lived by the $P_j(x,5,t)$ cohort between times t and t + 5, $m_{ij}(x,5)$ is the rate of movement from state i to state j between the ages of x and x + 5, and d represents the dead state. The PP functions are calculated from the linear relationship

$$PP_j(x,5) = 2.5\left[P_j(x,5,t) + P_j(x+5,5,t+5)\right] \tag{10.2}$$

Under the linear assumption, $P_j(x,5,t)$ can be viewed as being at age x + 2.5 at time t. Then $PP_j(x,5)$ can be seen as the number of person-years lived by the $P_j(x,5,t)$ persons between the ages of x + 2.5 and x + 7.5. In effect, the REAP procedure shifts age intervals upward by half an interval. The linear assumption was also used to change age intervals from (x,5) to (x + 2.5,5) for other life table functions.

Given a closed population and accurate counts, the four flow equations provide, for each age, the death rate for all marital statuses combined and three constraints on the five possible movements between living states. Specifically, interstate movements are possible from s to m (first marriages), from m to w (widowhoods), from m to v (divorces), from w to m (remarriage from widowed), and from v to m (remarriage from divorced). Population data alone thus leave two of the five rates unconstrained.

For the marital status model, the last two constraints can be found from available information. A widowhood arises when a married person dies, and U.S. mortality data by age, sex and marital status are available. Since husbands are typically 2–3 years older than their wives, wives' widowhood rates can be inferred from married male mortality. Here, we assume that husbands are 2.5 years older than their wives. Hence, for example, information on the mortality of married men aged 45–49 can be used to approximate the widowhood rate for married women aged 42.5–47.4.

The 2014 Kennedy-Ruggles (hereafter KR) study provides the final piece, the age-specific rates of divorce. Steven Ruggles was kind enough to provide me the 2008–2010 female divorce rates underlying Fig. 4 of the KR article. With m_{mw} found from married male mortality, m_{mv} derived from the KR divorce rates, and the state-specific death rates available from vital statistics data, let state-specific mortality differentials be described by

$$m_{jd}(x+2.5,5) = c_j(x+2.5,5)\, m_{sd}(x+2.5,5),\ j = m, w, v \tag{10.3}$$

where the c_j reflect the known differentials. Equation (10.1) then provide 4 equations with 4 unknowns: m_{sd} and the marriage rates m_{sm}, m_{wm}, and m_{vm}. The solutions are

$$\begin{aligned} m_{sd}(x+2.5,5) = \{&[P_s(x,5,t) + P_m(x,5,t) + P_w(x,5,t) + P_v(x,5,t)] \\ &-[P_s(x+5,5,t+5) + P_m(x+5,5,t+5) \\ &+ P_w(x+5,5,t+5) + P_v(x+5,5,t+5)]\}/PPT \end{aligned} \tag{10.4}$$

where $PPT = PP_s(x,5) + c_m(x + 2.5,5)\ PP_m(x,5) + c_w(x + 2.5,5)\ PP_w(x,5) + c_v(x + 2.5,5)\ PP_v(x,5)$. With m_{sd} known from Eq. (10.4), the three marriage rates are given by

$$m_{sm}(x+2.5,5) = \left[P_s(x,5,t) - P_s(x+5,5,t+5) - PP_s(x,5)\,m_{sd}(x,5)\right]/PP_s(x,5) \tag{10.5}$$

$$\begin{aligned} m_{wm}(x+2.5,5) = \Big[&P_w(x,5,t) - P_w(x+5,5,t+5) \\ &- PP_w(x,5)\,c_w(x+2.5,5)\,m_{sd}(x+2.5,5) \\ &+ PP_m(x,5)\,m_{mw}(x+2.5,5)\Big]/PP_w(x,5) \end{aligned} \tag{10.6}$$

$$\begin{aligned} m_{vm}(x+2.5,5) = \Big[&P_v(x,5,t) - P_v(x+5,5,t+5) \\ &- PP_v(x,5)\,c_v(x+2.5,5)\,m_{sd}(x+2.5,5) \\ &+ PP_m(x,5)\,m_{mv}(x+2.5,5)\Big]/PP_v(x,5) \end{aligned} \tag{10.7}$$

10.2.2 Data Assembly, Adjustment and Application

The present strategy uses available Census Bureau population and survey data and NCHS mortality data, augmented by KR divorce rates, to calculate occurrence/exposure rates and from them marital status life tables for U.S. Females during the 2000–2005 and 2005–2010 intervals. A fair amount of data management and approximation was required, and the details are described in the Appendix. However, some additional discussion is needed here concerning divorce data from the American Community Survey (ACS), which were used to produce the KR divorce rates.

The ACS is a major survey effort conducted by the U.S. Bureau of the Census (Elliott et al. 2010). The ACS asked whether the survey respondent was divorced within the past 12 months. Retrospective questions of that kind are known to be subject to potentially serious recall biases. In the case of divorce, reporting errors may also arise because the effective date of a divorce depends on state law, whose provisions may not be known to the respondent. The Census Bureau conducted several studies to evaluate the retrospective questions, and concluded that the ACS divorce data provided usable estimates. Nonetheless, the available evidence suggests that some caution is needed, since the ACS may overestimate the incidence of divorce (see Appendix).

Compared to the Schoen and Standish (2001) female divorce rates for 1995, the KR 2008–2010 rates were lower at ages below 35 and higher at ages above 35 (see Table 10.A1). At ages between 45 and 70, the KR rates exceeded 125 % of the 1995 divorce rates. That sizeable increase struck me as problematic. The KR rates at those ages were considerably greater than those seen at any time in the past, while crude rates of divorce had been declining since 2000 (see NCHS (2014a, b)). To control for possible overstatement at ages over 45, an alternate set of calculations was done with the KR divorce rates "capped" at 125 % of the 1995 rates. While arbitrary, that cap seems a reasonable upper bound.

Given the present inadequacy of the vital statistics on marriage and divorce, the ACS provides a valuable alternative source. At a minimum, one can say with considerable confidence that divorce rates from the ACS do not underestimate the incidence of divorce.

10.3 Results from the Marital Status Life Tables

Table 10.1 presents summary measures calculated from the marital status life tables for U.S. Females in 2000–2005 and 2005–2010. For each time interval, 2 sets of measures are shown: one derived from KR divorce rates and one where the KR rates are held to a maximum of 125 % of the 1995 rates. (Table 10.A1 shows those divorce rates along with the calculated 2000–2005 and 2005–2010 marriage rates.) Table 10.1 also provides comparable summary measures from 1988 and 1995 marital status life tables.

The proportion ever marrying (i.e. moving from state s to state m) has been declining. In 1988 and 1995, 87–88 % of women married at some point in their lives. In 2000–2005, that proportion fell to 84 %, and in 2005–2010 it declined to 80 %. In 2005–2010, women lived 45 % of their lives in the Never Married state, up from 39 % in 1988. The mean age at first marriage has increased since 1988, rising from 25.1 years in 1995 to 27.2 years in 2005–2010. That average age is well above the 20–21 year average age that prevailed some 50–60 years ago. Marriage has been retreating and singlehood advancing.

The proportion of marriages ending in divorce appears to have remained roughly constant, though there may have been a slight increase. In 2000–2005, 45–46 % of marriages ended in divorce, up from the 43 % found in 1988 and 1995. The capped KR rates suggest a possible decline back to 43 % in 2005–2010.

It is important to remember that the age-specific divorce rates show a fall in divorce at ages under 35 and a rise at higher ages. Aggregating to an overall measure of divorce is thus dependent on the weights given to the different ages. Kennedy and Ruggles (2014) saw a 25–40 % increase in divorce because their standardization procedure gave substantial weight to the higher ages. Here, the marriage, divorce, and mortality rates, by shaping the experience of the hypothetical life table cohort, generate their own weights, and reflect little change in the overall risk of divorce. I would argue that the life table proportion of marriages ending in divorce is a much more intuitively interpretable measure than a standardized divorce rate.

Other measures of marriage and marital stability are equivocal with respect to trends. There is little change in the number of marriages per person marrying, and

Table 10.1 Summary measures from marital status life tables for United States females, years 1988, 1995, 2000–2005, and 2005–2010

Measure	1988	1995	2000–2005		2005–2010	
			Interpolated KR rates	Interpolated & capped KR rates	KR Rates	Capped KR rates
1. Proportion ever marrying	.866	.878	.841	.841	.797	.797
2. Proportion of life never married	.391	.398	.429	.420	.449	.449
3. Mean age at first marriage	25.08	26.65	26.95	26.95	27.17	27.17
4. Number of marriages per person marrying	1.51	1.46	1.44	1.43	1.52	1.51
5. Proportion of marriages ending in widowhood	.393	.434	.356	.364	.356	.377
6. Proportion of marriages ending in divorce	.432	.425	.460	.449	.461	.428
7. Mean age at divorce	34.41	37.31	39.25	38.66	40.83	39.32
8. Average duration of a marriage	24.76	25.72	27.02	27.27	26.11	26.96
9. Proportion of life married	.412	.412	.407	.410	.396	.404
10. Remarriages of widows per widowhood	.063	.048	.038	.037	.036	.035
11. Proportion of life widowed	.100	.092	.068	.069	.068	.071
12. Remarriages from divorce per divorce	.723	.687	.630	.644	.717	.753
13. Mean age at remarriage from divorced	36.05	39.72	39.72	39.60	42.94	42.35
14. Average duration of a divorce	13.39	14.47	15.09	14.96	12.32	11.92
15. Proportion of life divorced	.108	.099	.105	.101	.086	.076

Source: Schoen and Weinick (1993); Schoen and Standish (2001); see text and Appendix for years 2000–2005 and 2005–2010

Notes: KR rates are the 2008–2010 divorce rates calculated by Kennedy and Ruggles (2014, Fig. 4). Capped KR rates are KR rates restricted to a maximum of 125 % of the 1995 marital status life table divorce rate (per Schoen and Standish 2001). Interpolated KR rates average the 1995 and KR rates, bringing them to the year 2002.5

Table 10.A1 Rates of First Marriage and Divorce, U.S. Females, 1995, 2000–2005 and 2005–2010

	REAP calculated first marriage rates		Divorce rates			
Age	2000–2005	2005–2010	KR 2008–2010	Capped KR	Interpolated KR to 2002.5	MSLT 1995
12.5–17.5	.00577	.00599	.0152	.0152	.02036	.02638
17.5–22.5	.04566	.04053	.0334	.0334	.04041	.04659
22.5–27.5	.09968	.08792	.0356	.0356	.03836	.03964
27.5–32.5	.09220	.08328	.0321	.0321	.03355	.03522
32.5–37.5	.06443	.05763	.0288	.0288	.02876	.02871
37.5–42.5	.03002	.02519	.0270	.0270	.02550	.02372
42.5–47.5	.01837	.01768	.0234	.0225	.02109	.01842
47.5–52.5	.01052	.01184	.0184	.0160	.01584	.01282
52.5–57.5	.00678	.00804	.0142	.0100	.01134	.00803
57.5–62.5	.00402	.00402	.0102	.0056	.00755	.00448
62.5–67.5	.00249	.00249	.0069	.0033	.00493	.00265
67.5–72.5	.00168	.00168	.0038	.0020	.00299	.00156
72.5–77.5	.00085	.00085	.0015	.0015	.0015	.00120
77.5–82.5	.00012	.00012	.0010	.0010	.0010	.00108
82.5–87.5	0	0	.0008	.0008	.0008	.00060
87.5+	0	0	.0002	.0002	.0002	.00029

Notes: KR rates are from Kennedy and Ruggles (2014). The MSLT 1995 divorce rates follow from Schoen and Standish (2001). Linear interpolation was used to produce the 2002.5 divorce rates from the 1995 MSLT rates and the KR 2008–2010 rates. The capped rates are no more than 125 % of the 1995 MSLT rates. For further discussion, see text and Appendix

the proportion of life lived married has held quite steady. The proportion of life lived in both the Widowed and Divorced states declined slightly. The mean age at divorce has risen, and was around age 40 in 2005–2010. The likelihood of remarriage from divorce remains high, though the measure was rather volatile. The mean age at remarriage after divorce has been rising, and was over age 42 in 2005–2010. The average duration of a marriage, an overall measure of stability, actually increased, but was only about 27 years.

10.4 Summary and Conclusions

Marital status life tables for American women in 2000–2005 and 2005–2010 were constructed using a new approach that combines vital statistics mortality data, census population and survey data, and recently available divorce data from the American Community Survey. Methodologically, the Rate Estimation from Adjacent Populations procedure makes greater use of population counts and allows multistate life tables to be calculated with less reliance on individual level data on movements between states. Substantively, the results provide a update on levels and trends in marriage and divorce.

The level of marriage has continued to decline, even though 80 % of women married in 2005–2010. The average age at first marriage has risen slightly, reaching 27.2 years of age. The probability that a marriage ends in divorce is 43–46 %, holding about constant when compared to 1988 and 1995.

Some countervailing signs regarding future trends can be noted. The fall in divorce rates at ages under 35 may indicate that marriage has become more stable for younger cohorts. A later age at marriage has long been associated with greater maturity and marital stability, and the continuing decline in the probability of ever marrying implies that persons who are less suited to marriage are less likely to marry. One may then ask why divorce rates have not fallen more. On balance, since the ACS data may overstate the likelihood of divorce, it is prudent to view divorce risks as remaining roughly level on a high plateau.

Assessments of changes in marriage and divorce patterns must also be qualified, because the available data do not offer the precision given by a well-functioning, contemporaneous vital statistics system. Issues of data quality and consistency have arisen, which required the use of numerous estimates, approximations, and adjustments.

Nonetheless, some tentative conclusions can be drawn. American marriage is still in retreat, as the proportion ever marrying is continuing to fall and the average age at first marriage is still rising. But marriage is not in free fall. Four out of five women marry, over half of marriages do not end in divorce, and a substantial majority of divorced women remarry. When–or whether—the retreat from marriage will end remains one of the great questions in contemporary demography.

Acknowledgements Assistance from Joshua Goldstein, Rose Kreider, Jamie Lewis, and Steven Ruggles is gratefully acknowledged.

Appendix: Data Assembly, Adjustment, and Application

Mortality Data

The mortality rates used in both the 2000–2005 and 2005–2010 marital status life tables were taken from the published United States Life Tables, 2005 (Arias et al. 2010, Table 3). Death rates for females aged x to x + n, were calculated as life table deaths, ${}_nd_x$, divided by life table person-years, ${}_nL_x$.

Life table person-year values were used to project the total female population in 2000 to 2005 and then to 2010. That was done to prevent distortions from in and out migration. The projected female populations, by age, were proportionally allocated to the four marital statuses using the data populations described below.

Mortality by marital status was obtained from National Vital Statistics Reports, Vol. 55, No. 19 (August 21, 2007), specifically from Table 25, "Number of deaths, death rates, and age-adjusted death rates for ages 15 and over, by marital status and sex: United States, 2004." Age 15 was taken as the minimum age at marriage. Husbands were assumed, on average, to be 2.5 years older than their wives, e.g. the death of a husband age 50 produced a widow aged 47.5. Five year age groups up to age 85 were generated from the published 10-year age groups up to age 75 by linear interpolation.

Divorce Data

The divorce data originated from the American Community Survey (ACS) conducted by the U.S. Bureau of the Census (Elliott et al. 2010). The survey inquired as to whether the survey respondent was divorced within the past 12 months. From that ACS data, Kennedy and Ruggles (2014) calculated age-specific female divorce rates for the years 2008–2010, and those rates are used for the 2005–2010 tables. The divorce rates for the 2000–2005 interval were found by linear interpolation to the year 2002.5 using the 1995 divorce rates underlying Schoen and Standish (2001) and the 2008–2010 KR rates.

In 2006, the Census Bureau conducted an evaluation of the ACS marital history questions (O'Connell et al. 2007). In Table 35, the Evaluation Report stated that 7.8 % of the 176 respondents who reported a divorce in the last 12 months did not actually have a divorce decree finalized during that time. The highest incidence of false reports, 13.5 %, was for persons aged 55–64. The Evaluation Report nonetheless concluded that the ACS produced reasonable estimates when compared to estimates from vital statistics. (The Evaluation Report did not come to a similar conclusion with respect to the retrospective marriage question.)

Elliott, Simmons, and Lewis (2010) also performed a review of results from the ACS retrospective marital history questions, reaching a similar conclusion with respect to the divorce data. Elliott et al. (2010) also reported some state level comparisons, including data for Delaware and New Hampshire, states they described as having high quality vital statistics that did not differ statistically from the ACS numbers. However, in 2007, 3215 divorce decrees were filed in Delaware, while the 2008 ACS estimated 4318 divorces to women. In New Hampshire, there were 4981 divorce decrees filed while the 2008 ACS estimated 5059 women divorcing in the previous 12 months. The ACS overestimate for New Hampshire is only 1.6 %, but for Delaware, it was 34.3 %. There is thus a real possibility that divorce is overreported in the ACS.

U.S. Population Data by Marital Status for 2000, 2005, and 2010

The Rate Estimation from Adjacent Populations (REAP) procedure requires counts of the female population, by marital status, at ages 0–4, 5–9, ..., 80–84, and 85+ for the years 2000, 2005, and 2010. The female population for all marital statuses combined for 2000 came from Census 2000, Summary File 1, Table 2, and for 2010 from the corresponding 2010 Summary File. For 2005, the aggregate populations were found from Current Population Survey, Annual Social and Economic Supplement, 2005, Table 1.1. However, the 2005 and 2010 population data were only used at ages 0–4 in 2005 and 2010, respectively. Populations at higher ages were found by 5-year survival, as described above.

The 2000, 2005, and 2010 female population by marital status, based on U.S. Census Bureau data, were taken from tables titled "Population, 5-year Age Groups, by Sex, Country, Marital Status, Age and Year" that appeared on the United Nations Economic Commission for Europe (UNECE) website, www.unece.org/pxweb/dialog/Print.asp?Matrix=005_GEPOPop5YearMaSta_r&timeid. For 2000, 10-year age groups at ages 55 and above were broken into 5-year age groups using linear interpolation and data from Kreider and Simmons (2003, Table 1). For 2005, linear interpolation was used augmented by data from Table 1, "Marital Status of the Population 15 Years and Over by Age and Sex: 2005", U.S. Census Bureau, Current Population Survey, Annual Social and Economic Supplement, 2005. For 2010, linear interpolation was again used on the UNECE data.

Finalizing the Rates of Interstate Transfer

The REAP procedure was applied to the 2000 and survived 2005 and 2010 populations to produce estimates of the 2000–2005 and 2005–2010 marriage rates. The initial estimates, subject to errors in the census enumerations, survey weights, interpolations, marital status reporting, and other assumptions, were not fully satisfactory. The estimated first marriage rates appeared quite accurate up to age 37.5, by which age the great majority of first marriages were entered. For higher ages, m_{sm} in both years was assumed to be (0.75) times the rates used in the 1995 U.S. Female marital status life tables of Schoen and Standish (2001), a reasonable approximation of the post-1995 decline in marriage (cf. the NCHS state level data in the table "Marriage Rates by State: 1990, 1995, and 1999–2011" downloaded 8/5/2014 from cdc.gov/nchs/data/dvs/ marriage_rates_90_95_99-11.pdf).

For widow remarriage, the REAP estimated rates were unusable. In both years those rates were set equal to (0.75) times the Schoen and Standish (2001) m_{wm} rates at every age. The error here is hard to assess, but widowhood plays a small role at the ages of principal interest here. The estimated remarriage from divorce rates in both years were quite reasonable up to age 47.5, after which they were assumed to be (0.75) times the m_{vm} rates underlying Schoen and Standish (2001). The bulk of remarriage from divorce events occur before age 47.5, so the error introduced by the fractional reduction should be minor.

Constructing the Marital Status Life Tables

The multistate life tables were calculated by the linear method (Schoen 1988, Chap. 4). Under age 15, values followed the U.S. Life Tables for 2005. Special procedures were needed for the open-ended 87.5+ age group. There, modified flow equations implemented matrix Eq. (4.21) of Schoen (1988, p. 70) to find the $PP_j(87.5+)$ values from known mortality and estimated marriage rates.

References

Arias, E., Rostron, B. L., & Tejada-Vera, B. (2010). *United States life tables, 2005. National vital statistics reports* (Vol. 58, No. 10). Hyattsville: National Center for Health Statistics.

Axinn, W. G., & Thornton, A. (2000). The transformation in the meaning of marriage. In L. J. Waite (Ed.), *The ties that bind* (pp. 147–165). New York: Aldine de Gruyter.

Copen, C. E., Daniels, K., Vespa, J., & Mosher, W. D. (2012). *First marriages in the United States: Data from the 2006–2010 national survey of family growth* (National Health Statistics Reports No. 49 (March 22)). Hyattsville: National Center for Health Statistics.

Elliott, D. B., Simmons, T., & Lewis, J. M. (2010). *Evaluation of the marital events items on the ACS*. U.S. Technical and Analytic Reports on the American Community Survey. Washington, DC: U.S. Census Bureau. Downloaded November 16, 2014, from https://www.census.gov/hhes/socdemo/marriage/data/acs/index.html

Goldstein, J. R. (1999). The leveling of divorce in the United States. *Demography, 36*, 409–414.

Kennedy, S., & Ruggles, S. (2014). Breaking up is hard to count: The rise of divorce in the United States, 1980–2010. *Demography, 51*, 587–598.

Kreider, R. M., & Simmons, T. (2003). *Marital status, 2000. Census 2000 brief (C2KBR30)*. Washington, DC: U.S. Census Bureau.

National Center for Health Statistics. (1996). Births, marriages, divorces, and deaths for 1995. *Monthly vital statistics report* (Vol. 44, No. 12). Hyattsville: Public Health Service

National Center for Health Statistics. (2014a). *Marriage rates by state*. Downloaded August 5, 2014, from website https://cdc.gov/nchs/data/dvs/marriage_rates_90_95_99-11.pdf

National Center for Health Statistics. (2014b). *National marriage and divorce trends*. Downloaded August 5, 2014, from website https://cdc.gov/nchs/nvss/marriage_divorce_tables.htm

O'Connell, M., Gooding, G., & Ericson, L. (2007). *2006 Evaluation report covering marital history* (American Community Survey Content test Report, p. 9). Washington, DC: U.S. Census Bureau.

Raley, R. K. (2000). Recent trends and differentials in marriage and cohabitation: The United States. In L. J. Waite (Ed.), *The ties that bind* (pp. 19–39). New York: Aldine de Gruyter.

Ruggles, S. (2015). *Patriarchy, power, and pay: The transformation of American families, 1800–2015*. Presidential Address at the Annual Meeting of the population Association of America in San Diego.

Schoen, R. (1988). *Modeling multigroup populations*. New York: Plenum.

Schoen, R. (2010). Gender competition and family change. *Genus, 66*, 95–120.

Schoen, R. (2015). Multistate transfer rate estimation from adjacent populations. *Population Research and Policy Review*, on line 21 September 2015.

Schoen, R., & Standish, N. (2001). The retrenchment of marriage: Results from marital status life tables for the United States, 1995. *Population and Development Review, 27*, 555–563.

Schoen, R., & Weinick, R. (1993). The slowing metabolism of marriage: Figures from 1988 U.S. marital status life tables. *Demography, 30*, 737–746.

Chapter 11
Emigration and The Stable Population Model: Migration Effects on the Demographic Structure of the Sending Country

Cristina Bradatan

11.1 Introduction

Mortality and fertility are often described as the main demographic phenomena, while migration is rather seen as a step-child. This is due to historical reasons (after all, demography started with *Bills of Mortality,* not Bills of Migration, while fundamental demographic models such as the *stable population* ignore migration), but also because, unlike fertility and mortality, migration data are not yet of a very high quality. The isolation of migration as a subject of study manifests itself even in demographic methods textbooks, where usually only a few pages are dedicated to mathematical models of migration.

Theories of migration do exist (Massey et al. 1993). In fact, migration theories might be even better than theories of fertility under low fertility conditions (Lutz 2006). In terms of models, the statistical ones tend to dominate the scene and be used on a regular basis. Although there are a number of models that use formal demography to model migration (Cerone 1987; Espenshade et al. 1982; Mitra 1983; Rogers 1990, 1995; Schmertmann 2012), these are not well integrated in applied research. While multistate population models are often used outside demography per se (Schoen 1988), this is not necessarily the case for migration.

When migration is integrated in formal demography models, either the rate of migration or the number of migrants is constant. Espenshade et al. (1982), for example, show that if a stable, below replacement fertility population receives a constant stream of immigrants that adopt the fertility and mortality of the host population, the result will be a stationary population. Mitra (1983, 1990) and Schmertmann (1992, 2012) use the same idea of a constant number of immigrants

C. Bradatan (✉)
Texas Tech University, 2500 Broadway, Lubbock, TX 79409, USA
e-mail: cristina.bradatan@ttu.edu

R. Schoen (ed.), *Dynamic Demographic Analysis*,
The Springer Series on Demographic Methods and Population Analysis 39,
DOI 10.1007/978-3-319-26603-9_11

entering into a below replacement stable population, extending the Espenshade et al. (1982) results by allowing the immigrant population to have a different schedule of mortality and fertility than the host population. Rogers' multistate work (1990, 1995), on the other hand, analyzes the situation in which migration rates between two or more regions/countries are constant (as are the fertility and mortality schedules in those countries) and highlights the effects of migration on the population structures.

The models reviewed above make some strong assumptions about the migrant population: either that the rates of migration stay constant, or that the number of immigrants does not change. In a real life situation, is it more probable to have a constant number or a constant rate of emigration? This is a difficult question to answer, not only because both number and rate of emigration tend to vary from one year to another, but also because good quality data to adequately measure the number/ rate of emigration for a country are lacking. While some countries keep some kind of statistics on emigration, most don't have reliable information.[1] The motivations and aspirations of potential migrants, the business cycle (Orrenius and Zavodny 2009), as well as the socio-economic institutions that link areas of immigration and emigration play an important role in shaping the direction and dimensions of migration flows (Massey 1999). In general, pull and push migration factors, as well as obstacles (such as restrictive migration policies in receiving or sending country or geographical distance), affect migration trends. Although migrants are a necessary addition to the labor markets of rich countries confronted with the problem of aging, other considerations (such as keeping the population ethnically homogenous) play an important role in the design and implementation of migration policies. Because migration streams are the result of so many factors, it is difficult to say whether the constant flow or the constant rate of emigration hypothesis is closer to reality. However, it is well established that age-specific migration rates display strong regularities (Rogers et al. (2005)). This makes a model with constant *emigration rates* more credible than one in which the number of emigrants remain constant for a long period of time.

One area in which migration models are often used is population projections, and they are tremendously important especially for countries with significant rates of immigration. In these cases, net migration is driven by the rates of immigration. The population structure of the sending country is often not taken into account and it is seen more like an unlimited reservoir of migrants. This might have to change – as recent events show, many traditional countries of emigration (e.g. Italy, Spain, Portugal) have moved toward below replacement levels of fertility, stopped sending emigrants, and in some cases, become themselves countries of immigration.

In this chapter I model the situation of a below replacement fertility population that also experiences emigration. The number and age structure of each of the two

[1] In Romania, the 2011 census included questions on people who used to be part of a household and emigrated.

populations formed (emigrants and stayers) is determined. Numerical calculations are then done for the case of Romania and Romanian emigrants to Spain.

11.2 Mathematical Description of the Model

Let's suppose that in country A there is a female human population of various ages a (a >= 0). The total number of this (female) population at time t is N(t); the number of women age a at time t is N(a,t); women's childbearing age is limited to [α, β], the death rate at age a is m(a), w(a) is the age-specific fertility rate, and the annual number of female births, at time t is B(t). Then:

$$B(t) = \int_{\alpha}^{\beta} N(x,t)\, w(x) dx \tag{1.1}$$

11.2.1 Constant Age Schedule of Emigration Rates

We suppose that the rate of emigration for all ages is constant in time (o(a), a >= 0) and the fertility and mortality schedules of the total population (stayers and emigrants) are the same and do not change in time for neither of them. The total population (stayers + emigrants) age a at time t is N(a,t); constant rate of growth for population total population is γ. Stayer population age a at time t is S(a,t); emigrant population age a at time t is E(a,t). Then:

$$N(a,t) = B(t-a) \exp\left(-\int_0^a m(x)dx\right) \tag{1.2}$$

If the population is stable, then in the long term:

$$B(t) = Be^{\gamma t}$$

where B = B(0). A stable population with below replacement fertility has γ <0. Hence the total population age a at time t is:

$$N(a,t) = Be^{\gamma(t-a)} \exp\left(-\int_0^a m(x)dx\right) \tag{1.3}$$

The stayer population age a at time t is:

$$S(a,t) = Be^{\gamma(t-a)} \exp\left(-\int_0^a (m(x) + o(x))\, d\right)$$

and total stayer population is:

$$S(t) = Be^{\gamma t} \int_{0}^{\infty\infty} \exp\left(-\gamma a - \int_{0}^{a} (m(x) + o(x))\, dx\right) da \tag{1.4}$$

Basically emigration acts as a 'surmortality' in reducing the size of the stayer population.

The emigrant population at time t is:

$$E(t) = N_0 e^{\gamma t} - S(t) \tag{1.5}$$

where N_0 is the total population at time 0.

11.3 Case Study: Romanian Emigration to Spain

As noted before, one of the problems in studying migration using formal demography is the quality of data. If a country has large numbers of immigrants, many of these immigrants tend to be undocumented, and they often do not appear in the official data. In the US, for example, it is estimated that 12–15 million immigrants are currently undocumented, a sizeable amount even for a country where historically 7–10 % of population has been foreign born.

The particular case I focus on, Romanian migration to Spain after 2007, has three advantages: (1) a significant migration flow; (2) very low fertility in the sending country; and (3) decent quality data.

1. *Significant migration flow.* The stock of Romanian immigrants in Spain during 2007–2014 stayed between 600,000 and 800,000, representing 3–4 % of the total Romanian population.
2. *Very low fertility.* ATFR of 1.3–1.4 is an unusual occurrence in a sending country. Usually sending countries have a young population structure and, because of that, the effect of emigration on the demographic future of the country is limited. This is not at all the case in Romania, where the ratio between retirees and workers is now up to 1.2.
3. *Decent quality data.* There are two main reasons why the data in this case are decent. First of all, on January 1, 2007, Romania became a member of the European Union. Although Spain imposed some work restrictions (until 2014), integration in the European Union made it easier for Romanians to get legal work status in Spain. The second reason is the Spanish health system: a person can have access to this system regardless of their legal status if they can prove they have a Spanish address. This creates an incentive for immigrants – legal and illegal – to register with the local authorities.

Fig. 11.1 Age specific migration rates, Romanians to Spain, females, 2008

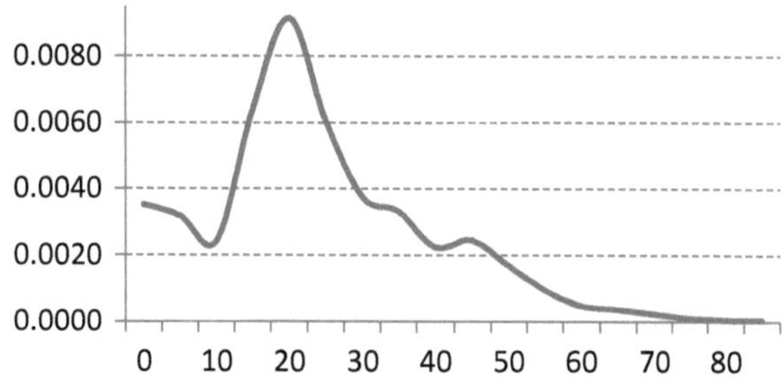

For the purposes of this study, I use data 2008 data from the Institutul National de Statistica (Romania) and the Instituto Nacional de Estatistica (Spain) to calculate how the population structure in Romania and Romanians in Spain will look in the future. The assumptions are that there is no immigration into Romania, and that emigrants do not change their mortality or fertility while in Spain. Both these assumptions are supported by observed information. Immigration to Romania is very small. In 2014, 0.9 % of Romanian population was foreign born, many of them classified as such due to border changes rather than immigration (Eurostat 2014). Regarding fertility, immigrant women in Spain who are not of African origin tend to have low levels of fertility, similar to their country of origin (Castro-Martin and Rosero-Bixby 2011). Due to data restrictions, I also assume that emigrants from Romania go only to Spain, and do not return. This assumption makes the rate of emigration from Romania lower than it is in reality, but unfortunately there are no good estimates of the number and age schedule of Romanians who leave or re-enter the country.

In 2008, 61,262 Romanian emigrants went to Spain, of which 30,387 were females. In the following, I consider 2008 as time 0, and focus only on female population. Figure 11.1 shows the age pattern of migration rates in 2008. This pattern is similar to the one presented by Rogers et al. (2005) as general for age specific migration rates.

In Table 11.1, I calculate the intrinsic growth rate for the Romanian population without taking into account emigration and using the 2008 fertility and mortality rates using Coale's iterative process (Preston et al. 2001).

11.3.1 The Below Replacement Level Stable Population with Constant Rate of Emigration

I assume that the age specific migration rates will remain the same as in 2008. In Fig. 11.2, the dashed line shows the age structure of population *without emigration*, while the dark columns show the age structure of the female population with continuous migration for 100 years. Table 11.2 shows the dependency ratio and percentage of people age 65+ and age 0–14 in both populations at three points in time: 0, 100 and 150.

Table 11.1 The intrinsic growth rate, Romania, 2008

Age group	nLx/l0	nFx	nLx/L0*nFx	First iteration (Coale)	y(γ1)	Second iteration (Coale)	y(γ2)	Third iteration (Coale)	y(γ3)
15	4.68	0.0187	0.0874	1.3228	0.1157	1.3189	0.1153	1.3190	0.1153
20	4.93	0.0330	0.1624	1.4329	0.2326	1.4275	0.2318	1.4276	0.2318
25	4.97	0.0403	0.2006	1.5521	0.3114	1.5449	0.3100	1.5451	0.3100
30	4.97	0.0279	0.1386	1.6812	0.2329	1.6721	0.2317	1.6723	0.2317
35	4.97	0.0100	0.0494	1.8211	0.0900	1.8097	0.0894	1.8100	0.0894
40	4.96	0.0021	0.0106	1.9727	0.0209	1.9586	0.0207	1.9590	0.0207
45	4.93	0.0001	0.0005	2.1368	0.0010	2.1198	0.0010	2.1202	0.0010
		NRR	0.6495		1.0045		0.9999		1.0000
		γ0	−0.01599	γ1	−0.01582	γ2	−0.01582	γ3	−0.01582

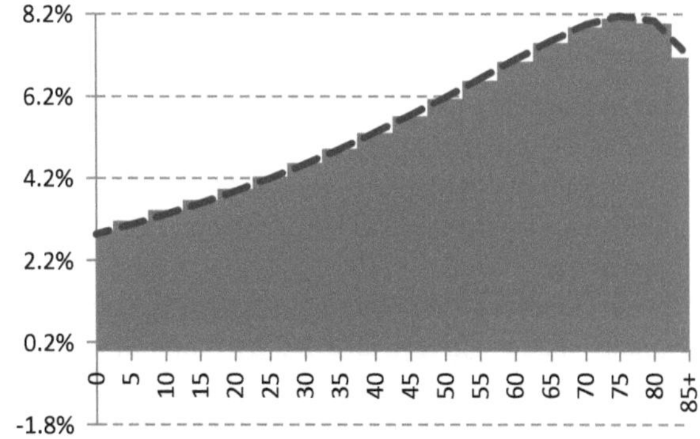

Fig. 11.2 Female population structure, stable population with (*shaded area*) and without migration (*dashed line*)

Table 11.2 Some summary measures for years 0, 100 and 150, Romania

		Without emigration	With emigration
0	Total female population	10,586,266	10,555,879
	Number of female emigrants, Year 0	30,387	
	Dependency ratio	50.1 %	50.2 %
	%65+	17.86 %	17.91 %
	%0–14	15.51 %	15.51 %
100	Total female population	4,059,747	3,907,732
	Number of females living in Spain	152,015	
	Dependency ratio	92.5 %	92.4 %
	%65+	38.9 %	38.5 %
	%0–14	9.2 %	9.5 %
150	Total female population	1,840,653	1,771,731
	Number of females living in Spain	68,922	
	Dependency ratio	92.5 %	92.4 %
	%65+	38.9 %	38.5 %
	%0–14	9.2 %	9.5 %

Emigration to Spain starts in Year 0 (2008); the percentage emigrating is 0.3 %. The number of emigrants, and their descendants, eventually reaches 3.9 % in the stable population, as both populations (with and without emigration) decrease in size significantly.

One of the interesting results of this exercise is that emigration, in the long run, lowers the dependency ratio in the in the sending population. Long term emigration depletes the shrinking stayer population of its older age segments while having very little effect on the 0–14 segment. As time goes by, in the below replacement population, the shrinking birth cohorts represent a larger segment of the population with emigration than in the case without emigration. Note that the number of emigrants decreases significantly from year 100 to Year 150.

11.4 Conclusions

This chapter offer new insights into the demographic behavior of a below replacement population that experiences emigration. While other studies look to the effects of immigration on stable populations, this chapter analyzes what happens if the sending country itself has below replacement levels of fertility and constant rates of emigration.

A first result worth mentioning is that the age structure of the sending population is slightly but positively affected by emigration in the long term. While the birth cohorts shrink due to the below replacement fertility, emigration also shrinks the older cohorts, leading to a slightly more balanced population structure. A second interesting result from the model is that the size of the population of Romania with emigration is not much smaller than the size of the population without emigration. Emigration acts as a 'surmortality', but at current levels has only a modest effect, stabilizing at under 4 % of the population.

References

Cerone, P. (1987). On stable population theory with immigration. *Demography, 24*, 431–438.

Espenshade, T. J., Bouvier, L. F., & Arthur, W. B. (1982). Immigration and the stable population model. *Demography, 19*, 125–133.

Eurostat. (2014). *Migration and migrant population statistics*, Retrieved on May 21, 2015, from http://ec.europa.eu/eurostat/statistics-explained/index.php/Migration_and_migrant_population_statistics#Foreign_and_foreign-born_population

Lutz, W. (2006). Toward building a comprehensive migration projections framework. In N. Howe & R. Jackson (Eds.), *Long-term immigration projection methods: Current practice and how to improve it* (WP#2005-3). Chestnut Hill: Center for Retirement Research at Boston College.

Martin, T. C., & Rosero-Bixby, L. (2011). Maternidades y fronteras la fecundidad de las mujeres inmigrantes en España. *Revista Internacional de Sociología, 1*, 105–131.

Massey, D. S. (1999). International migration at the dawn of the twenty-first century: The role of the state. *Population and Development Review, 25*(2), 303–322.

Massey, D. S., Arango, J., Hugo, G., Kouaouci, A., Adela, P., & Edward Taylor, J. (1993). Theories of international migration: A review and appraisal. *Population and Development Review, 19*(3), 431–466.

Mitra, S. (1983). Generalization of the immigration and the stable population model. *Demography, 20*, 111–115.

Mitra, S. (1990). Immigration, below-replacement fertility, and long-term national population trends. *Demography, 29*(4), 595–612.

Orrenius, P., & Zavodny, M. (2009). *Tied to the business cycle: How immigrants fare in good and bad economic times*, Migration Policy Institute. Retrieved on May 3, 2014, from www.migrationpolicy.org

Preston, S. H., Heuveline, P., & Guillot, M. (2001). *Demography. Measuring and modeling population processes*. Malden: Blackwell Publishing.

Rogers, A. (1990). The multistate stable population model with immigration. *Mathematical Population Studies, 2*, 313–324.

Rogers, A. (1995). *Multiregional demography. Principles, methods, and extensions*. New York: Wiley.

Rogers, A., Castro, L. J., & Lea, M. (2005). Model migration schedules: Three alternative linear parameter estimation methods. *Mathematical Population Studies, 12*, 17–38.
Schmertmann, C. (1992). Immigrants' ages and the structure of stationary populations with below-replacement fertility. *Demography, 29*(4), 595–612.
Schmertmann, C. (2012). Stationary populations with below-replacement fertility. *Demographic Research, 26*(14), 319–330.
Schoen, R. (1988). Practical uses of multistate population models. *Annual Review of Sociology, 14*, 341–361.

Chapter 12
Exploring Stable Population Concepts from the Perspective of Cohort Change Ratios: Estimating the Time to Stability and Intrinsic r from Initial Information and Components of Change

David A. Swanson, Lucky M. Tedrow, and Jack Baker

12.1 Introduction

Cohort Change Ratios (CCRs) have a long history of use in demography, dating back to at least Hardy and Wyatt (1911). Under the rubric of "Census Survival Ratios," they have been used to estimate adult mortality (Swanson and Tedrow 2012; United Nations 2002) and under the rubric of the "Hamilton-Perry" method, they are used to make population projections (Hamilton and Perry 1962; Smith et al. 2013; Swanson and Tayman 2013, 2014; Swanson and Tedrow 2013; Swanson et al. 2010). However, CCRs appear to have been largely overlooked in regard to examining the concept of a stable population (Caswell 2001; Coale 1972; Dublin and Lotka 1925; Lotka 1907; Preston et al. 2001; Schoen 2006; United Nations 1968).

The authors are grateful to a number of people for comments and suggestions, including Hiram Beltran-Sanchez, Stan Drezek, Barry Edmonston, Victor M. Garcia-Guerrero, David Hamiter, Richard Verdugo, Robert Schoen, Webb Sprague, and Jeff Tayman.

D.A. Swanson (✉)
Department of Sociology, University of California Riverside, Riverside, CA 92521, USA
e-mail: dswanson@ucr.edu

L.M. Tedrow
Department of Sociology, Western Washington University, Bellingham, WA 98225, USA
e-mail: Lucky.Tedrow@wwu.edu

J. Baker
Geospatial and population Studies, University of New Mexico, Albuquerque, NM 87106, USA
e-mail: kali@unm.edu

R. Schoen (ed.), *Dynamic Demographic Analysis*,
The Springer Series on Demographic Methods and Population Analysis 39,
DOI 10.1007/978-3-319-26603-9_12

We believe that it is worthwhile to examine CCRs in regard to the concept of a stable population because they contain information about both migration and mortality. As a means of exploiting this information, we explore CCRs as a tool for examining the transient dynamics of a population as it moves toward the stable equivalent that is captured in most formal demographic models based on asymptotic population dynamics. We employ a Leslie Matrix framework with an invariant set of CCRs and a regression-based approach to model trajectories toward stability.

The usual approach to generating a stable population is the use of a constant set of fertility and mortality rates applied to an arbitrarily chosen age distribution (Caswell 2001; Coale 1972; Dublin and Lotka 1925; Lotka 1907; Preston et al. 2001). When a given population is subjected to constant fertility and mortality rates, it will eventually reach stability and have a constant rate of growth (Caswell 2001: 79–92). This constant rate of change is known by several names, but in this chapter we use the term "intrinsic r," which is denoted in this study by "***r***." It also is the case that ergodicity stipulates that the initial age distribution is "forgotten" by the time the population in question reaches stability (Caswell 2001: 79–92, 386–397). Because CCRs are invariant and always positive, the Leslie Matrix framework we use represents a process will lead to a stable population that is ergodic.

Preston et al. (2001), among others, observe that a stable population also will result if a constant set of migration rates is included with sets of constant fertility and mortality rates. A number of papers have extended stable population models to include migration, and have examined its impact on long-term population stability. Espenshade (1986: 249), for example, states: "When migration is recognized, it is often to note that migration rates can be incorporated into survival rates so that no substantial modifications of the stable model are required." Sivamurthy (1982) also considers net-migration within the standard stable population model in the same manner. We find that this approach is both similar and dissimilar to the Hamilton-Perry model we employ in this chapter. It is similar in that net-migration schedules are components of the CCRs we use in the sub-diagonal of the Leslie matrix; it is dissimilar in that the approach developed by Sivamurthy (1982) and applied by Espenshade (1986) allows only for net in-migration, while ours allows for both net in-migration and net-outmigration.

It is important to acknowledge that there are analytic solutions to some of the questions we are asking. For example, the eigenvalues and eigenvectors of the CCR Leslie matrix can be computed and analyzed to yield "damping ratios," which provide a basis for estimating the time to stability (Caswell 2001: 95–97). Similarly, Caswell (2001: 74–75) provides an analytic basis for estimating ***r*** that can be used with CCRs. In Sect. 12.8, we use these analytic approaches to develop estimates of time to stability and ***r***, which are compared to those found using the CCR Leslie Matrix approach. We also acknowledge that while a CCR approach has not before been used, it has been proven that any population subject to a constant set of positive rates (such as CCRs) will converge to stability (Alho 2008; Cohen 1979a; Espenshade 1986; Mitra and Cerone 1986). This is consistent with the Perron-Frebonius and ergodicity theorems (Caswell 2001: 79–87, 369–370). While we know that analytic solutions exist that should work with the CCR approach and expect a population subject to a constant set of CCRs to converge to stability, in point of fact, however, the CCR approach has not yet been examined. Moreover, neither the Perron-Frebonius nor ergodicity theorems actually produce a stable population. Hence, we believe that the CCR approach is worth exploring.

In addition to using CCRs as a new way to examine the concept of population stability, we also use a measure that, like CCRs, has been employed by demographers (and others) for a long time, but appears to have been overlooked in regard to population stability. This is the Index of Dissimilarity (Hobbs 2004: 157–158), which we use as a measure of both stability itself and the distance to stability. In our employment of the Index of Dissimilarity, we refer to it as the "Stability Index" (***S***). Its application is useful here because it is a bounded measure that is easy to interpret and it has characteristics not found in existing measures of the distance to stability.

We use a CCR Leslie Matrix framework in conjunction with a series of regression models to estimate the number of years to stability at selected levels of ***S***, which represent not only stability itself, but "quasi-stability" points on the way to stability. As we discuss later, we find that these regression models work reasonably well in providing an estimate of the time both to these selected points of quasi-stability and stability itself for a given population within the CCR Leslie Matrix framework. Importantly, these models provide information on the role of the components of change in determining the time to stability, as well as the initial conditions (as measured by the initial Stability Index). Continuing the use of regression analysis, we also find that a regression model works reasonably well in estimating the intrinsic rate of increase (***r***) from the initial rate of increase (IRI)

Including this Sect. (12.1), this chapter is composed of nine sections. In the next Sect. (12.2), we discuss the CCR method while in Sect. 12.3 we briefly discuss stable population concepts. The Stability Index and the CCR approach are discussed in Sect. 12.4. Section 12.5 discusses the data we employ along with the Leslie Matrix we use to implement this approach. Section 12.6 describes the regression models used in the estimation of time to selected points of quasi-stability and stability while Sect. 12.7 describes the estimation of ***r*** from the initial rate of increase (IRI). Section 12.8 provides a comparison of the estimates of time to stability and ***r*** found using the CCR approach with results found using analytical methods for the same populations. Section 12.9 concludes this chapter with a discussion.

12.2 Cohort Change Ratios

Because we use a constant set of CCRs to project a population to stability, we discuss them in conjunction with the Hamilton-Perry method. The Hamilton-Perry Method is a variant of the cohort-component method that has far less intensive input data requirements. Instead of mortality, fertility, migration, and total population data, which are required by the cohort-component method, the Hamilton-Perry method requires data only from two census counts (or estimates) that provide population data by age (Hamilton and Perry 1962; Smith et al. 2013; Swanson and Tayman 2013; Swanson et al. 2010).

The Hamilton-Perry method moves a population by age (and sex) from time t to time $t+k$ using CCRs typically computed from data in the two most recent censuses.[1] It consists of two steps. The first uses existing data to develop CCRs and the second applies the CCRs to the cohorts of the launch year population to move them into the future. The second step can be repeated infinitely, with the projected population serving as the launch population for the next projection cycle. The formula for the first step, the development of a CCR is:

$$_{n}CCR_{x,i} = {}_{n}P_{x,i,t} / {}_{n}P_{x-k,i,t-k} \tag{12.1}$$

where

$_{n}P_{x,i,t}$ is the population aged x to $x+n$ in area i at the most recent of the two points in time for which the data are available(t),
$_{n}P_{x-k,i,t-k}$ is the population aged $x-k$ to $x-k+n$ in area i at the earlier of the two points in time for which the data are available (t-k),
k is the number of years between the two points in time for which the population data are available in area i and it needs to be consistent with the age groups ($_{n}P_{x}$) used for the population in question and not greater than 10.

The basic formula for the second step, moving the cohorts of a population into the future is:

$$_{n}P_{x+k,i,t+k} = ({}_{n}CCR_{x,i}) * ({}_{n}P_{x,i,t}) \tag{12.2}$$

where

$_{n}P_{x+k,i,t+k}$ is the population aged $x+k$ to $x+k+n$ in area i at time $t+k$

[1]The input data used to generate cohort change ratios need to be separated by a time interval that is consistent with the age groups used in the input data. For example, if the data are in 5 year age groups (up to the terminal, open-ended age group), the time interval should be either 5 years or 10 years. If the data are in 10-year age groups then the time interval will need to be 10 years. If the data are in single-year age groups, then the time interval should be 1 year. Fortunately, most population data are provided in 5-year age groups.

Although we do not provide a proof here, it is easy to show that using CCRs to move a population through time is consistent with the fundamental demographic equation. This consistency is important for two reasons. First, as noted by Land (1986) any quantitative approach to forecasting is constrained to satisfy various mathematical identities, and a demographic approach should ideally satisfy demographic accounting identities, which are summarized in the fundamental demographic equation. The second reason is based on the argument by Vaupel and Yashin (1985) that a demographic forecasting method needs to be consistent with the fundamental demographic equation in order to minimize the potential errors associated with hidden heterogeneity.

$${}_{n}CCR_{x,i} = {}_{n}P_{x,i,t}/{}_{n}P_{x-k,i,t-k}$$

${}_{n}P_{x,i,t}$ is the population aged x to x + n in area i at the most recent point in time for which the data are available (t),

k is the number of years between the two points in time for which the population data used to construct the CCRs were available. This time interval becomes the length of the forecast cycle and must be consistent with the age groups (${}_{n}P_{x}$) used for the population in question and should not be greater than 10.

The CCRs reflect differential net undercount error and both the effect of mortality and migration. CCRs can be less than one (1.00) or greater than one (1.00). In the absence of differential net undercount error, the following observations hold: (1) in any age group where a CCR is greater than one, net in-migration is occurring; (2) in young ages (i.e., 20–24, 25–29, and 30–34) where mortality rates are low, CCRs less than one generally indicate net out-migration; and (3) at older age groups where mortality is high, CCRs less than one generally provide a picture of cohort survival rates. Thus, in the absence of differential net undercount error, CCRs and the combined effects of mortality and migration change with age, with mortality becoming a dominant component of a CCR at older ages (e.g., 60–64, 65–69, and 75+).[2]

Given the nature of the CCRs, 5–9 is the youngest age group for CCRs can be calculated if there are 5 years between the points in time that the data are assembled. If there are 10 years between the data points, then 10–14 is the youngest age group for which CCRs can be calculated. To project the population aged 0–4 (and 5–9) one can use the Child Woman Ratio (CWR), or more generally a "Child Adult Ratio" (CAR). It does not require any data beyond what is available in the decennial census. There are different ways to develop a CAR (Hamilton and Perry 1962; Smith et al. 2013: 176–180; Swanson and Tayman 2013; Swanson et al. 2010). As we discuss in Sect. 12.5, we do not use "CARs" because we use employ age-specific fertility rates to generate the number in the youngest age group, which given our five-year data structure is 0–4.

CCRs for the oldest open-ended age group differ slightly from the CCRs for the age groups up to the oldest open-ended age group and for which a CAR is not required. If, for example, there are 5 years between the points in time for which the data are assembled ($k = 5$) and the final closed age group is 70–74, with 75+ as the terminal open-ended age group, then calculations for the ${}_{\infty}CCR_{75,i,t}$ require the summation of the appropriate age groups to get the population age 70+ at time $t - k$, which is then used as the denominator in finding the CCR for those aged 75+:

[2]If one has a life table, the CCRs for a given population could be compared to their corresponding survival rates and the effects of migration could be separated from the effects of mortality. This would be similar to using a life table to estimate net migration by age using the Forward Life Table Survival Method. Again, an important assumption is that differential net undercount by age is absent or at least very minimal.

Table 12.1 A 2010 Hamilton-Perry projection of Austria using 2000–2005 CCRs & fertility data[a]

	2000	2005	CCR	2010 Forecast
Total population: 0–4 years	416,996	373,688	Mid-Point ASFR[b]	412,804
Total population: 5–9 years	474,442	424,606	1.01825	380,508
Total population: 10–14 years	468,613	482,338	1.01664	431,673
Total population: 15–19 years	486,243	478,246	1.02056	492,253
Total population: 20–24 years	470,256	500,377	1.02907	492,148
Total population: 25–29 years	570,381	485,939	1.03335	517,065
Total population: 30–34 years	704,616	578,844	1.01484	493,149
Total population: 35–39 years	715,158	706,090	1.00209	580,055
Total population: 40–44 years	615,389	712,287	0.99599	703,255
Total population: 45–49 years	521,659	609,722	0.99079	705,728
Total population: 50–54 years	499,456	513,512	0.98438	600,200
Total population: 55–59 years	494,015	486,579	0.97422	500,273
Total population: 60–64 years	419,019	475,689	0.96290	468,529
Total population: 65–69 years	344,843	395,638	0.94420	449,146
Total population: 70–74 years	331,663	312,967	0.90756	359,067
Total population: 75 years and over	580,664	648,169	0.71046	682,846
Total population	8,113,413	8,184,691	1.00879	8,268,696

[a]Source data: US Census Bureau's International Data Base (http://www.census.gov/population/international/data/idWmformationGateway.php)
[b]The age-specific fertility rates in the source data are female dominant and for a single year They were averaged to represent 2002.5 and adjusted to the total population (both males and females) and to represent a 5-year period to correspond with the forecast cycle
The final values are, by age group

<= 19	20–24	25–29	30–34	35–39	40–44	>= 45
0.037	0.211625	0.26725	0.174125	0.067875	0.01375	0.001375

$$_\infty CCR_{75+,i,t} = {}_\infty P_{75,i,t} / {}_\infty P_{70+,i,t-k} \quad (12.3)$$

The formula for projecting the population 75+ of area i for the year t + k is:

$$_\infty P_{75,imt+k} = ({}_\infty CCR_{75,i,t}) * ({}_\infty P_{70,i,t}) \quad (12.4)$$

Table 12.1 provides an illustrative example of the Hamilton-Perry Method for Austria, which uses data from the US Census Bureau's International Data Base for 2000 and 2005 to generate a 2010 population projection of the population by age for both sexes combined.

Table 12.1 shows that launching from a total population of 8184.691 in 2005, the Hamilton-Perry Method generates a 2010 total population of 8,268,696 for Austria using the 2000-2005 CCRs and a midpoint (2002.5) set of age-specific fertility rates. The increase largely reflects Austria's net in-migration among young adults and their children (all of the CCRs from age 5–9 to age 35–39 exceed 1.000).

12.3 A Stable Population: A Brief Overview of the Traditional Approach

A stable population has an invariable relative age structure and a constant rate of growth. That is, the proportion of people in each age group remains constant over time and the population as a whole has a constant rate of increase (Coale 1972; Dublin and Lotka 1925; Lotka 1907; Preston et al. 2001). As mentioned earlier, an important feature of the stable population model is ergodicity, whereby over time a population "forgets" its initial age distribution as it convergences on stability (Coale 1972; Cohen 1979a; Preston et al. 2001). There is both a strong and weak form of the ergodicity theorem (Caswell 2001: 79–92 & 386–387; Cohen 1979a). We use ergodicity as a general guide for part of our analysis.

Alfred J. Lotka is generally credited with formulating the idea of a stable population and exploring many of its important features, including the finding that in the absence of migration, a population subject to constant fertility and mortality rates would eventually have a constant rate of natural increase (Dublin and Lotka 1925; Lotka 1907). Continuing the analytical tradition established by Lotka, many researchers have examined the idea of a stable population and refined its underlying theory and extended its applications (Alho and Spencer 2005; Arthur 1981; Arthur and Vaupel 1984; Bacaër 2011; Bennett and Horuchi 1984; Caswell 2001; Coale 1972; Cohen 1979a; Kim and Sykes 1976; Le Bras 2008; Pollard et al. 1974; Popoff and Judson 2004; Preston et al. 2001; Preston and Coale 1982; Rogers 1985; Schoen 1988, 2006; United Nations 1968). Much of this research has, however, been confined to examining a population not affected by migration. Preston et al. (2001) and others (Espenshade 1986; Sivamurthy 1982) have suggested that this is an un-necessarily restrictive assumption. Nonetheless, other than a few exceptions, such as Espenshade et al. (1982), Rogers (1985, 1995), and (Rogers et al. 2010), this restriction appears to remain a governing force in the examination of stable population ideas. It is useful to note that even the approach for dealing with migration developed by Sivamurthy (1982) and used by Espenshade (1986) is limited in its application because it requires that a population have only net in-migration at all ages, a condition not always found in human populations.

Another restrictive assumption that has governed much of the work on stable populations is defined by the so-called "two-sex" problem (Pollak 1986; Preston and Coale 1982). In this problem (which evidently stems from Lotka's 1907 formulation of a stable population), only one sex (virtually always women) was examined in the context of a stable population because of problems reconciling the numbers of births resulting from including both sexes. However, as Preston et al. (2001) show a "female-dominant" approach to fertility offers a convenient way around this problem, one that has been employed in different ways by others (Barclay 1958: 216–222; Keyfitz and Flieger 1968). Yet another somewhat restrictive idea associated with the traditional approach is that if one is using a discrete approach, such as found in this chapter, a discrete version of Lotka's equation is required. Caswell (2001: 197), however, observes that the best way to implement a discrete version of Lotka's continuous equation is to use a discrete-time model rather than attempt to write discrete versions of Lotka's equation. This is the approach we follow, as discussed in the next section.

12.4 A Stable Population: The CCR Approach and the Index of Stability

The CCR approach simply takes the cohort change ratios found at a given point in time and holds them constant until the population reaches stability. In terms of our implementation of this approach within the Leslie Matrix framework, this also means we hold the initial ASFRs constant as well.

To determine when a population has reached stability, the well-known "Index of Dissimilarity" is employed as an "Index of Stability" (***S***).[3] The index is defined as:

$$S = \left\{ 0.5 * \sum \left| \left({}_nP_x / \sum {}_nP_x \right)_{t+y} - \left({}_nP_x / \sum {}_nP_x \right)_t \right| \right\}. \quad (12.5)$$

[3]Often, the Index of Dissimilarity is expressed as a percentage, whereby the formula shown in Eq. (12.5) is multiplied by 100. In our use of this Index, we define "zero" to six significant digits. That is, when ***S*** is equal to "0.000000," we define this stability. If fewer or more significant digits were used, the point at which stability is reached would, of course, be different.

It is worthwhile to note here that Keyfitz and Flieger (1968: 23 and 24–41) display a "dissimilarity" score between a current population age distribution and the age distribution for the corresponding stable population. The index is the sum of positive differences between the two distributions. This index is only one simple step from the Index of Dissimilarity. However, even so, it is neither employed by Keyfitz and Flieger (1968) to define a stable population nor used to estimate time to stability. However, Keyfitz (1968: 47) does use it to define the distance to stability and other measures of this distance are found in Caswell (2001), Cohen (1979b), Schoen (2006), Schoen and Kim (1991) and Tuljapurkar (1982).

Also, as noted in the text and described by Keyfitz (1968: 47), the Index of Dissimilarity could be used in conjunction with the relative age distribution at stability and at the initial launch point. As an example of this use, the highest value found in the 62 country data set is for Hong Kong, which has a Dissimilarity Index of .399073; the lowest is found for Guatemala, with a Dissimilarity Index of .05001. Thus, Hong Kong's age distribution at origin is furthest from its stable age distribution while Guatemala's is closest. As would be expected, Hong Kong's time to stability (740 years) is much longer than Guatemala's (250 years). These two respective indices also provide an easy-to-interpret measure of how different the initial population age structure is from the age structure at stability. For Hong Kong, 39.91 % of the initial population needs to be re-allocated to match its relative age distribution at stability while for Guatemala only 5 % needs to be reallocated. Due to the specific dynamics underlying a country's path to stability, the Index is not the sole determinant of time to stability, however. For example, Hong Kong does not take the longest time to reach stability of the 62 countries (Singapore does, at 890 years) and Guatemala does not take the shortest time to reach stability (El Salvador does, at 225 years).

where

p = population
y = number of years between census counts/projection cycles
x = age
n = width of the age group (in years)
t = year

S compares the relative age distribution at one point in time (t + y) with the relative age distribution at the preceding point in time (t) within the forecast cycle (the forecast cycle that we employ is 5 years) and measures the proportion of one population that would have to be re-allocated to match the relative age distribution of the other. ***S*** ranges from 0 to one (1); a score of zero means that there is no difference between the two relative age distributions and no re-allocation is needed, which is the minimum re-allocation that can take place. A score of one (1) means that maximum re-allocation is required for the two relative age distributions to match. A score of one (1) can be interpreted in several ways, but a common interpretation is all of the numbers by age in one population would have to be re-allocated in order to match the distribution of the numbers by age in the comparison population. Since we are dealing with the same population as viewed at two successive points in time, this leads to viewing a score of one (1) as an indication that all of the numbers by age at time t would have to be reallocated to match the numbers by age of the same population at the preceding point in time in terms of the forecast cycle.

S exploits the idea that when a population is stable, the sum of the differences between the relative size of corresponding age groups at time t + y and time t is zero (which we have operationalized as ***S*** = 0.000000). Thus, at a point when the sum of the differences across all of the corresponding age groups is zero between the time point at the end of a five-year forecast cycle and the preceding time point of the five-year forecast cycle, the population has reached stability. The advantage of using the Index of Dissimilarity as ***S*** is that it provides a bounded measure (between zero and 1) and has a clear interpretation. As mentioned earlier, this index can be used both to define stability and provide a measure of the distance to stability and we use it here in both regards. It could, of course, be used in conjunction with the traditional approach, but this appears not to be the case in that our search of the literature found nothing in regard to the use of the Dissimilarity Index to either define stability (see e.g., Caswell 2001; Preston et al. 2001; Schoen 2006) or measure the distance to stability (see, e.g., Caswell 2001; Schoen 2006; Schoen and Kim 1991; Tuljapurkar 1982). We believe that ***S*** possesses several desirable characteristics not found in other measures. First, in regard to stability itself, it is a summary measure of population age structure. Second, as a measure of the distance to stability, it only requires information on two successive current age structures, unlike, say, the Kullback Index, which requires information on the current age structure and the age structure at stability (Schoen 2006; Schoen and Kim 1991; Tuljapurkar 1982). In addition, the Dissimilarity Index can also be calculated between an initial age structure of a population (or the population's age structure at any point on the path to stability) and its age structure at stability, which makes it conceptually similar to the

Kullback Index. This use of the Dissimilarity Index is described by Keyfitz (1968: 47). Thus, we find that there are four useful features of the Index of Dissimilarity in regard to the concept of population stability. First, it provides a summary measure of relative age structure at origin, an important aspect of initial conditions. Second, it provides a measure of stability itself in that when ***S*** = zero (in the context of using the CCR or some other approach within a Leslie Matrix framework) a population has converged to stability. Third, by looking at ***S*** at any given point on the path to stability, we get an idea of the distance to stability in that we can see how far it is from zero. Fourth, by computing an Index of Dissimilarity between the age structure at origin (or any other point on the path to stability) and the age structure at stability, we can see how much of the initial age structure must be "re-allocated" in order to match the age structure at stability. These four features are found neither in any other single measure of stability nor any other single measure of the distance to stability.

Using ***S*** in conjunction with the CCR approach is a natural fit because the latter is implemented using a chain of fixed forecast cycles, which in the case of moving to stability, ends when successive age distributions are proportionately equal and ***S*** = zero. We generate a chain of fixed forecast cycles by using a Leslie Matrix (Caswell 2001: 8–34), which we illustrate here using our example, Austria. The CCRs and the ASFRs for Austria shown in Table 12.1 are held constant from the launch year (2005) to a year where ***S*** = zero (relative to the preceding year in the five-year projection cycle). This occurs at the year 2485.

Figure 12.1 provides the change in ***S*** from 2005 to 2485 for Austria as it proceeds to stability. It shows that the path to stability is nearly monotonic and definitely not linear. It initially declines rapidly to the point where S is approximately equal to .005, but the change in ***S*** slows substantially around the year 2185, which is 180 years from the launch year. From there to 2485, ***S*** moves incrementally to zero as can be seen in Fig. 12.1.

With some variations, the path to stability shown for Austria in Fig. 12.1 is generally found for all of the other 61 countries we use in this analysis. For many, the path is fully monotonic, others, nearly monotonic, but all are non-linear. This is consistent with findings elsewhere (Nair and Nair 2010; Schoen and Kim 1991). There is an initial and rapid decline in ***S***, the Index of Stability, which at some point slows. From the point at which it slows, it moves very slowly until stability is reached. As such, taking into account the slightly non-monotonic nature of the initial part in which ***S*** declines rapidly, these paths generally fit the form of "long-tailed" negative exponential distributions, where those showing monotonic decline would be better fits than those showing decline that is not precisely monotonic.

12.5 Data and Methods

As illustrated in the example for Austria, applying a constant set of CCRs and ASFRs to a given population will yield a stable population. We pursue this idea by applying this approach to 62 countries taken from the US Census Bureau's

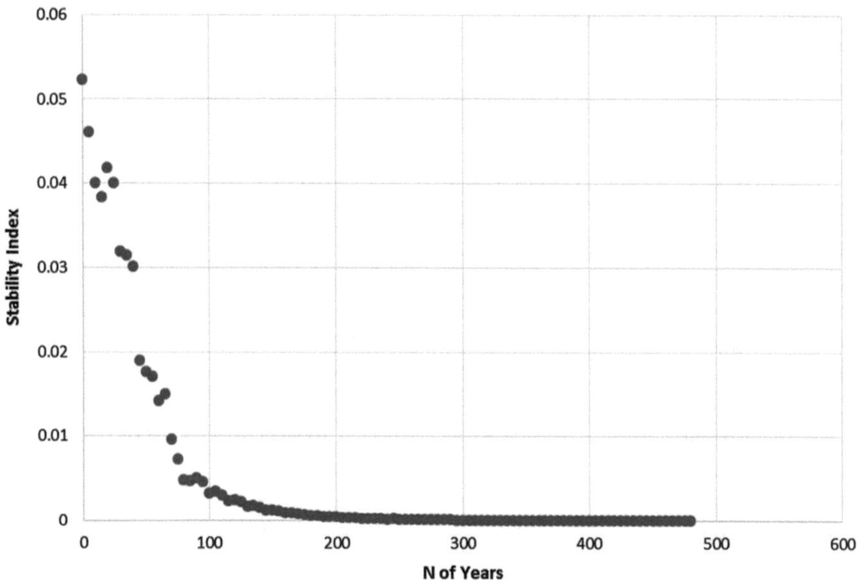

Fig. 12.1 Austria: path to stability

International Data Base. These countries were selected using two major criteria: First, the United Nations (2008) identifies them as having "reasonably reliable data;" and, second, their launch year populations are greater than 500,000. The data in the International Data Base are provided on an annual basis, so clearly they represent estimates informed by census and register information. These data allowed us to select a five-year forecast cycle, which is consistent with the five-year age groups we use (0–4, 5–9, 10–14, . . ., 70–74, 75+).

Exhibit 12.1 provides a list of these 62 countries along with their initial population counts and the region of the world in which they are found.

For these countries we used data from the early part of the twenty first century to develop the input data needed to perform the projections. For those countries which have census counts in years ending in one and six (e.g., Australia, Canada, Fiji, Ireland, the United Kingdom), we used data for 2001 and 2006; for countries which have census counts in years ending in zero or with excellent population registers, we used data for 2000 and 2005 (e.g., Austria, Cuba, Finland, Poland, the United States). We organized the population data into 16 age groups, 0–4, 5–9, 10–14, . . ., 70–74, and 75+. For these countries, we also obtained ASFRs for the same years for which we obtained the population data. In both cases, we selected data from these past points in time in order to ensure that the input data were, in fact, "reasonably reliable" in that these data could have been informed by subsequent census counts and administrative data (e.g., 2006 and 2011 for countries such as Australia, Canada, Fiji, Ireland, and the United Kingdom; or 2010 for countries such as Austria, Cuba, Finland, Poland, and the United States).

Exhibit 12.1 List of 62 countries

N	Country	Region	IDB population 2000[a]	IDB population 2005[a]
1	Canada[a]	N. America	31,376,736	32,656,679
2	Costa Rica	N. America	3,882,581	4,208,691
3	Cuba	N. America	11,071,849	11,198,439
4	El Salvador	N. America	5,849,822	5,956,221
5	Guatemala	N. America	11,085,025	12,182,548
6	Jamaica[a]	N. America	3,837,878	4,089,964
7	USA	N. America	282,162,411	295,516,599
8	Chile	S. America	15,174,571	15,979,150
9	Uruguay	S. America	3,219,793	3,264,911
10	Venezuela	S. America	23,492,753	25,269,177
11	Armenia	Asia	3,100,045	3,084,084
12	Azerbaijan	Asia	8,463,076	8,825,439
13	Hong Kong[a]	Asia	6,714,968	6,955,186
14	Georgia	Asia	4,818,805	4,790,009
15	Israel	Asia	6,114,570	6,742,915
16	Japan	Asia	126,775,612	127,715,356
17	Kazakhstan	Asia	15,687,251	16,122,665
18	Kyrgyzstan	Asia	4,937,128	5,164,248
19	Saudi Arabia	Asia	21,311,904	23,642,207
20	Singapore[a]	Asia	4,169,481	4,713,561
21	Tajikistan	Asia	6,229,697	6,814,791
22	Turkmenistan	Asia	4,385,485	4,664,155
23	Uzbekistan	Asia	25,041,821	26,539,888
24	Albania	Europe	3,158,352	3,024,533
25	Austria	Europe	8,113,413	8,184,691
26	Belarus	Europe	10,033,392	9,806,452
27	Belgium	Europe	10,263,618	10,364,388
28	Bosnia-Herzgovnia	Europe	3,805,512	3,893,097
29	Bulgaria	Europe	7,818,495	7,450,349
30	Croatia	Europe	4,410,830	4,495,904
31	Czech Republic	Europe	10,268,899	10,266,923
32	Denmark	Europe	5,337,416	5,432,335
33	Estonia	Europe	1,379,835	1,332,893
34	Finland	Europe	5,168,595	5,223,442
35	France	Europe	61,255,363	63,059,742
36	Germany	Europe	82,183,670	82,439,417
37	Greece	Europe	10,559,110	10,668,354
38	Hungary	Europe	10,147,425	10,057,624
39	Ireland[a]	Europe	3,872,700	4,309,024
40	Italy	Europe	57,784,373	59,037,808

(continued)

Exhibit 12.1 (continued)

N	Country	Region	IDB population 2000[a]	IDB population 2005[a]
41	Latvia	Europe	2,376,178	2,290,237
42	Lithuania	Europe	3,654,387	3,596,617
43	Macedonia/Former Yugoslavia	Europe	2,014,512	2,045,262
44	Moldova	Europe	4,180,215	3,948,261
45	Montenegro	Europe	732,302	699,259
46	Netherlands	Europe	15,930,181	16,299,097
47	Norway	Europe	4,492,400	4,624,875
48	Poland	Europe	38,654,164	38,557,964
49	Portugal	Europe	10,335,597	10,568,212
50	Romania	Europe	22,447,353	22,197,164
51	Russian Federation	Europe	147,053,966	143,319,518
52	Serbia	Europe	7,604,335	7,502,197
53	Slovakia	Europe	5,400,320	5,431,363
54	Slovenia	Europe	2,010,557	2,011,070
55	Spain	Europe	40,589,004	43,704,367
56	Sweden	Europe	8,924,354	9,082,561
57	Switzerland	Europe	7,277,250	7,448,224
58	Ukraine	Europe	49,005,222	46,959,420
59	United Kingdom[a]	Europe	59,374,727	60,846,809
60	Australia[a]	Oceania	19,294,257	20,489,472
61	Fiji[a]	Oceania	810,728	843,945
62	New Zealand[a]	Oceania	3,837,878	4,089,964

[a]For Australia, Canada, Fiji, Hong Kong, Ireland, Jamaica New Zealand, Singapore, & The United Kingdom, the years selected are 2001 and 2006, not 2000 and 2005, respectively

The population data were used to generate the (constant) set of CCRs that was applied to the most recent launch year (2005 or 2006) to take the country in question to stability. The ASFRs were averaged between the 2 years. Because they related only to the female population in each age group, they were "deflated" so that they applied to the total population (both males and females) and then multiplied by five to match the five-year cycle used in the projection sequence.

The 16 age groups yielded a 16×16 Leslie Matrix. The CCRs are found in the major diagonal and the ASFRs in the first row, elements 5 through 10 (which correspond to age groups 20–24, 25–29, ..., 45–49). Exhibit 12.2 shows the layout of this matrix for Austria.[4]

[4]The Leslie Matrix was implemented as a "macro" in Excel using Excel's coding language, VBA. The code as well as a "template" excel file with instructions on how to implement the Leslie Matrix are available on request from the authors. Also available from the authors are the files for all 62 countries as well as the summary file containing life expectancy at birth, the total fertility rate, and the mean CCR for ages 20–24, 25–29, and 30–34.

Exhibit 12.2 The Leslie matrix for Austria

0	0	0	0.037	0.211625	0.26725	0.174125	0.067875	0.01375	0.001375	0	0	0	0	0	0
1.018249576	0	0	0	0	0	0	0	0	0	0	0	0	0	0	0
0	1.016642709	0	0	0	0	0	0	0	0	0	0	0	0	0	0
0	0	1.020556408	0	0	0	0	0	0	0	0	0	0	0	0	0
0	0	0	1.02906777	0	0	0	0	0	0	0	0	0	0	0	0
0	0	0	0	1.03335	0	0	0	0	0	0	0	0	0	0	0
0	0	0	0	0	1.014837	0	0	0	0	0	0	0	0	0	0
0	0	0	0	0	0	1.002092	0	0	0	0	0	0	0	0	0
0	0	0	0	0	0	0	0.995985503	0	0	0	0	0	0	0	0
0	0	0	0	0	0	0	0	0.990791	0	0	0	0	0	0	0
0	0	0	0	0	0	0	0	0	0.984383	0	0	0	0	0	0
0	0	0	0	0	0	0	0	0	0	0.974218	0	0	0	0	0
0	0	0	0	0	0	0	0	0	0	0	0.962904	0	0	0	0
0	0	0	0	0	0	0	0	0	0	0	0	0.944201	0	0	0
0	0	0	0	0	0	0	0	0	0	0	0	0	0.907564	0	0
0	0	0	0	0	0	0	0	0	0	0	0	0	0	0.710457	0

Exhibit 12.3 Summary statistics for the variables used in the study

Variable	Mean	Std. dev.	Maximum	Minimum
Initial S	0.04506	0.01545	0.08378	0.01919
e_0	75.04	4.80	81.90	63.90
Mean CCR20-34	1.00	0.06	1.29	0.84
TFR	1.7804	0.6430	4.2298	0.9078
Initial r	0.00419	0.00840	0.02569	−0.01114
Intrinsic r (***r***)	−0.00468	0.01078	0.01995	−0.02518
N of years to				
S = .01	75.83	35.53	162.66	22.84
S = .005	102.82	31.20	216.06	45.27
S = .001	173.05	48.55	340.00	91.76
S = .0005	204.24	56.76	395.00	103.48
S = zero	489.92	140.91	890.00	225.00

The population data used to calculate CCRs were also used to calculate the initial Stability Index for each of the 62 countries. We also calculated a measure of net migration from the CCRs. This measure is the mean of the CCRs for age groups 20–24, 25–29, and 30–34. We selected these ages because they are closely associated with the ages at which adult migration is most likely to occur at the national level. The CCRs for these age groups also include mortality, but the mortality effects at these age groups are minimal.

In addition to the fertility data and the population data needed to develop CCRs, we acquired life expectancy data from the Census Bureau's International Data Base for the 62 countries used in this study. We did this to have a complete set of indicators for all three of the components of population change.

Exhibit 12.3 provides summary statistics for these measures as well as the average times to stability (when ***S*** = zero, which recall we have operationalized as ***S*** = 0.000000) and the selected points of quasi-stability used in the study, ***S*** = .01, ***S*** = .005, ***S*** = .001, and ***S*** = .0005. The selection of these points has no substantive significance beyond the fact that our visual inspections of the graphs showing the paths to stability for all 62 countries suggested that they generally encompass

There are different ways in which the Leslie Matrix could be implemented in terms of the constant ASFRs and CCRs. For example, once one developed the ASFRs for both sexes (as we have done) and set them up in a forecast cycle (which in our case here is for a 5 year period), the ASFRs could then be adjusted for infant mortality. One also could determine the mid-cycle populations of child-bearing ages (15–19, 20–24, ..., 45–49) and then apply the either the unadjusted ASFRS or mortality-adjusted ASFRs to them. We implemented the ASFRs without an adjustment for mortality and applied them to the population at the beginning of the forecast cycle. In the long run to stability, the different implementations are not likely to create substantial differences in the time to stability, but they could make a difference if one were attempting to develop realistic forecasts with much shorter horizons (e.g., 10 years, 20 years and even 50 years).

portions of the path to stability that in terms of time are rapid (.01), somewhat less rapid (005), slow, (.001) and very slow (.0005). An example of this can be seen in Fig. 12.1.

12.6 Time to Stability

As the title suggests, in this section we are primarily interested in examining the time it takes for a population to reach stability, given its initial conditions and its components of change. However, we also are interested in an issue raised by ergodicity; namely, that at some point on the path to stability, the initial conditions are "forgotten" (Caswell 2001). Hence, in this section, we also explore the point(s) on the path to stability at which this occurs.

Given our earlier work (Swanson and Tedrow 2013) in exploring CCRs and the time to stability, we elected to again use regression as the major tool of inquiry. It is well-suited to our task for several reasons: (1) regression models can be specified both in accordance with ergodicity and with our assumption that as a component of population change, migration should be examined along with births and deaths; (2) regression models use empirical data; and (3) characteristics of the models (e.g., the coefficient of variation, statistical tests of inference, and standardized regression coefficients) support analysis in regard to ergodicity. That is, we construct and examine regression models using a combination of variables representing initial conditions and the components of change as predictors of the time to selected quasi-stable points on the path to stability as well as to stability. In this regard, one would expect, for example, that the initial Stability Index would affect the time to a point on the path to stability, but after that point it would no longer have an effect – it would be "forgotten" before stability was reached. Swanson and Tedrow (2013), in fact, found support for this in that the initial Stability Index served reasonably well as a predictor of the time to $\boldsymbol{S} = .01$, but not to $\boldsymbol{S} =$ zero. A related question is the role played by the components of change in determining time to stability. Clearly, they play a role, but what is the relative importance of fertility vs. mortality, vs. migration? This question has only been partially answered, and generally only in the context of fertility variation (Kim and Schoen 1993a; Coale 1972; Liaw 1980). Again, we use regression analysis to explore this issue. We explored a number of regression models in terms of the time to stability using the NCSS statistical analysis system. What resulted is a model in which fertility, migration, and mortality all play a role, but the initial conditions (in the form of the initial Stability Index) do not. The model is provided below as Eq. (12.6), along with its characteristics:

Estimated N of Years to $\boldsymbol{S} =$ zero =

$$\underset{p=.007}{-824.79} + \underset{p=.0242}{\left(6.73^{*}e_0\right)} + \underset{p=.0001}{\left(927.84^{*}\text{MEAN\_CCR\_20\_34}\right)} - \underset{p=.0004}{\left(69.82^{*}\text{TFR}\right)} \quad (12.6)$$

n = 62

$R^2 = .597$
adj $R^2 = .576$

where

e_0 = life expectancy at birth (an index of mortality)
MEAN_CCR_20-34 = Mean of the CCRs, Age 20–24, 25–29 & 30–34 (an index of migration that is positively related to net in-migration: as it increases, so does net in-migration)
TFR = Total Fertility Rate (an Index of Fertility)
and the p values ($\alpha = .05$) are found below the intercept term and each of the three regression coefficients.

In the model shown as Eq. (12.6), life expectancy is positively related to the time to stability, as is net in-migration, while fertility is inversely related to the time to stability. Since life expectancy is inversely related to mortality, we can see that: (1) as fertility and mortality increase, the time to stability declines; and (2) as net in-migration increases, the time increases.[5] In advance of generating this equation, we had no expectations in terms of the signs and magnitudes of the regression coefficients, and, therefore, we had no expectations regarding the effects of these variables beyond the idea that the initial Stability Index would not be likely to play a role and that the components of change would play roles. We return to this point in our discussion of Exhibit 12.4 and again in the last section.

Because of the different scales at which the predictor variables are measured, we examine standardized regression coefficients to get an idea of the relative importance of these three components of change. The standardized versions of the coefficients found in Eq. (12.6) are, respectively, for the measures of life expectancy, migration, and fertility, .2376, .4286, and −.3327. These values suggest that in terms

[5]Given that the path to stability is non-linear, we also explored regression models in which the time (number of years) to stability was transformed using natural logarithms. However, we found that other than the change in the regression coefficients to accommodate the transformation, these models were not substantially different than their non-transformed counterparts. For example, the model that corresponds to the provided in Eq. (12.6) has an R^2 of .61 and an adjusted R^2 of .59 and the rank-order of the standardized coefficients is the same as found for the model shown in Exhibit 12.3 for time to $\boldsymbol{S}$ = zero, which are those associated with Eq. (12.5). It is useful to note here that the NCSS regression procedure employs Huber's method when skewed residuals are encountered. As such, it is a robust approach and its results will vary from those found using OLS methods which do not employ this method when skewed residuals are encountered.

It is worthwhile to note that some of the effects of the predictor variables found in Eq. (12.6), may also be non-linear on their own and interactive. We have not explored these possibilities here, but they may prove useful in future work.

It also is worthwhile to mention work by Preston (1986) in which he found that there is a close approximation between the intrinsic growth rate of a population and the mean of age-specific growth rates below age T, the mean length of a generation. He concluded therefore that where a disparity exists between the intrinsic growth rate and the actual growth rate of a population (whether or not net migration is included in both rates), it must be attributable to an unusual growth rate of the population block above age T.

Exhibit 12.4 The effect of the initial *S* score and the components of change on selected points on the path to stability

Standardized coefficient[b]					
Variable	Time to $S = .01$	Time to $S = .005$	Time to $S = .001$	Time to $S = .0005$	Time to $S =$ zero
Initial S	.6175	.4787	.2297	N/A[a]	N/A[a]
Mean CCR20–34	.1263	.2815	.4895	.4727	.4286
TFR	−.2246	−.2296	−.3338	−.4306	−.3327
e_0	N/A[a]	N/A[a]	N/A[a]	N/A[a]	.2376
ADJ R^2	.52	.41	.50	.51	.58

[a]Not statistically significant ($\alpha = .05$)
[b]The coefficients shown are for models for which only the statistically significant predictor variables are present. Models were, of course, constructed in which non-significant variables were present, but when such models were found, they were re-run without the non-significant variables, the results of which are shown here

of the time to stability, the level of net in-migration plays the largest role, fertility the second largest, and life expectancy, the least. They also suggest that the time to stability is longer for a population with low mortality, low fertility and high net in-migration than it is for a population with high mortality, high fertility, and low net in-migration. While we do not show the full results, the former description fits Singapore very well ($e_0 = 81.7$, MeanCCR20-34 = 1.287, and TFR = 0.908) which takes 890 years to reach stability; and the latter description fits El Salvador ($e_0 = 71.8$, MeanCCR20-34 = 0.869, and TFR = 2.73), which takes only 225 years to reach stability.

To examine the question in regard to the effect of initial conditions on the path to stability, we examined regression models using the components of change in conjunction with the initial Stability Index as predictors of the time to $S = .01$, $S = .005$, $S = .001$, $S = .0005$, and $S =$ zero. We summarize the results of this investigation in the form of Exhibit 12.4.

As can be seen in Exhibit 12.4, the initial conditions (in the form of the initial Stability Index) have an effect all the way to the time when $S = .001$ and play the largest role in terms of the times to $S = .01$ and $S = .005$, respectively. As we move from $S = .01$, to $S = .005$, to $S = .001$, we reach the last point where this predictor variable is statistically significant and we can see that it declines steadily to this point (from .6175 at $S = .01$ to .4787 at $S = .005$, to .2297 at $S = .001$). By the time we reach the point of quasi-stability where $S = .0005$, the initial value of *S* is no longer statistically significant and it remains so to the point of stability when $S =$ zero. Both migration (in the form of Mean CCR for age groups 20–24, 25–29 & 30–34) and fertility (in the form of the Total Fertility Rate) have an effect throughout the entire path, with migration having less of an effect initially (at $S = .01$) then having a larger effect than fertility from $S = .005$ all the way to when $S =$ zero. Mortality, in the form of e_0, is not statistically significant on the path to stability when fertility and migration are present until the point where $S =$ zero, at which time it has the smallest effect of the three predictor variables (.2376). As was the case with the discussion of our expectations regarding Eq. (12.6), we had no firm expectations regarding the regression coefficients found in Exhibit 12.4 other than the following: (1) the initial Stability Index would have an effect up to some point of quasi-stability but not to stability, per the concept of ergodicity; and (2) it seemed likely to us that all else being equal, fertility would have an inverse relationship with the time to stability, as would mortality, all else being equal.

12.7 Estimating Intrinsic r

A number of methods exists for estimating intrinsic r, which, recall is denoted as $\boldsymbol{r}$ by us (Barclay 1958: 216–222; Coale 1957, 1972; Dublin and Lotka 1925; Keyfitz and Flieger 1968; Lotka 1907; McCann 1973; Pressat 2009: 318–328; Preston et al. 2001:138–170; United Nations 1968), but we not aware of the direct use of regression analysis in which the initial rate of increase in a given population is used a predictor variable.[6] We note in regard to our use of regression analysis that analytic methods are preferable when relationships are understood. However, as Barclay (1958: 216) observes the determination of a non-stationary population is a complex task and the literature does not reveal a direct relationship between the initial rate of increase (which, recall we denote by IRI) in a given population to $\boldsymbol{r}$ (Barclay 1958; Coale 1957, 1972; Dublin and Lotka 1925; Keyfitz and Flieger 1968; Lotka 1907; McCann 1973; Pressat 2009; Preston et al. 2001) As an initial exploration of this relationship, and given the results yielded from employing regression to estimate the time to stability for a given population, we, therefore, employ regression analysis.

In earlier work, Swanson and Tedrow (2013) used data for 67 countries found in Keyfitz and Flieger (1968) in a "proof of concept" test. These 67 cases represent are the most recent entries for national and ethnic populations in Keyfitz and Flieger (1968); they also were used by McCann (1973) in constructing a quadratic regression model to estimate mean generation length, which he then employed to estimate $\boldsymbol{r}$ in conjunction with the natural logarithm of the net reproduction rate. The independent variable is the natural rate of increase (denoted here by NRI), which Keyfitz and Flieger (1968) found by subtracting the crude death rate from the crude birth rate for these 67 populations. The dependent variable is the intrinsic rate of increase, $\boldsymbol{r}$, found by Keyfitz and Flieger (1968) for these same 67 populations. Swanson and Tedrow (2013) found that a simple bivariate regression equation worked very well in estimating $\boldsymbol{r}$ from NRI for these 67 countries:

$$\text{Estimated Intrinsic rate of increase}, \boldsymbol{r} = \underset{p=.0222}{-1.1719} + \underset{p<.0001}{(1.0532 * \text{NRI})} \tag{12.7}$$

[6]While it appears that regression analysis has not been used to estimate intrinsic r from an initial r, Bourgeois-Pichat employed it to estimate intrinsic r from the proportional age distribution of a given population (see Keyfitz and Flieger 1968: 40).

n = 67
$r^2 = .8992$

The results strongly support the idea that ***r*** can be estimated from NRI using linear regression. The coefficient of determination is high ($r^2 = .8992$) and the both the intercept and slope coefficient are statistically significant (given $\alpha = .05$) at $p = .0222$ and $p < .0001$, respectively.

Given these results for the 67 countries taken from Keyfitz and Flieger (1968), we now turn our attention to the same 62 country data set used to generate the regression model for estimating time to stability from the score of the initial Stability Index. Here we do not use the "Natural Rate of Increase (NRI), as found in the data provided by Keyfitz and Flieger (1968), but, instead the Initial Rate of Increase (IRI). As discussed earlier, the former takes into account only the difference between the crude birth rate and crude death rate while the latter takes into account all three of the components of change, births, deaths, and migration.

$$\text{Estimated Intrinsic rate of increase, } \boldsymbol{r} = \underset{p<.0001}{-0.0096} + \underset{p<.0001}{(1.778 * \text{IRI})} \tag{12.8}$$

n = 62
$r^2 = .881$

As was the case for using NRI as a predictor variable for the 67 country data set taken from Keyfitz and Flieger (1968), we find that a simple bivariate regression model works well for predicting ***r*** from NRI using our 62 country data set: the coefficient of determination is high ($r^2 = .881$) and both the intercept and slope coefficient are statistically significant (given $\alpha = .05$) at $p < .0001$ and $p < .0001$, respectively. Taking into account the differences between NRI and IRI, it appears that the regression approach to estimating intrinsic r from initial measures of population change is reasonably robust.

Given our results for estimating time to stability, it is a natural question to ask what role the components of change play in estimating ***r***. To answer this question, we constructed a multiple regression model using IRI and the components of change as predictor variables. The results are found below with the model shown as Eq. (12.9).

Estimated Intrinsic rate of increase, ***r*** =

$$\underset{p<.0001}{-.1227} + \underset{p=.0151}{(.2922^*\text{IRI})} + \underset{p=0032}{(.0002^*e_0)} + \underset{p<.0001}{(.0767^*\text{MEAN\_CCR\_20\_34})} + \underset{p=.0158}{(.0128^*\text{TFR})} \tag{12.9}$$

n = 62
$R^2 = .952$
Adj $R^2 = .948$

The model shown as Eq. (12.9) shows that the components of change play a role along with IRI in determining ***r***. Because the regression coefficients are all positive, we can see that each of the components has a positive relationship with ***r***. By looking at the standardized regression coefficients for the model shown in Eq. (12.9), we obtain an idea of their relative importance. In order of size, we find that fertility plays the largest role, with a standardized coefficient of 0.7345 for the variable TFR; migration has the second largest standardized coefficient, with .4805 for the variable, MeanCCR20-34; IRI has the third largest standardized coefficient at .2366, while the smallest effect is found for mortality, for which the standardized coefficient for e_0 is .1088.

12.8 Comparison of CCR-Based Estimates of Time to Stability and *r* with Estimates Found Using the Analytic Approach

Estimates of time to stability and the intrinsic growth rate found using the simulation and regression approaches utilized here should correspond directly with those arrived at by analytic solutions. To explore their similarity, we computed time to stability using the damping ratio within a matrix model, as described by Caswell (2001: 95–97). All calculations were conducted in the "**R**" software package (www.Rproject.org), using the PopBio package (Stubben and Milligan 2007). Within a matrix model framework, the dominant eigenvalue of a square projection matrix (such as what we employ, one based on age-specific fertility rates and cohort change ratios) is the equivalent of the Euler-Lotka growth rate (Caswell 2001; Sykes 1969). This solution holds at stability, such that the ratio of the dominant eigenvalue to the absolute value of the second largest eigenvalue provides a measure of the percentage rate of convergence of the population on stability for each time-step in a population projection. In formulaic terms:

$$\rho = \frac{\lambda_1}{|\lambda_2|} \tag{12.10}$$

The damping ratio may be used to approximate the time to stability in an asymptotic model (Caswell 2001) as:

$$t_x = \frac{\log(x)}{\log(\rho)} \tag{12.11}$$

From this relationship, the number of years required for a population to reach convergence, which by convention is a point in time where x = 10 (Caswell 2001) may be estimated as:

$$\textit{time to convergence} = 5 * t_x \tag{12.12}$$

This time should match to a high degree of precision the number of years required in each simulation.

This method provides an estimate of time to convergence that is widely used in population ecology (Caswell 2001; Rogers-Bennett and Leaf 2006), but differs from those traditionally presented in human demographic studies (Kim and Schoen 1993a, b; Schoen 2006; Schoen and Kim 1991; Tuljapurkar 1982; Cohen 1979b). The damping ratio is specifically chosen here for comparison with the results of our simulation—as it provides a direct measure of the number of years required to achieve stability that can be compared to those estimated using projection. As such, it provides a straightforward basis of comparison in the correspondence between the results of an asymptotic model and those based on demographic projection reported here. The implications of the damping ratio for studies of population convergence should be similar to alternative approaches (Kim and Schoen 1993b) because all convergence measures (as well as patterns of fluctuation in age-structure or growth rate during the process of convergence) ultimately depend upon the relationship between the largest and second-largest eigenvalues of a projection matrix (Keyfitz 1977).

These asymptotic estimates were compared to those arrived at via the damping ratio measure and the results of this analysis is presented in Exhibit 12.5. On average, small differences characterized discrepancies between analytic and simulated solutions for time to convergence. While the presence of some outlying values is clearly observable in the difference between mean and median and coefficient of variation measures for the CCR and analytic solutions in terms of years required to converge, on average these differences are 5 years or less in numeric terms. In percentage terms, the algebraic differences suggest that the CCR Leslie matrix-based approach provides a lower estimate of the time to convergence than does the analytic solution (mean = −4.22 %, median = −2.92 %). Absolute differences are less than 7 % on average between the two sets when the mean is used

Exhibit 12.5 Summary of the comparison of estimates of time to stability found using the Leslie Matrix approach and an analytic approach for the 62 countries[a]

Estimate of time to stability (S = zero)					
	CCR Leslie Matrix approach	Analytic approach	Algebraic difference	Percent difference	
	(1)	(2)	(2) − (1)	Algebraic	Absolute
Mean	97.85	102.84	− 5.01	− 4.22	6.99
Median	92.50	93.69	− 2.64	− 2.92	5.91
Std Dev	27.97	33.09	N/A	N/A	N/A
C.V.	0.29	0.32	N/A	N/A	N/A

[a]The time to stability for the CCR approach as shown in Exhibit 12.3 was divided by 5 (the width of the age groups and the length of the forecast interval used in the CCR approach was five years) to compare the time to stability found using the analytic approach. For example, the mean time to stability found in Exhibit 12.3 is 489.92, which is equal to mean shown here (97.85) multiplied by 5

Exhibit 12.6 Summary of the comparison of estimates of r found using the Leslie Matrix approach and an analytic approach for the 62 countries

Estimate of ***r***					
	CCR Leslie Matrix approach	Analytic approach	Algebraic difference	Percent difference	
	(1)	(2)	(2) − (1)	Algebraic	Absolute
Mean	−0.0050	−0.0047	0.00030	−4.05 %	8.67 %
Median	−0.0047	−0.0045	0.00020	−3.20 %	4.49 %
Std Dev	0.0106	0.0105	N/A	N/A	N/A
C.V.	−2.12	−2.23	N/A	N/A	N/A

(mean = 6.99 %) and under 6 % when the median is considered (median = 5.91 %). These relatively small differences likely stem from rounding issues associated with imprecision in floating point arithmetic in the Excel software package and from the arbitrary use of "10" as the converging scale in Eq. (12.11) (Caswell 2001). Overall, these results suggest a strong correspondence between the estimates of time to convergence associated with both approaches. The correlation between the two is extremely high (r = 0.96).

To get an idea of the consistency between the CCR approach to estimating ***r*** and the analytic approach, the latter was estimated using a method suggested by Caswell (2001: 74–75) in which the natural logarithm of the ratio of each population age group at stability to its corresponding age group at the launch point is summed across all age groups using the proportion of each age group at origin as a weight in the summation process.

Moving on to the estimates of ***r***, Exhibit 12.6 provides a summary of the comparisons across the 62 countries found using the CCR approach and the analytic approach. As can be seen in Exhibit 12.6, there is close agreement between the two approaches. In terms of measures of centrality, the mean of the 62 values of ***r*** estimated using the CCR approach is -0.0050 while the mean for the 62 values of ***r*** estimated using the analytic approach is −0.0047, an algebraic difference of 0.00030 (subtracting the former from the latter). The mean algebraic percent difference is −4.05 % and the mean absolute percent difference is 8.67 %. The median of the 62 values of ***r*** estimated using the CCR approach is −0.0047 while the mean for the 62 values of ***r*** estimated using the analytic approach is −0.0045, an algebraic difference of 0.00020 (subtracting the former from the latter). The median algebraic percent difference is −3.20 % and the median absolute percent difference is 4.49 %.

In terms of dispersion, Exhibit 12.6 shows that the standard deviation of the estimates of r found using the CCR approach is 0.0106 while the standard deviation for the estimates found using the analytic approach is 0.0105. Given the close correspondence between the means found using the two approaches and their standard deviations, it is not surprising that the coefficients of variation are similar, −2.12 for the former and −2.23 for the latter.

12.9 Discussion

In terms of our findings regarding the time to stability, initial conditions are forgotten as a population moves to stability, which is consistent with ergodicity. However, given the results of our analysis (as summarized in Exhibit 12.4), it is clear that initial conditions (as represented by the initial Stability Index at the launch year) play a role well into the path to stability, up to and including, the quasi-stable point where $S = .001$. The average time to reach this point (across all 62 countries) is 173 years (with a standard deviation of 48.6). In this context, it is useful to note that the average time to reach stability is 490 years (with a standard deviation of 141 years). Thus, it appears that the initial conditions play a role in the path to stability up to the time that, on average, a country is approximately one-third of the way to stability. Further, as can be seen in Exhibit 12.4, the effects of initial conditions as measured by S at the launch year diminish as a country moves from its launch year to the year of quasi-stability where $S = .001$.

In terms of the components of change, fertility and net in-migration play a role all the way from launch to stability, which is when **S** =0.000000. As seen in the standardized coefficients found in Exhibit 12.4, the effect of fertility increases from launch to the point of quasi-stability where $S = .0005$, when it reaches value of -.4306. It then diminishes to -.4010 at the point of stability where $S =$ zero. In terms of net in-migration, the standardized coefficient increases from launch to the point of where $S = .001$, when it reaches a value of .4895 and then diminishes to .4727 when $S = .005$ and finally, to .3884 when $S =$ zero.

Life expectancy, our indicator for the mortality component, does not play a role until some point between $S = .0005$ and stability (when $S =$ zero). As can be seen in Exhibit 12.4, the coefficient for this variable is not statistically significant until $S =$ zero, where it has a value of .2325. Since the average time to $S = .0005$ is 204 years (with a standard deviation of 57 years), and the average time to stability is 490 years (with a standard deviation of 141 years), it appears that it takes a long time for the effects of mortality to come into play, on average.

Our analysis suggests that the initial value of S is the most important determinant on the path to stability up to the point when $S = .005$, which, on average, takes 103 years to achieve for our set of 62 countries (with a standard deviation of 31 years). The standardized coefficient for Initial S is much larger than those for fertility and net in-migration at both $S = .01$ and $S = .005$. However, by the time the point of $S = .001$ is reached, the initial value of S becomes the least important determinant in that its standardized coefficient (.2297) is exceeded (in the absolute sense) by both the standardized coefficient for net in-migration (.4895) and the standardized coefficient for fertility (−.3338). At the time when $S = .005$ is reached, the initial value of S is no longer statistically significant and the effects of fertility and net in-migration are about equal (in the absolute sense), with standardized coefficients of .4727 and −.4306, respectively. At this same point, the effect of mortality has not yet come into play.

Although we did not show the results of all of the regressions that were constructed, we did find that the initial rate of population change does not play a role in that this variable was not statistically significant in any of the multiple regressions. Similarly, no initial conditions (e.g., proportion of the population aged 0–4, proportion of the population aged 75 years and over, the difference between the proportion aged 0–4 and the proportion aged 75+) other than the initial value of **S** were statistically significant in any of the regressions we constructed. These findings serve to complement those generated by the traditional approach to examining the path to stability in which only fertility and mortality are considered as components of change (Cohen 1979a, b; Keyfitz 1974; Kim and Schoen 1993a; Schoen 2006; Schoen and Kim 1991; Tuljapurkar 1982).

The results obtained here are most comparable to those of Kim and Schoen (1993a), who use an alternative measure of the rate of convergence and relate it to variation in the net-maternity function in a standard birth-death model. Kim and Schoen (1993a) and Schoen (2006) argue that if the second largest eigenvalue is real (the solution will not hold if it is complex), then there is a "constant, ultimate force of convergence" given by:

$$\mathbf{h}^* = \mathbf{1} - \left\{\left(\left|\boldsymbol{\lambda}_2\right|/\boldsymbol{\lambda}_1\right|\right)\right\}^2 \qquad (12.13)$$

This measure, of course, suggests a negative exponential convergence upon stability. It differs from the damping ratio approach employed in this paper and is relevant to analyzing determinants of the rate of convergence whereas our use of the damping ratio (which would give a geometric approach to a predefined level of stability in which the ratio of the largest to second largest eigenvalue approaches 10) is utilized to compare results of a simulation approach to an asymptotic population model.

On this basis, Kim and Schoen (1993a) further suggested that if the net-maternity function in a standard birth-death model of population dynamics is parameterized in the fashion suggested by Keyfitz (1977) – using a normal curve - then the moments of the distribution of Lotka's stable net maternity function may be used to analyze variation in the speed of convergence as:

$$\mathbf{h}^* \approx \mathbf{1}\text{–}\mathbf{exp}\left[-\mathbf{4n}\boldsymbol{\pi}^2\boldsymbol{\sigma}^2/\boldsymbol{\mu}^3\right] \qquad (12.14)$$

Specifically, Kim and Schoen (1993a) suggest that the rate of convergence measured by h* should be inversely proportional to the mean of the stable net maternity function. Greater variability in the stable net maternity function and a lower age at mean childbearing, under the model of Kim and Schoen (1993a) convergence should occur more rapidly. This idea corresponds to findings of Coale (1972), who also suggested that greater variance in age at reproduction would converge upon stability more rapidly. Utilizing data from 177 populations originally analyzed by Keyfitz and Flieger (1968) and Kim and Schoen (1993a) illustrated that a strong relationship between the h* shown in Eq. (12.14) and the h* measure calculated using the first and second largest eigenvalues as in Eq. (12.13) (R^2 of 0.98). They also found that alternative measures of convergence, such as population entropy (Tuljapurkar 1982) also co-vary in the same direction with net-maternity functions.

In this study, we find that levels of TFR are negatively related to the years required to reach convergence upon stability in demographic projections. Kim and Schoen (1993a) found that the level of net maternity did not affect their results—only the mean age at childbearing. Since NRR and TFR should be positively related (because NRR is simply the TFR schedule discounted for survivorship), we might have expected to find more similar results. In this study, the effect of increasing levels of the TFR is negative. For each one unit increase in TFR (holding constant the other variables), a reduction in time to stability of approximately 68 years is observed. For each one unit increase in the average CCR of the 20–34 year old age intervals (holding constant the other variables), we observe a 928 year increase in the time to stability. This suggests that net in-migration has a much stronger and more pronounced effect upon variation in the time required to reach stability than does fertility in a births-deaths only model such as those examined by Coale (1972) or Kim and Schoen (1993a). This is not surprising, given that shifts in fertility are known to have more pronounced effects on age-structure than those in mortality when births-deaths only models are considered (Coale and Trussell 1974; Caswell 2001). However, we are surprised by the orders of magnitude difference observed in this study when migration effects are included. To our knowledge, this finding is a novel one in the literature on stable population dynamics.

We believe that these findings in regard to the role played by initial conditions (i.e., age structure) as measured by ***S*** in regard to the path to stability are novel. While the effect of varying initial age structures can be seen in simulations (e.g., Caswell 2001: 12–13), it appears that they have not been quantified. As such, our regression analyses provide a starting point for developing an analytical description of this process. Also useful as a point of departure for further analysis is the finding that the mean length of time to stability using the CCR approach with our 62 country dataset is 489.92 years. This length of time likely reflects at least partially the fact that CCRs can be greater than 1.0, owing to the effect of migration. Continuing this line, our analysis suggests that the time to stability is increased as net in-migration increases. The preceding points regarding migration also lead to the realization that the age structure of a stable population found using the CCR approach can look very different than that found using the traditional approach. Due to the effects of migration being incorporated into the CCRs, the former may have, for example, more people in a given age group than found in a preceding age group, something not found in the latter.

In regard to migration, it is important to note that because CCRs are always greater than zero and can encompass both net in-migration and net out-migration, they can be used with a Leslie Matrix with assurance that a given population will converge to stability. This is not the case with other approaches that have looked at accommodating migration as part of the process to convergence in that they only allowed for net in-migration in order to provide assurance that a given population would converge (e.g., Espenshade 1986; Sivamurthy 1982). As such, the CCR Leslie Matrix approach is more general in regard to accommodating migration. While we believe that our use of CCRs provides new insights and capabilities, we also note that nothing extraordinary is found in their use in regard to the mathematical foundation of the Leslie Matrix approach. That is CCRs are always greater than zero. Thus, when one is using CCRs as the process to stability in the context of a Leslie Matrix, the matrix is "positive," which means the population in question will converge (Caswell 2001: 79; Schoen 2006: 26–29).

Turning to the topic of estimating intrinsic r (***r***) from initial conditions, we confirm the finding of Swanson and Tedrow (2013) that ***r*** can be estimated from initial measures of population change. The model (Eq. (12.7)) they constructed using the natural rate of increase (NRI) from 67 populations selected from Keyfitz and Flieger (1968) is similar to the model constructed using the initial rate of increase (IRI) from the 62 populations used here (Eq. (12.8)). In the earlier model, the coefficient of determination was .899 and in the current model it is .881. Given that the model found in Eq. (12.7) uses NRI and that the one found in Eq. (12.8) uses IRI, the regression coefficients in both models are somewhat different at 1.0532 and 1.778, respectively. When the individual components of change are added as predictor variables, the model for estimating ***r*** is improved even more in that R^2 is .952. We provided an idea of the relative importance of the predictor variables by looking at their standardized coefficients. The highest one, .7345, is found for the fertility variable, TFR, while the second highest one, .4805, is found for the net in-migration variable, MeanCCR20-34. The standardized coefficient for IRI is the third highest at .2366 while the lowest, .1086, is found for the mortality variable, which we operationalized as e_0.[7]

As was the case regarding our findings on the path to stability, we believe that our findings regarding the estimation of ***r*** from a measure of initial population change (i.e., both NRI and IRI) complement: (1) those generated by the more traditional approach to examining the path to stability in which only fertility and mortality are considered as components of change (Cohen 1979a, b; Keyfitz 1974; Kim and

[7]Based on comments by Barry Edmonston, we also constructed models for estimating r using the initial Stability Index (***S***). In the model in which all three components of change were included along with initial r, we found that e_0 was not statistically significant. We then eliminated this predictor variable and re-ran the model with the other two components of change, Initial ***S***, and initial r, and found a model with an adjusted R^2 of .948 and predictor variables that were all statistically significant. These results suggest that the difference between the initial age distribution and the stable age distribution may be a factor in the difference between initial r and ***r***, a suggestion provided by Barry Edmonston. This idea may also account for the difference between the regression model for estimating ***r*** from initial r that was constructed using the Keyfitz and Flieger (1968) data and the model for estimating r from initial r that was constructed using the Census Bureau's International Data Base. That is, the differences found between initial population age structures and the stable ones for each of the 67 populations taken from Keyfitz and Flieger (1968), on the one hand, may vary from the differences found for each of the 62 populations taken from the US Census Bureau's International Data Base, on the other. This is a topic for future research.

Schoen 1993a; Preston 1986; Schoen and Kim 1991; Tuljapurkar 1982); and (2): a continuous approach (as opposed to our discrete approach using CCRs), such as found in Schoen (2006: 71). Importantly, we also find that estimates of ***r*** and time to stability generated by the CCR Leslie Matrix approach are consistent with estimates developed from the analytic approach.[8]

In summary, Cohort Change Ratios (CCRs) appear to be useful as a tool for examining the idea of a stable population. The consistency found between CCR-based estimates of both time to stability and ***r*** and those using the analytic approach suggest that the former is consistent with the theoretical foundation of stable population theory. Given this consistency, a major benefit of the CCR approach is the ability to easily deal with both sexes and all of the components of change, including migration. In this regard, a recent analysis of the path to stability for India that used the traditional framework by looking at the reproduction and survivorship of females would have been more realistic if it had employed the CCR approach, which would have accommodated all of the components of change (including migration) in regard to both males and females (Nair and Nair 2010). In addition, the CCR approach could have easily been implemented within the Leslie Matrix framework.

In conjunction with regression models and variables representing fertility, migration, and mortality, we believe that our examination of the CCR approach and the use of the ***S*** Index has yielded some useful insights on the effect of variables on the path to stability, insights not fully available from existing analytic methods, but yet consistent with ergodicity.[9] It is worthwhile to note again here that ergodicity

[8]There may be approaches other than the one we employ (Caswell 2001: 95–97) to compare with the estimate of time to stability generated from our CCR approach using regression and the Stability Index (***S***). For example, it may be possible to substitute a variation of the Kullback Distance (Nair and Nair 2010; Schoen 2006: 29–33) for the Stability Index as an independent variable in a regression model such as we employed. With appropriate modifications, the Kullback Distance potentially could be used with cohort change ratios and its results compared with both those generated by the CCR method and the analytic approach we used. It is useful to note that the Kullback Distance declines monotonically during the process of convergence (Schoen 2006: 31), which is similar to the behavior of ***S***, where the initial decline may not be monotonic, but becomes so at some point and overall, is monotonic or nearly so. Also, like ***S***, the Kullback Distance possesses a number of desirable properties (Schoen 2006: 31). However, the Kullback Distance also may generate different values than the method we used and, as such, yield different summary statistics in a comparison with the CCR approach. Similarly, there variations on the analytic approach we used to estimate ***r***, which is taken from Caswell (2001: 74–75). Descriptions of variations that potentially could be used can be found in Barclay (1958: 216–222), Coale (1957, 1972), Dublin and Lotka (1925), Keyfitz and Flieger (1968), Lotka (1907), Pressat (2009: 318–328), Preston et al. (2001:138–170), and United Nations (1968). Again, as we noted in regard to time to stability, these approaches may generate different values of ***r*** than the method we used and, as such, yield different summary statistics in a comparison with the values of ***r*** generated by the CCR approach.

[9]Caswell (2001: 572) notes that demographers have addressed the "two-sex" problem since the 1940s, but that much of the literature focuses on the "consistency" problem: how to make estimates of intrinsic r based on male and female life tables agree. Although he notes that those studies that deal with demographic dynamics in any detail have focused on models lacking age structure, examples of studies using age can be found in Schoen (1988).

(in either the form of the strong or weak theorems) states that initial conditions are forgotten and that "vital statistics"(which can be generalized to include CCRs) are the determinants of the stable age structure. Our findings suggest that the initial Stability Index plays a role about one-third of the way on the temporal path to stability while fertility and migration play a role along the entire path and mortality only does so toward the end of the temporal path to stability. This finding could lead to a refinement of the concept of ergodicity.

It is important to note that if different definitions were used in place of those we used to operationalize the predictor variables for initial conditions and the components of population change, it is likely that the regression models resulting from them would vary from ours. It may also be the case that if different age groupings (e.g., 0–4, 5–9,..., 85–89, 90+) and forecast cycle lengths (e.g., 10 years), the results would be different. However, it is likely the case that there would be findings common to them as well and these common findings would serve to point the way to increased understanding of the path to stability.

In terms of future research, one area might be the use of different points of quasi-stability. Recall that in this study we use $\boldsymbol{S} = .01$, $\boldsymbol{S} = .005$, $\boldsymbol{S} = .001$, and $\boldsymbol{S} = .0005$ because they generally encompass portions of the path to stability that in terms of time are rapid (.01), somewhat less rapid (005), slow, (.001) and very slow (.0005). It may be the case that different points yield different insights.

Another area for future research is to examine CCRs in conjunction with ideas promulgated by Keyfitz (1974) for examining stable processes across two (or more) interacting populations, ideas explored by, among others, Keyfitz (1980), Kim and Schoen (1993b), and Liaw (1980). Because it can deal with both sexes and migration quite handily, the CCR approach appears to be more tractable in regard to examining the path to stability in such populations. Another area, which we mentioned earlier, would be to develop formal statements like the strong and weak forms of the ergodicity theorem that specify the effect of both initial conditions and all three of the components of population change on the path to stability. Yet another area could involve decomposing CCRs into their survivorship and migration components and examining the effects of these two components of change directly.

In conclusion, we know that regression models are generally not as satisfying as analytical expressions in regard to describing relationships. It would be much more elegant to express the time to stability in terms of an analytic expression that incorporates the initial Stability Index (and possibly other information about initial conditions) and components of change than it is to express the relationship in the form of a regression model.[10] The same can be said about the relationship between the initial rate of increase in a given population and its intrinsic rate of increase. However, we also note that regression analysis has already been successfully employed in conjunction with stable population analysis, to include the Bourgeois-

[10]In regard to the usefulness of empirical findings, we note that in discussing the exploration of Kim and Sykes (1976) on stable population concepts, Cohen (1979a: 286) observed that their numerical experiments uncovered empirical regularities that invite theoretical explanation.

Pichat method for estimating intrinsic r from the proportional age distribution of a given population (Keyfitz and Flieger 1968: 49; United Nations 1968), McCann's (1973) method for estimating mean generation length from a trial value of the intrinsic rate of increase, and the generation of model life table families and from them, stable populations (Coale and Demeny 1966).

References

Alho, J. (2008). Migration, fertility, and aging in stable populations. *Demography, 45*(3), 641–650.

Alho, J., & Spencer, B. (2005). *Statistical demography and forecasting*. New York: Springer.

Arthur, W. B. (1981). Why a population converges to stability. *The American Mathematical Monthly, 86*(8), 557–563.

Arthur, W. B., & Vaupel, J. (1984). Some relationships in population dynamics. *Population Index, 50*(2), 214–226.

Bacaër, N. (2011). *A short history of population dynamics*. Dordrecht: Springer.

Barclay, R. (1958). *Techniques of population analysis*. New York: Wiley.

Bennett, N., & Horuchi, S. (1984). Mortality estimation from registered deaths in less developed countries. *Demography, 21*(2), 217–233.

Caswell, H. (2001). *Matrix population models: Construction, analysis, and interpretation* (2nd ed.). Sunderland: Sinauer Associates, Inc.

Coale, A. J. (1957). A new method for calculating Lotka's r – The intrinsic rate of growth in a stable population. *Population Studies, 11*, 92–94.

Coale, A. J. (1972). *The growth and structure of human populations: A mathematical investigation*. Princeton: Princeton University Press.

Coale, A. J., & Demeny, P. (1966). *Regional model life tables and stable populations*. Princeton: Princeton University Press.

Coale, A. J., & Trussell, J. (1974). Model fertility schedules: Variations in the age structure of childbearing in human populations. *Population Index, 40*(2), 185–258.

Cohen, J. (1979a). Ergodic theorems in demography. *Bulletin of the American Mathematical Society, 1*(2), 275–295.

Cohen, J. (1979b). The cumulative distance from an observed to a stable population age structure. *SIAM Journal of Applied Mathematics, 36*, 169–175.

Dublin, L., & Lotka, A. (1925). On the true rate of natural increase: As exemplified by the population of the United States, 1920. *Journal of the American Statistical Association, 20*, 305–339.

Espenshade, T. (1986). Population dynamics with immigration and low fertility. *Population and Development Review, 12*, 248–261.

Espenshade, T., Bouvier, L., & Arthur, W. B. (1982). Immigration and the stable population model. *Demography, 19*, 125–133.

Hamilton, C. H., & Perry, J. (1962). A short method for projecting population by age from one decennial census to another. *Social Forces, 41*, 163–170.

Hardy, G. F., & Wyatt, F. B. (1911). Report of the actuaries in relation to the scheme of insurance against sickness, disablement, &c., embodied in the national insurance bill. *Journal of the Institute of Actuaries XLV*, 406–443.

Hobbs, F. (2004). Age and sex composition. In J. Siegel, & D. A. Swanson (Eds.), *The methods and materials of demography* (2nd ed., pp. 125–173). San Diego: Elsevier Academic Press.

Keyfitz, N. (1968). *Introduction to the mathematics of population*. Reading: Addison-Wesley.

Keyfitz, N. (1974). A general condition for stability in demographic processes. *Canadian Studies in Population, 1*, 29–35.

Keyfitz, N. (1977). *Introduction to the mathematics of population*. New York: Addison-Wesley.

Keyfitz, N. (1980). Multistate demography and its data: A comment. *Environment and Planning A, 12*, 615–622.

Keyfitz, N., & Flieger, W. (1968). *World population: An analysis of vital data*. Chicago: University of Chicago Press.

Kim, Y., & Schoen, R. (1993a). On the intrinsic force of convergence to stability. *Mathematical Population Studies, 4*(2), 89–102.

Kim, Y., & Schoen, R. (1993b). Crossovers that link populations with the same vital rates. *Mathematical Population Studies, 4*(1), 1–19.

Kim, Y., & Sykes, Z. (1976). An experimental study of weak ergodicity in human populations. *Theoretical Population Biology, 10*, 150–172.

Land, K. (1986). Methods for national population forecasts: A review. *Journal of the American Statistical Association, 81*, 888–901.

Le Bras, H. (2008). *The nature of demography*. Princeton: Princeton University Press.

Liaw, K. L. (1980). Multistate dynamics: The convergence of an age-by-region population system. *Environment and Planning A, 12*, 589–613.

Lotka, A. J. (1907). Relation between birth rates and death rates. *Science (New Series), 26*(653), 21–22.

McCann, J. (1973). A more accurate short method of approximating Lotka's r. *Demography, 10*(4), 567–570.

Mitra, S., & Cerone, P. (1986). Migration and stability. *Genus, 42*(1-2), 1–12.

Nair, S. B., & Nair, P. S. (2010). Momentum of population growth in India. In S. Nangia, N. C. Jana, & R. B. Bhagat, (Eds.), *State of natural and human resources in India, Part 2* (pp. 399–424). New Delhi: Concept Publishing Company, Ltd.

Pollak, R. (1986). A re-formulation of the two-sex problem. *Demography, 23*, 247–259.

Pollard, A., Yusuf, F., & Pollard, G. (1974). *Demographic techniques* (3rd ed.). New York: Pergamon Press.

Popoff, C., & Judson, D. (2004). Some methods of estimation for statistically underdeveloped areas. In J. Siegel & D. A. Swanson (Eds.), *The methods and materials of demography* (2nd ed., pp. 603–641). San Diego: Elsevier Academic Press.

Pressat, R. (2009). Demographic analysis: Projections on natality, fertility and replacement (2nd Paperback Printing). New Brunswick: Aldine Transaction.

Preston, S. (1986). The relation between actual and intrinsic growth rates. *Population Studies, 40*(3), 343–351.

Preston, S., & Coale, A. J. (1982). Age structure, growth, attrition, and accession: A new synthesis. *Population Index, 48*, 217–259.

Preston, S., Heuveline, P., & Guillot, M. (2001). *Demography: Measuring and modeling population processes*. Malden: Blackwell Publishing.

Rogers, A. (1985). *Regional population projection models* (Vol. 4, Scientific Geography Series). Beverly Hills: Sage Publications.

Rogers, A. (1995). *Multiregional demography: Principles, methods, and extensions*. New York: Wiley.

Rogers, A., Little, J., & Raymer, J. (2010). *The indirect estimation of migration*. Dordrecht: Springer.

Rogers-Bennett, L., & Leaf, R. (2006). Elasticity analysis of size-based red and white abalone matrix models: Management and conservation. *Ecological Applications, 16*(1), 213–224.

Schoen, R. (1988). *Modeling multigroup populations*. New York: Plenum Press.

Schoen, R. (2006). *Dynamic population models*. Dordrecht: Springer.

Schoen, R., & Kim, Y. (1991). Movement toward stability as a fundamental principle of population dynamics. *Demography, 28*(3), 455–466.

Sivamurthy, M. (1982). *Growth and structure of human population in the presence of migration*. London: Academic.

Smith, S., Tayman, J., & Swanson, D. A. (2013). *A practitioner's guide to state and local population projections*. Dordrecht: Springer.

Stubben, C., & Milligan, B. (2007). Estimating and analyzing demographic models using the popbio package in R. *Journal of Statistical Software, 22*(11), 1–23.
Swanson, D. A., & Tayman, J. (2013). *Subnational population estimates*. Dordrecht: Springer.
Swanson, D. A., & Tayman, J. (2014). Measuring uncertainty in population forecasts: A new approach (pp. 203–215). In M. Marsili & G. Capacci (Eds.), *Proceedings of the 6th EUROSTAT/UNECE work session on demographic projections National Institute of Statistics*, Rome, Italy.
Swanson, D. A., & Tedrow, L. (2012). Using cohort change ratios to estimate life expectancy in populations with negligible migration: A new approach. *Canadian Studies in Population, 39*, 83–90.
Swanson, D. A., & Tedrow, L. (2013). Exploring stable population concepts from the perspective of cohort change ratios. *The Open Demography Journal, 6*, 1–17.
Swanson, D. A., Schlottmann, A., & Schmidt, R. (2010). Forecasting the population of census tracts by age and sex: An example of the Hamilton-Perry method in action. *Population Research and Policy Review, 29*(1), 47–63.
Sykes, Z. M. (1969). Stochastic versions of the matrix model of population dynamics. *Journal of the American Statistical Association, 64*(325), 111–130.
Tuljapurkar, S. (1982). Why use population entropy? It determines the rate of convergence. *Journal of Mathematical Biology, 13*, 325–337.
United Nations. (1968). *The concept of a stable population: Applications to the study of populations with incomplete demographic statistics* (Population Studies No. 39). New York: United Nations, Department of Economic and Social Affairs.
United Nations. (2002). *Methods for estimating adult mortality*. New York: Population Division, United Nations.
United Nations. (2008). *2006 Demographic yearbook*. New York: Department of Economic and Social Affairs, United Nations.
Vaupel, J., & Yashin, A. (1985). Heterogeneity's ruses: Some surprising effects of selection on population dynamics. *The American Statistician, 39*(August), 176–185.

Part V
The Dynamics of Population Size and Structure

Chapter 13
Estimating the Demographic Dynamic of Small Areas with the Kalman Filter

Manuel Ordorica-Mellado and Víctor M. García-Guerrero

13.1 Introduction

The demand for demographic data at the micro-regional level has increased due to the need to deal more effectively with the needs of specific population sectors and to make decisions in the government and private sectors. In countries such as Mexico, for example, there are many small localities. According to the 2010 Mexican Census, there are 189,000 localities with fewer than 2,500 inhabitants, which are home to 23 % of the Mexican population. Many of them belong to the poorest parts of Mexico, where a significant number of indigenous people live in isolated, scattered areas. This pattern of isolation and dispersion challenges economic, social and demographic planning. The population living in this kind of community experiences the most intense socioeconomic lags in Mexico together with extreme poverty and marginalization. Better demographic estimates for such small areas would make it possible to produce more specific information that would help local governments allocate resources to help diminish this inequality.

Making demographic estimates and projections for small geographic areas, however, is no easy task. The estimation toolkits of many demographers contain a range of methods that can be placed into three categories: statistical models, mathematical models and sample surveys (Swanson and Tayman 2012, p. 3). Demographic projections and estimates at the national and state level are calculated using the cohort component method. However, this methodology has not been possible to replicate in small areas due to the lack of information, but mainly because the indicators have a high level of variability. Estimates in small level areas are

M. Ordorica-Mellado (✉) • V.M. García-Guerrero
Center of Demographic, Urban and Environmental Studies, El Colegio de México A.C., Camino al Ajusco No. 20, Col. Pedregal de Sta. Teresa, C.P., 10740 Mexico, DF, Mexico
e-mail: mordori@colmex.mx; vmgarcia@colmex.mx

R. Schoen (ed.), *Dynamic Demographic Analysis*,
The Springer Series on Demographic Methods and Population Analysis 39,
DOI 10.1007/978-3-319-26603-9_13

therefore calculated using mathematical models and, more recently, they have been calculated using spatial statistical techniques and remote sensing.

Most methodological improvements in demographic estimates and projections have been made at the national level or for large areas. It is hard to find new methodological approaches for small areas; as stated above, this is due to the lack of demographic information and the data quality for this geographical size. At best, information is only available on the total population, and it is also extremely difficult to obtain data on mortality, fertility, and migration (both internal and international) at the level of small areas.

When information is available for small areas, birth rates, death rates and growth rates are extremely sensitive to small changes in both the number of events and/or their exposures. This is why many demographic techniques yield good results for large population aggregates.

In this chapter, we propose a new technique for estimating and projecting the population of small areas. In particular, we propose a heuristic method for small areas where housing hardly ever grows vertically but rather horizontally. This means that the population first expands its territory as the population grows before it expands vertically (as large metropolises do nowadays). This particularity is found in many ancient populations (such as the Maya and the Aztecs) although unfortunately we do not have empirical information on either their territorial expansion or their demographic dynamic. In Mexico, we found a modern population with that housing characteristic located in a rural area within Mexico City, called Villa Milpa Alta, in the municipality of Milpa Alta. It allows us to explore the new estimation technique proposed using empirical information from two sources, the Census and satellite images.

The aim of this chapter is to present a heuristic methodology to estimate the population for small areas. The method is a derivation of the Kalman filter, used to estimate the population in a hamlet with an area of 2.5 km^2 (Villa Milpa Alta). There are no data on births, deaths or migration. The only data available is on the population as a whole, by large age groups and sex. The methodology proposed would make it possible to combine different sources of information, such as the total population, obtained from the Census with information from satellite images.

The chapter is structured as follows. The following section contains a brief description of the methods used to estimate the population in small areas. This is followed by a description of the proposed methodology and the way it is applied to the case of Villa Milpa Alta. The chapter ends with some final remarks and some suggestions for future directions of research.

13.2 Preliminaries

There are three kinds of methods to estimate the demographic dynamic for small areas: (1) mathematical models, (2) those based on symptomatic variables (statistical models) and (3) sample surveys (Cavenaghi 2012; Swanson and Tayman

2012). Estimates based on mathematical models are useful for small areas where there are no vital statistics. The most commonly used mathematical models are:

1. The linear function, the simplest procedure, assumes that the annual population increase from a recent period will be repeated in the future. The exponential function, based on the assumption that population growth is exponential, applies when fertility and mortality levels remain constant. The logistic function assumes that the population grows quickly in the beginning, then it reaches a peak, after which the growth rate decreases. The logistic proportion estimation can also be used to project small populations (Arriaga 2001).
2. The relative increments method is based on the determination of the population's absolute growth as a proportion of a larger area corresponding to the population of every one of the lesser areas (Madeira and Simões 1972). The cohort ratio method is an adaptation of the cohort-component method (CCM) for smaller areas. Cohorts' growth rates are calculated on the basis of a larger projection area by the CCM and applied to smaller cohorts (Duchesne 1987). The differential method estimates the population by sex and age from a set of subareas and then adjusts them to the larger population (Gonzalez and Torres 2012).
3. Other mathematical functions have been used to make population estimates and projections, such as the geometric function, the Gompertz function, the modified exponential model and so on. It is also common to find applications using simple linear regression models, nonlinear regression models and multiple linear regression models (Pittenger 1976).

The second set of methods is based on symptomatic variables that attempt to update estimates based on variables associated with demographic change, which are of extremely good quality. Bay (1998) identifies symptomatic variables with statistical information related to changes in the number of inhabitants. For several decades, researchers have used a number of variables related to the population increase. These indicators are regularly gathered by public and private institutions, mainly for administrative purposes. This is the case, for example, for tax declarations, school enrollment, water intakes, light plugs in households, and number of voters. The accuracy and reliability of the estimates depends on the relationship between these variables and the size of the population. A useful tool for establishing the number of people in small areas has been constructed on the basis of indirect indicators of population size, using the regression method, whose dependent variable is the population, while independent variables are, for example, school enrollment and births and deaths.

Howe (2004) defines symptomatic variables as a set of available data that are related in any way to population changes. These methods take into count variables that should satisfy two requirements: having a high correlation with the population level and having permanent records in order to provide sufficient information to calibrate the models. The most common symptomatic variable methods are:

1. Distribution by apportionment, which assumes that the ratio between the population of every small area and the total population equals the corresponding symptomatic variable (Jardim 2001).

2. Proportional distribution, based on the assumption that the population varies in the same proportion as the symptomatic variable (Jardim 2001).
3. The method of vital rates, proposed by Donald Bogue (1950), uses vital statistics information on births and deaths in every small area for the base year and the year concerned for the estimation. It is assumed that a reliable estimate of the population of small areas is obtained through the ratio between local rates and the larger area.
4. The census-ratio method, which considers the rates of occurrence of symptomatic variables, assuming that the local population has a rate of change proportional to the larger area. The rate difference method assumes that the rate of growth of the symptomatic variable for a lesser area equals that registered in a larger area (Chaves Esquivel 2001).
5. The ratio correlation method assumes that the evolution of the population is correlated with the variation of the set of symptomatic variables (Schmitt and Crosetti 1954).

Finally, there are other methods: the rate correlation method, based on an exponential approximation (Chaves Esquivel 2001) and the difference correlation method, which is a variant of the ratio correlation. The difference is that population variations are taken from the differences rather than the ratios (O'Hare 1976).

All these methods are based on census information, sociodemographic statistics or surveys and on symptomatic indicator variables. Conversely, the one presented in this chapter is based on census data and geographic information obtained from satellite images, based on the evolution of the habitable area, which provides a good indication of the demographic dynamic. The remote sensing information is independent from the other sources, and can also be combined with information using symptomatic variable methods. It is important to recognize that most of the geographic information generated using satellite images is used, and that geospatial systems have been refined over time. One advantage of the method proposed in this work is that the estimation of the population is based on information that is entirely different from that normally used by demographers: the size of the geographic area under study.

The Kalman filter can also be extremely useful for estimating the population in ancient times, where the only source of information is the area where people lived. That area and its population density can shed light on the number of people living in that site.

13.3 Methodology

A filter is a device used to remove unwanted elements from certain mixtures, like water, and the term has been used in regard to electronic signals. The idea is to separate the "signal" from "noise". Noise is what statisticians call random error. When we turn on the radio, there is good signal and there is noise. An appropriate filter is one that allows us to listen without interference. The filter is based on concepts from the Least Squares Analysis of Stochastic Processes in Bayesian Statistics and Dynamical Systems.

The Kalman Filter (KF) is a modern version of the Least Squares Method (LSM). The main difference between both techniques is that in the LSM the parameters are fixed while in the KF they are stochastic processes, which means that the parameters are a succession of random variables changing with time.

In November 1958, when Rudolf Emil Kálmán[1] was traveling from Princeton to Baltimore by train, he thought about an application of the State Variables to the Wiener Filter which is used to estimate a desired random process by linear time-invariant filtering of an observed noisy process. (Grewal and Andrews 1993, p. 13) The Wiener filter minimizes the mean square error between the estimated random process and the desired process. At first, Kalman's idea was met with skepticism but now it has shown its usefulness.

The Kalman Filter is a recursive least squares estimator that is unbiased under Gaussian noise. It has a lot of potential applications in many fields, from aeronautics to economics. Actually, in 1960 the Kalman Filter (from here just "the filter") was considered to estimate the Apollo's trajectory to the moon (Grewal and Andrews 1993, p. 14). It has been applied to problems related to the aerospace industry and to analyze underwater radar signals and has recently been used in the area of economics. Many papers about the filter are published in engineering journals, meaning that many statisticians and demographers are unaware of the usefulness of this methodology. Nevertheless, the filter has a great potential in applications, due to its similarity with linear models and the analysis of Time Series. It is a modern way to discuss Least Squares Theory. The advantage is that the parameters of the Kalman Filter are stochastic processes; in other words, they vary over time. In the conventional regression model, parameters are fixed. In this chapter we propose an application of the filter to the field of demographic dynamics over time.

In summary, the Kalman filter is a set of equations to obtain an efficient recursive solution by the least squares method. This solution allows us to calculate a linear, unbiased and optimal estimator of the state of a process at each point in time (t) based on the information available at time $t-1$, and updated with the additional available information at time t.

Thus, for δ and γ in $\mathbb{R}$, the Kalman Filter is defined by the following two equations:

$$N'_{t+n} = \Phi N_t + W_t, \tag{13.1}$$

[1]Rudolf Emil Kálmán was born in Budapest, Hungary in May 19th, 1930. His family immigrated to the U.S. during World War II. He is an American electrical engineer, mathematician and inventor. For his work, president Obama rewarded him with the National Medal of Science in 2009.

$$S_t = \delta + \gamma N_t + V_t', \tag{13.2}$$

where N_t' is the first population estimate and V_t' is an error term (we are using the prime symbol to denote the order of the estimates in the process). Eq. (13.1) is the transition equation and Eq. (13.2) is the observation equation. In the second equation, δ is the initial surface value and γ is the per capita surface's growth rate. Notice that clearing N_t in Eq. (13.2) we can get the second population's estimate:

$$N''_{t+n} = \alpha + \beta S_{t+n} + V_{t+n} \tag{13.3}$$

where $\alpha = -\delta/\gamma$, $\beta = 1/\gamma$ and $V_{t+n} = -V_t'$. Demographically, the parameter Φ represents the growth rate of the geographic area between times t and $t+n$; β represents the relation between the population and the geographic area at moment t; α is the intercept, which in this case is zero because we have taken the values of the abscissa and the intercept relative to their respective mean, N_t represents the population at time t and S_t represents the geographic area at moment t. W_t and V_t are the errors of the equations which follow a normal distribution, $W_t \sim N\left(0, \sigma_W^2\right)$ and $V_t \sim N\left(0, \sigma_V^2\right)$, where σ_W^2 and σ_V^2 are the variance of the population estimation from the demographic growth rate of the surface and from the relation between population and surface, respectively.

Let

$$\widehat{N}_{t+n} = \omega N'_{t+n} + (1-\omega)\, N''_{t+n} \tag{13.4}$$

where N'_{t+n} and N''_{t+n} are two independent estimations of the population and therefore, the variance of the sum is the sum of variances, because the covariance is equal to 0. The value of ω represents the variance's weight in the estimation of the linear regression between the population and the surface with respect to the sum of the variances of both estimations. Calculating the variance of (13.4) we obtain:

$$Var\left(\widehat{N}_{t+n}\right) = \omega^2 Var\left(N'_{t+n}\right) + (1-\omega)^2 Var\left(N''_{t+n}\right) \tag{13.5}$$

Differentiating the last equation with respect to ω and equating it to zero, we find that the optimal value of ω is:

$$\omega = \frac{Var\left(N''_{t+n}\right)}{Var\left(N'_{t+n}\right) + Var\left(N''_{t+n}\right)}.$$

When $\omega = 1$, the estimated value of N_{t+n} (denoted as $\widehat{N}_{t+n}$) is N'_{t+n} and when $\omega = 0$ $\widehat{N}_{t+n} = N''_{t+n}$. The real value of ω varies between 0 and 1, depending on the value of the variances from different population estimates. The value of ω is estimated from the value of the variances of both population estimates.

13.4 Application

For this study, population data were taken from the Population Censuses of 1970, 1990, 2000 and 2010, in addition to the Population Counts of 1995 and 2005, and satellite images of the area of Villa Milpa Alta, for 1978, 1997, 2000 and 2010, measured in square kilometers. In 1978, the area was 1.04 km^2, in 1997 it was 1.32, in 2000, it was 2.40 and in 2010 it was 2.55 (see Map 13.A.1). The information was interpolated from the built area available for obtaining the estimated areas in the census years and in the years when population counts were carried out, with the aim of drawing a correlation between the census population and the surface. Information at the locality level has been available since 1970. According to the census, the total population was 13,347 inhabitants in 1990; 13,655 in 1995; 16,536 in 2000; 17,957 in 2005 and 18,274 in 2010. Although a population census was taken in 1980, it was of poor quality and no reliable data were published at the locality level. Information on the area and population from 1970 to 2005 is given below. The purpose of the study is to obtain an estimate of the 2010 population on the basis of the information from 2005, in order to make a linear regression between population and area, and then use the data on the estimated area for 2010 (Table 13.1).

In order to analyze the relation between the two variables, a linear correlation was drawn between the areas of each year minus the mean of the period and the population of each year minus the mean of the period, yielding a coefficient approximately equal to 1. The deviations regarding the mean of both variables were taken in order to have an intercept equal to zero. The regression equation obtained is as follows:

$$N_t - \overline{N} = 4,034.43 \left(S_t - \overline{S}\right), \tag{13.7}$$

Table 13.1 Milpa Alta village: surface and population

Year	Surface[a]	Population
1970	0.7077	9,451
1990	1.2150	13,347
1995	1.2875	13,655
2000	2.4047	16,536
2005	2.4767	17,957
2010	2.5487	18,274

Source: Census and Population Counts from the National Institute of Statistics and Geography (INEGI 2015a) and own calculations based on Satellite Images from INEGI (2015b) and National Commission of Territorial Studies

[a]Interpolated from the observed data for years 1970, 1978, 1997, 2000 and 2010

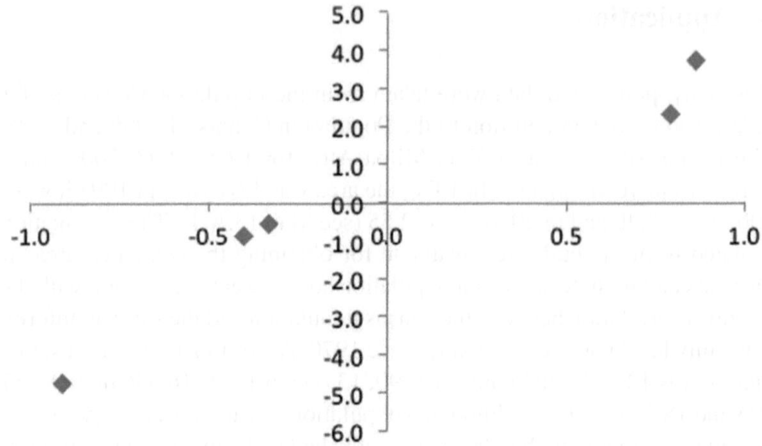

Fig. 13.1 Population deviation with respect to their mean (y) and surface deviation with respect to their mean (x), 1970, 1990, 1995, 2000, 2005 (Source: Own calculations based on Population Census 2010 and Population Count of 2005 (INEGI 2015a))

where N_t and S_t are the population and surface at time t, respectively, and $\overline{N}$ and $\overline{S}$ are the 1970–2010 mean population and the mean surface over the 1970–2010 period, respectively.

A high correlation is observed between both variables. The correlation coefficient is 0.96 and the coefficient of determination $R^2 = 0.93$. An analysis of the information shows that the larger the population, the larger the area. It is important to note that we have only five points in time (Fig. 13.1).

We chose to apply the Kalman Filter methodology in Villa Milpa Alta because the location is in a very small rural area, where almost all the inhabitants live in one-story houses. The expansion of the built area gives us an idea of population growth. The few two-story houses are in the downtown area of the locality, where the church and town hall are located. This locality is near the downtown area of Mexico City, so it was possible to verify the surface extension, and the construction types.

13.5 Results and Conclusions

In order to estimate the 2010 population, we calculated it in two ways. We first calculated the 2010 population projecting the census population from 2005 to 2010, based on the surface growth rate between 2005 and 2010, and did not use the population growth rate between 2000 and 2005. It is assumed that the surface growth rate is constant in the 5-year period from 2005 to 2010. The surface growth rate is a better indicator than the population growth rate for short-term projections.

The second way to calculate the population takes into account the relation between the population and the surface. Based on the area estimated in 2010, we obtained a population estimate for 2010. The mean between both estimations is

Table 13.2 Population of Villa Milpa Alta observed and estimated

T	N_t^O	r_S	N_t''	S_t	N_t'	$\widehat{N}_t$
2005	17,957	0.006	17,957	2.48	17,957	
2006			18,060			
2007			18,164			
2008			18,268			
2009			18,373			
2010	18,274		18,479	2.55	17,943	18,211

Source: Own calculations based on Population Census 2010 and Population Count of 2005 (INEGI 2015a)

taken and it is assumed that the variance of the estimate obtained from area increase estimate equals the variance obtained from the ratio between population and area, which was calculated on the basis of a regression model. In this case we started the simulation with $\omega = 0.5$. Once an estimate of the census population of 2010 was obtained, a new value of ω was calculated using that new information (Table 13.2).

N_t^O is the observed census population in year t, $r_S = \ln(S_{t+n}/S_t)/n$ is the surface annualized increasing rate, N'' is the estimated population using the surface increase rate, where $N''_{t+n} = N''_{t+n-1}\exp(r_S)$, S_t, is the observed surface in year t in square kilometers, N'_t is the estimated population based on the linear relation between population and surface (according to Eq. (13.7)) and $\widehat{N}_t$ is the estimated population according to Kalman Filter based Eq. (13.4).

The appropriate value of ω, estimated from the population census, i.e., the "real" (observed) population, can be obtained from $N^O_{t+n} = \omega N'_{t+n} + (1-\omega)N''_{t+n}$. Clearing the value of ω, the following is obtained:

$$\omega = \frac{N^O_{t+n} - N''_{t+n}}{N'_{t+n} - N''_{t+n}}.$$

Thus, in our example, ω equals 0.62, which means that the population from census 2010 (the real one) is closer to the estimated population for 2010 from the area increase rate between 2005 and 2010. In other words, that value of ω means that the new population estimate takes into account 62 % of the estimate based on the surface growth rate and 38 % of the estimate of the population from the regression between surface and population. As was shown above, the population estimate from the surface growth rates is calculated with data of the period 2005–2010, while the population estimate from the regression model is calculated with data for the period 1970–2010. That means that the earlier information has a lot of noise that is filtered out by the Kalman procedure. The difference between the observed population in 2010 and the Kalman estimate is 63 persons (18,274–18,211). That means an error of 0.34 %, which is very reasonable in statistical terms. Since Villa Milpa Alta is a place with a slow demographic dynamic, we will assume that census undercount is not significant. Therefore, we will assume it as the standard for evaluating estimates. The last value of ω could be used for a later estimation. Every time we have new information, it will be possible to adjust the model.

This method can be used when there is a positive relation between population and area; meaning that as the population grows, the housing area also increases. A limitation of this method is found when the population decreases and there are unoccupied houses. It is straightforward to use this method when houses are just one-story high, because they provide a good estimate of the number of people living in them. Nevertheless, it is possible to apply this method to areas with taller buildings, but instead of using the area, we will use the volume or height of the buildings.

The method proposed here allows us to proxy the demographic dynamics of certain ethnic groups living in isolated or mountainous areas. The study of these populations is very important in the sense that some of them are almost extinct while the remainder live in poverty. In this context, the study of demographic methods that help us improve estimates of the demographic dynamics of very small areas will help enable decision makers to ensure that public policies accurately target sparse populations that usually live in poverty. Thus, dynamic demographic analysis finds a huge applied research area that will ultimately help increase the wellbeing of many of the world's people.

Appendix 13.A

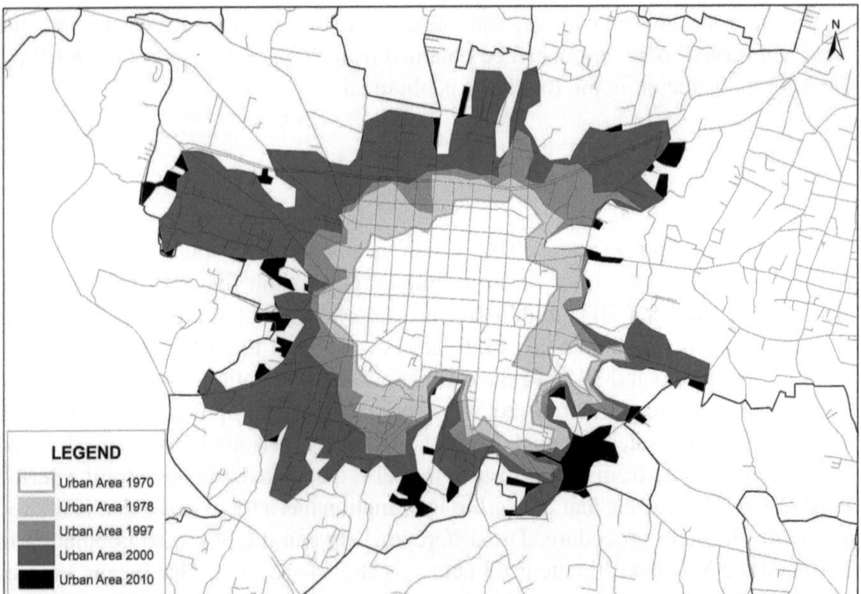

Map 13.A.1 Villa Milpa Alta. Stages of Growth of the Urban Area (Source: National Commission of Territorial Studies for 1970 and 1978, and INEGI (2015b) for 1997, 2000 and 2010)

References

Arriaga, E. (2001). *Population analysis with microcomputers* [Spanish title El análisis de la población con microcomputadoras]. Cordoba-Argentina: National University of Córdoba.

Bay, G. (1998). *The use of symptomatic variables in population estimation of small areas* [Spanish title "El uso de variables sintomáticas en la estimación de la población de áreas menores"], *Notas de Población* No. 67/68. Santiago-Chile, CELADE, pp. 181–208. Available at http://www.cepal.cl/publicaciones/xml/1/5431/LCG2048_p7.pdf

Bogue, D. J. (1950). A technique for making extensive population estimates. *Journal of the American Statistical Association, 45*(250), 149–163.

Cavenaghi, S. (2012). *Population estimates and projections in Latin America* [in Spanish Estimaciones y proyecciones de población en América Latina: desafíos de una agenda pendiente] Serie E- Investigaciones N.° 2/ALAP, Rio de Janeiro-Brazil.

Chaves Esquivel, E. (2001). *Symptomatic variables in population estimates at the Canton level in Costa Rica* [Spanish title "Variables sintomáticas en las estimaciones poblacionales a nivel cantonal en Costa Rica"] *Notas de Población* No 71. Santiago, CELADE, pp. 51–72. Available at http://www.cepal.cl/publicaciones/xml/3/7223/LCG2114_p3.pdf

Duchesne, L. (1987). *Method of population projections by sex and age for intermediate and small areas. Method of cohorts ratio* [Spanish title Método de proyecciones de población por sexo y edad para áreas intermedias y menores, Método de relación de cohortes] Santiago: CELADE.

Grewal, M. S., & Andrews, A. P. (1993). *Kalman filtering: Theory and practice*. Prentice-Hall: Englewood Cliffs.

Howe, A. (2004). *Assessing the accuracy of Australia's small area population estimates, 2001*. Australian Population Association: Canberra. http://www.apa.org.au/upload/2004-5C_Howe.pdf.

Jardim, M. L. (2001). *Use of symptomatic variables to estimate the population spatial distribution. Application to Rio Grande du Sul's Municipalities* [Spanish title "Uso de variables sintomáticas para estimar la distribución espacial de población. Aplicación a los municipios de Rio Grande do Sul, Brasil"], *Notas de Población* No 71. Santiago, CELADE, pp. 21–50. Available at http://www.cepal.cl/publicaciones/xml/3/7223/LCG2114_p2.pdf

Madeira, J. L., & Simões, C. C. S. (1972). Estimativas preliminares da populaçaõ urbana e rural, Segundo as unidades da Federaçaõ, 1960/1980: por uma nova metodologia. *Revista Brasileira de Estadística*, Río de Janeiro, ABE-IBGE, V. 33(129), 3–11.

National Institute of Statistics and Geography (INEGI). (2015a). National census and population counts. Available at http://www.inegi.org.mx/est/contenidos/proyectos/ccpv/default.aspx

National Institute of Statistics and Geography (INEGI). (2015b). *Urban geostatistical cartography*. Available at http://www.inegi.org.mx/geo/contenidos/urbana/default.aspx

O'Hare, W. (1976), Report on a multiple regression method for making population estimates. *Demography*, Maryland, *13*(3), 369–379.

Pittenger, D. B. (1976). *Population forecasting*. Cambridge, MA: Population Studies Division, Office of Program Planning and Fiscal Management, State of Washington, Ballinger Publishing Company.

Gonzalez, L., & Torres, E. (2012). Capítulo 4. Estimaciones de población en áreas menores en América Latina: revisión de métodos utilizados" in Cavenaghi (org.) Estimaciones y Proyecciones de Población en América Latina, Serie e-Investigaciones No. 2, ALAP, Rio de Janeiro, Brazil.

Swanson, D., & Tayman J. (2012). *Subnational population estimates* (The Springer series on demographic methods and population analysis 31). Netherlands, Springer.

Schmitt, R. C., & Crossetti, A. H. (1954). Accuracy of the ratio-correlation method for estimating postcensal population. *Land Economics, 30*(3), 279–281.

Chapter 14
Are the Pension Systems of Low Fertility Populations Sustainable?

Nan Li

Among the world's 198 countries and areas with more than 100,000 population in 2010, 75 had a total fertility below replacement level (2.1 children per woman) in 2005–2010, and the number of such countries is projected to reach 137 in 2045–2050 (United Nations 2013). The decline in fertility levels to below replacement level is a common trend that will lead to population decline and reduce the pressure on the environment and on natural resources. On the other hand, below-replacement fertility will also cause population aging and higher pension burdens of pay-as-you-go (PAYGO) systems (Lee et al. 2003). To compensate the effect of mortality decline, raising the retirement age has been gradually adopted as a solution by more and more countries. How to deal with the effect of low fertility, however, is still a challenge to most countries at low level of fertility, and this challenge will spread to other countries quickly.

Facing the challenge of long-term and wide-spread low fertility, considerations are naturally given to funded pension systems (World Bank 1994), in which current workers save to a fund that will pay their pensions in the future. Obviously, funded systems are costly and risky. Thus, a decision on whether to establish a funded pension system would require comparing the benefits and costs of a funded system to that of a PAYGO system. Such a comparison, however, is difficult to make, because PAYGO systems are built for populations in a certain period but funded systems are based on cohorts' lifecycles. Connecting cohort savings to period consumption usually requires complex models that can only be made for countries

Views expressed in this chapter are those of the author's and do not necessarily reflect those of the United Nations.

N. Li (✉)
United Nations Population Division, United Nations, New York, NY 10017, USA
e-mail: li32@un.org

R. Schoen (ed.), *Dynamic Demographic Analysis*,
The Springer Series on Demographic Methods and Population Analysis 39,
DOI 10.1007/978-3-319-26603-9_14

with reliable data on pension systems (Anderson et al. 2001; Borsch-Supan 2001). To address this issue, a time-based cohort old-age dependency ratio is proposed, which requires only demographic data. Further, comparing a time-based cohort old-age dependency ratio with a period ratio could indicate the difference between the pension burdens of a funded and a PAYGO system. Furthermore, such comparisons can be done for all the countries and areas of the world.

14.1 The Time-Based Cohort Old-Age Dependency Ratio

The period old-age dependency ratio can be defined as the quotient of old-age population aged 65 and over to working population aged 20–64. These ages may differ between countries and change over time, but the main results of this chapter will not be affected.

Denoting the population aged $[x,x+5)$ in year t by $P(x,t)$, and excluding centenarians for the reason to be indicated soon, the period old-age dependency ratio can be expressed as

$$P_{ODR}(t) = \frac{\sum_{x=65}^{95} P(x,t)}{\sum_{x=20}^{60} P(x,t)}. \tag{14.1}$$

The period dependency could be realised through PAYGO pension systems in more developed countries, or through informal kinship networks in less developed nations.

On the other hand, defining the period age-specific old-age dependency ratio as

$$P_{ASODR}(x,t) = \frac{P(x,t)}{\sum_{y=20}^{60} P(y,t)}, x = 65, 70, \ldots, \tag{14.2}$$

$P_{ODR}(t)$ can also be written as

$$P_{ODR}(t) = \sum_{x=65}^{95} P_{ASODR}(x,t). \tag{14.3}$$

Decomposing $P_{ODR}(t)$ into age-specific components leads to proposing a time-based cohort measure below.

At first glance, cohort dependencies are lifecycle measures, and therefore cannot be compared with period dependency at a certain time. But, if an old-age group can be viewed to depend on the working-age populations in the same period, it could

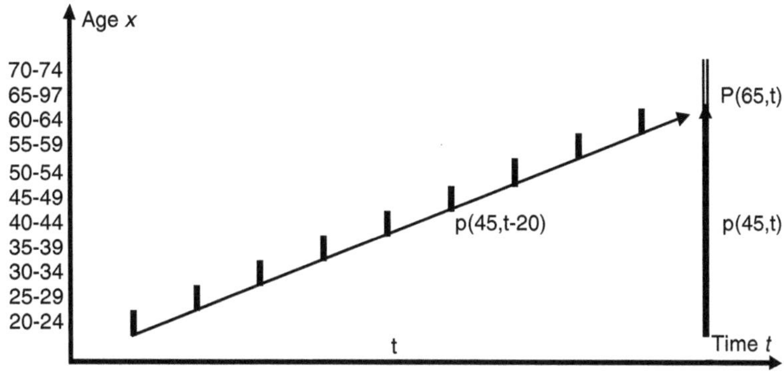

Fig. 14.1 Conceptual framework of the time-based cohort and period old-age dependency ratios

also be seen to depend on its own working times in earlier years. Arranging old-age groups to depend on the working-age populations in the same cohorts, a certain time can be assigned to the old-age dependency ratio for cohorts, which is called the *time-based cohort old-age dependency ratio*.

Assuming that, at time *t*, the old-age population in a certain age group depends on the person-years of themselves at working ages in earlier times through savings and asset accumulation, the cohort age-specific old-age dependency ratio can be defined as

$$C_{ASODR}(x,t) = \frac{P(x,t)}{\sum_{y=20}^{60} P(y, t-(x-y))}, x = 65, 70, \ldots, \qquad (14.4)$$

Figure 14.1 shows the case of how the old-age population aged 65–69 at time *t* are supported by populations aged 45–49 in the same time (vertical arrow), or by populations aged 45–49 in the same cohort but 20 years earlier (oblique arrow). The population aged 65–69 years at time *t*, *P(65,t)*, could be supported by the working-age population in the same period (vertical arrow); they may also be supported by the working ages of their own at earlier times (oblique arrow). Changing the '65' to '*x*' and '45' to '*x*-20' and drawing oblique arrows parallel to that in Fig. 14.1 illustrates how the old-age populations in other age groups are supported by working age populations, in the same cohort. Summing up such supports over all the ages of 65 and older will lead to the time-based cohort old-age dependency ratio, as can be seen below. There is a fundamental difference between (14.2) and (14.4). The denominator of (14.2) is the number of working-age population in a certain period, and is therefore constant with regard to *x*, the starting age of an old-age group. On the other hand, the denominator of (14.4) is the working-age person-years of a cohort at age *x* in the certain period, and therefore changes with age *x*.

Noting that $C_{ASODR}(x,t)$ represents the number of old-age population aged $[x,x+5)$ at time t per one working-age person in the same cohort, the sum of $C_{ASODR}(x,t)$ over age x indicates the total number of old-age population at time t per one working-age person in the corresponding cohorts. Consequently, similar to the procedure of computing the period old-dependency ratio, the time-based cohort old-age dependency ratio is obtained as the sum of the cohort age-specific old-age dependency ratios:

$$C_{ODR}(t) = \sum_{x=65}^{95} C_{ASODR}(x,t), \tag{14.5}$$

where the referred time, t, is naturally introduced.

Since $C_{ODR}(t)$ indicates the number of old-age population shouldered on every working-age person in corresponding cohorts, it is comparable to $P_{ODR}(t)$ that indicates the number of old-age population shouldered on every working-age person in the same period. Furthermore, computing $C_{ODR}(t)$ and $P_{ODR}(t)$ requires data only on historical and projected populations by age.

Computing $P_{ODR}(t)$ needs data on population only at year t. Calculating $C_{ODR}(t)$, however, requires population data from year t to 75 years before (when those aged 95 years in year t were 20 years old). If the end age were chosen as 105 instead of 100 years, for example, it would require population data 80 years before t. Nonetheless, computing $C_{ODR}(t)$ and comparing it with $P_{ODR}(t)$ can be done for all the countries of the world, using data from the World Population Prospects (e.g., United Nations 2013).

Looking at (14.4), it is clear that the effect of birth size is eliminated, because both the numerator and the denominator are from the same cohort. Thus, $C_{ODR}(t)$ is invariable to fertility change, which is an attractive feature with regard to the prospect of low fertility situations. Detailed properties of $C_{ODR}(t)$, and its relationships with $P_{ODR}(t)$, are discussed below.

14.2 Relationships Between $P_{ODR}(t)$ and $C_{ODR}(t)$

14.2.1 Stable Populations

Real populations are not stable. Nonetheless, stable populations are usually good approximations of real populations, and therefore the relationships found in stable populations are often approximately true in real populations.

With zero migration and unchanging mortality and fertility rates, the population would converge to a stable population (Pollard 1973), of which the age structure is constant over time and can be written as $L(x)\exp(-r\overline{x})$, where $L(x)$ and $\overline{x}$ represent the person-years and a middle age in $[x,x+5)$, respectively, and r indicates the growth rate. Thus, in the long run the $P_{ODR}(t)$ would converge to the old-age dependency ratio of the corresponding stable population, namely P^S_{ODR}

$$\mathop{Lim}_{t\to\infty} P_{ODR}(t) = P^S_{ODR} = \frac{\sum_{x=65}^{95} L(x)\exp(-r\overline{x})}{\sum_{x=20}^{60} L(x)\exp(-r\overline{x})}. \tag{14.6}$$

Since the value of r is usually small, (14.6) can be approximated as

$$\begin{aligned} P^S_{ODR} &\approx \frac{\sum_{x=65}^{95} L(x)}{\sum_{x=20}^{60} L(x)} - r\frac{\sum_{x=65}^{95}\overline{x}L(x)\sum_{x=20}^{60}L(x) - \sum_{x=65}^{95}L(x)\sum_{x=20}^{60}\overline{x}L(x)}{\left[\sum_{x=20}^{60}L(x)\right]^2} \\ &= \frac{\sum_{x=65}^{95} L(x)}{\sum_{x=20}^{60} L(x)} - r\left\{\frac{\sum_{x=65}^{95}\overline{x}L(x)}{\sum_{x=65}^{95}L(x)}\frac{\sum_{x=65}^{95}L(x)}{\sum_{x=20}^{60}L(x)} - \frac{\sum_{x=20}^{60}\overline{x}L(x)}{\sum_{x=20}^{60}L(x)}\frac{\sum_{x=65}^{95}L(x)}{\sum_{x=20}^{60}L(x)}\right\} \end{aligned} \tag{14.7}$$

Noticing that $\sum_{x=65}^{95}\overline{x}L(x)/\sum_{x=65}^{95}L(x)$ and $\sum_{x=20}^{60}\overline{x}L(x)/\sum_{x=20}^{60}L(x)$ are the average ages of the working-age and old-age population, which we denote as μ_w and μ_o respectively, (14.7) can be written as

$$P^S_{ODR} \approx C^S_{ODR} - C^S_{ODR}\cdot(\mu_o - \mu_w)\cdot r, \tag{14.8}$$

where C^S_{ODR} represents cohort old-age dependency ratio of the corresponding stationary population. Further, because (see Pollard 1973)

$$r \approx \frac{Log(NRR)}{\mu_b}, \tag{14.9}$$

where μ_b represents the average age of delivering birth and NRR is the net reproduction rate that reflects fertility level. Accordingly, (14.8) is expressed as

$$P^S_{ODR} \approx C^S_{ODR} - \frac{\mu_o - \mu_w}{\mu_b}C^S_{ODR}\cdot\log(NRR). \tag{14.10}$$

Equation (14.10) leads to the basic conclusion of this chapter as is shown in (14.11),

$$P^S_{ODR} \begin{cases} < C^S_{ODR}, & NRR > 1, \\ = C^S_{ODR}, & NRR = 1, \\ > C^S_{ODR}, & NRR < 1. \end{cases} \tag{14.11}$$

which is that, when fertility level is higher than replacement (*NRR* > *1*), in the long run the period old-age dependency ratio would be smaller than the time-based cohort old-age dependency ratio; when fertility level is below replacement (*NRR* < *1*), the period old-age dependency ratio would be bigger than the time-based cohort old-age dependency ratio; and in stationary populations *(NRR* = *1)* the two ratios would be the same.

14.2.2 Real Populations

Relaxing the constant fertility assumption: reducing the fertility level will shrink the working-age population and hence raise the period old-age dependency ratio, and vice versa. On the other hand, changes of fertility level will not affect the cohort dependency ratio.

The effect of mortality decline can also be discussed approximately: survivors at old ages have accumulated more effects of mortality decline than have working-age individuals. Therefore, mortality decline would raise both $P_{ODR}(t)$ and $C_{ODR}(t)$. Furthermore, mortality decline affects the numerators of $P_{ODR}(t)$ and $C_{ODR}(t)$ in the same way. Thus, although more recent mortality decline affects the denominator of $P_{ODR}(t)$ while earlier mortality decline affects the denominator of $C_{ODR}(t)$, overall the gap between $P_{ODR}(t)$ and $C_{ODR}(t)$ would not change dramatically.

For most countries, the effect of migration is negligible compared to that of mortality and fertility change. For countries with significant immigration, $P_{ODR}(t)$ will be reduced immediately if the majority of immigrants are in working ages. But in the long run immigration will raise $P_{ODR}(t)$, if immigrants stay in the host country after retirement (Schmertmann 1992).

Entirely different from the effects on $P_{ODR}(t)$, immigration always raises $C_{ODR}(t)$, regardless of its time trend and age pattern. This is because that, the effect on $C_{ODR}(t)$ from one immigrant, at the starting point of working age or younger, is equivalent to an increase of fertility, and therefore is zero. On the other hand, the effect of one immigrant, older than the starting point of working age, is equivalent to a decrease of mortality, and therefore is to increase $C_{ODR}(t)$. Putting the zero and increase effects together, the total effect is to increase $C_{ODR}(t)$.

14.3 An Illustration

As is indicated above, when fertility level is higher than replacement, there is $P^S_{ODR} < C^S_{ODR}$; and fertility decline to below replacement leads to $P^S_{ODR} > C^S_{ODR}$. To illustrate this in real populations, the data of Italy back to 1900 is used as an example (Human Mortality Database 2013). Data on populations after 1950, and especially after 2010, for Italy and other countries in this chapter, are collected from the 2012 revision of the World Population Prospects (United Nations 2013).

The total fertility (TF) of Italy was higher than replacement level before 1975. As the stable-population analysis predicts, we observe in Fig. 14.2 that $P_{ODR}(t) < C_{ODR}(t)$ up to 2020, 45 years later than 1975. Note that 45 years is about the middle age of the working population. The reversal, from $P_{ODR}(t) < C_{ODR}(t)$ to $P_{ODR}(t) > C_{ODR}(t)$ at 45 years later than 1975, is a result of fertility decline crossing the replacement level at 1975, and indicates a common trend that would occur in other countries. This reversal could perhaps explain why PAYGO systems were widely adopted among more developed countries when fertility levels were high. This reversal should also cause considerations about establishing funded pension systems, especially for developing countries, where the traditional kinship networks would become insufficient owing to below-replacement fertility, and where pension systems are yet to be established.

The TF for low-fertility countries are assumed in World Population Prospects (United Nations 2013) to increase in the long run. If the future levels of TF are not rising significantly, Italian $P_{ODR}(t)$ in the future would be higher than that in

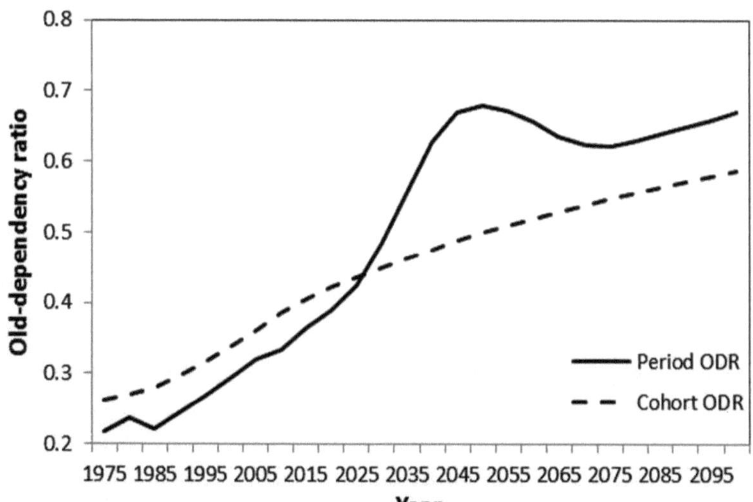

Fig. 14.2 Period and time-based cohort old-age dependency ratios, Italy

Fig. 14.2. On the other hand, as is also indicated, the increase in $C_{ODR}(t)$ is mainly due to the decline of mortality at old ages, and it could be reduced only by raising the starting age of old population, which is 65 in this chapter.

14.4 Examples and Discussion

Population and fertility data of China, Japan, and the Republic of Korea are collected from the 2012 revision of the World Population Prospects to provide examples. Japan and the Republic of Korea are selected because of their low fertility levels; China is chosen due to the sizes of its population and economy.

In Fig. 14.3, the curves before 2010 display observations (or estimation for China), and after 2010 represent projections. The projected increases of total fertility, in the medium variant of the United Nations (2013), reflect the belief that societies will act to raise fertility. Projections of the Government of Japan and the Republic of Korea (personal communication) are also displayed in Fig. 14.3. Obviously, the Governments are not as optimistic as the United Nations. For low-fertility countries, subtle rises of total fertility were observed in recent years, as can be seen from the data of Japan and the Republic of Korea in Fig. 14.3. These small rises are also the empirical basis of projecting the long-term increase of total fertility. Whether such small rises reflect robust trend or random fluctuation, however, is yet an open question. In fact, a more recent investigation (Goldstein et al. 2013) found

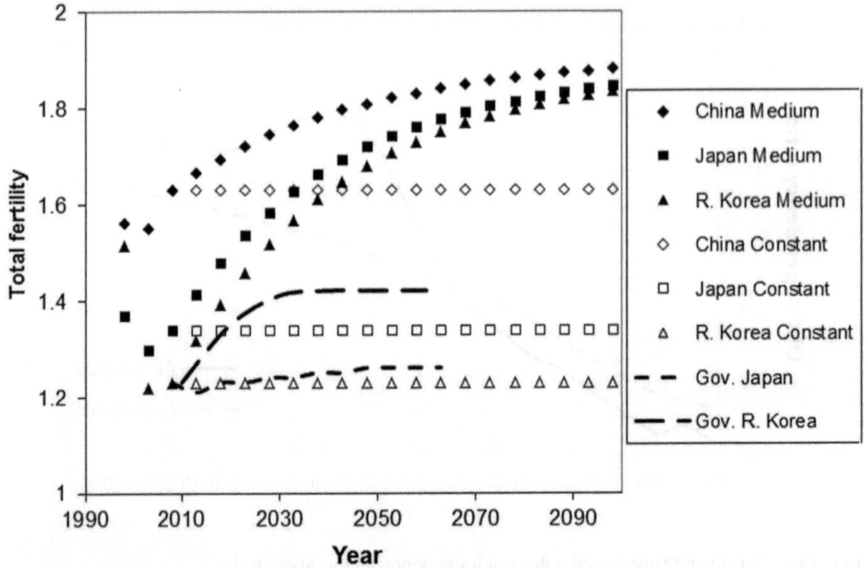

Fig. 14.3 Total fertility of China, Japan and the Republic of Korea

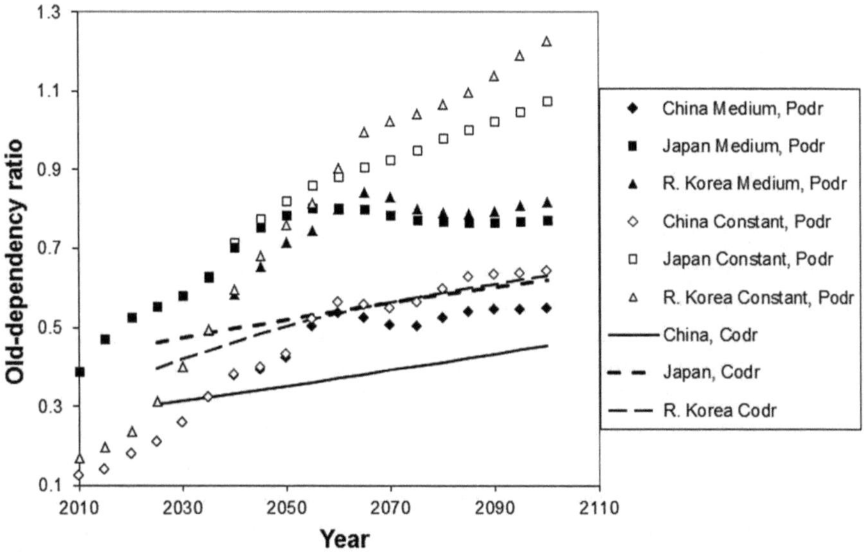

Fig. 14.4 Old-age dependency ratios of China, Japan, and the Republic of Korea

declines of fertility in European countries. Fertility changes were highly uncertain in history, and it should be so in the future. For this reason, the variants of constant-fertility at the 2005–2010 levels (United Nations 2013) are also used as references.

In Fig. 14.4, period old-age dependency ratios are displayed in diamonds (China), squares (Japan), and triangles (the Republic of Korea), with solid black characters representing medium fertility and white-filled characters denoting constant fertility. Cohort time-based old-age dependency ratios are shown as curves, of which solid, short dash, and long dash stand for China, Japan, and the Republic of Korea, respectively.

As can be seen in Fig. 14.4, in 2010, the $P_{ODR}(t)$ of Japan was about 0.4, much higher than that of China and the Republic of Korea, because the levels of fertility and mortality of Japan had been lower than the other two countries in a number of decades before 2010. The historical lower levels of mortality and fertility had made the PAYGO burden of Japan more than twice that of China and the Republic of Korea.

After 2010, it can be seen that, in the first 40 years, the values of $P_{ODR}(t)$ are projected to rise in the same way, regardless of rising or constant fertility. This is because rising fertility can only affect $P_{ODR}(t)$ after about 40 years, when the boom in births reaches the middle of working ages. The alarming information is that in 2050, 40 years after 2010, the $P_{ODR}(t)$ of China is projected to be about 0.4, the level of Japan in 2010. In 2050, the $P_{ODR}(t)$ of Japan and the Republic of Korea are projected to be around 0.7, a level that calls into question the maintainability of PAYGO systems.

After 2050, the change of $P_{ODR}(t)$ would depend on fertility level. If total fertility rises as in the case of the medium variant (United Nations 2013), then $P_{ODR}(t)$ would level off around 2050. If total fertility could not be raised in the future, then, as the constant fertility scenarios show, $P_{ODR}(t)$ would increase by an additional 20 % for China, and surge to much higher than unity for Japan and the Republic of Korea. It is also interesting to note that, a hump in $P_{ODR}(t)$ occurs for China around 2060, even if fertility remains constant after 2010. This is because fertility decline will cause waves in births, a phenomenon analysed by Schoen and Kim (1996).

How to increase fertility in low-fertility countries is not actually known, either in theory or in practice. What else, then, might help? An obvious answer is to look at funded systems, where the pension burden is measured by $C_{ODR}(t)$. Using data that begins in 1950, the earliest year at which the $C_{ODR}(t)$ can be computed is 2025. As Fig. 14.4 shows, the $C_{ODR}(t)$ of Japan and the Republic of Korea would be higher than that of China. The reason is that mortality levels of Japan and the Republic of Korea were historically lower than that of China. Nonetheless, the values of $C_{ODR}(t)$ in 2025 for Japan would still be lower than the corresponding $P_{ODR}(t)$, and therefore the pension burdens of funded system would be lower than that of PAYGO system. For China and the Republic of Korea, $C_{ODR}(t)$ are projected to be higher than $P_{ODR}(t)$ in 2025 but lower than 0.4, and the reversal would come soon.

For the years after 2025, the $C_{ODR}(t)$ will be independent from fertility change, and would remain constant if there were no mortality decline. Under the mortality declines projected by the United Nations (United Nations 2013), the increases in $C_{ODR}(t)$ are trivial comparing to that of $P_{ODR}(t)$ (Fig. 14.4). In 2050, for instance, the levels of $C_{ODR}(t)$ would be lower than that of $P_{ODR}(t)$ by 17 %, 34 %, and 30 % for China, Japan, and the Republic of Korea, respectively. Moreover, for the years after 2050, the reductions of pension burden from using the funded system would be larger, if total fertility remained below the medium variants of the United Nations project. For instance, if total fertility remained constant in the future, then, in 2100 the reductions of pension burden would be 30 %, 42 % and 49 % for China, Japan, and the Republic of Korea, respectively.

14.5 Summary

For low-fertility populations, the pension burdens of funded system are significantly lower than that of the PAYGO system. Accordingly, funded systems of low-fertility populations are sustainable. This is not hard to understand. In the situation of below-replacement fertility, workers and retirees are both declining over time; and therefore the pension burdens for current workers to support future retirees (funded system) would be lighter than that for the current workers to support current retirees (PAYGO system).

Compared to PAYGO systems, however, funded systems require reliable financial systems to accumulate and manage the assets on which the elderly depend. To better understand this requirement, Lee and Mason (2011) provided comprehensive concepts and measures. To meet this requirement is not easy, but not unrealistic either. In fact, funded systems have long been used as a component of pension programs in many countries. For example, the RRSP of Canada and 401 K plans of the United States are funded systems. Theoretical discussions about mixed pension systems are also available (e.g., De Santis 2003). Moreover, funded systems predominate in a number of countries, such as Chile and Mexico.

References

Anderson, M., Tuljapurkar, S., & Li, N. (2001). How accurate are demographic projections used in forecasting pension expenditure? In T. Boeri et al. (Eds.), *Pensions: More information, less ideology* (pp. 9–29). Boston: Kluwer Academic Publisher.

Borsch-Supan, A. (2001). What we know and what we do not know about the willingness to provide self-financed old-age insurance. In T. Boeri et al. (Eds.), *Pensions: More information, less ideology* (pp. 113–136). Boston: Kluwer Academic Publisher.

De Santis, G. (2003). The demography of an equitable and stable intergenerational transfer system. *Population-E, 58*, 587–622.

Goldstein, J. R., Kreyenfeld, M., Jasilioniene, A., & Orsal, D. K. (2013). Fertility reactions to the "Great Recession" in Europe: Recent evidence from order-specific data. *Demographic Research, 29*, 85–104.

Human Mortality Database. (2013). University of California, Berkeley (USA), and Max Planck Institute for Demographic Research (Germany). Available at www.mortality.org or www.humanmortality.de. Data downloaded on April 2013.

Lee, R., & Mason, A. (2011). *Population aging and the generational economy, a global perspective*. Cheltenham: Edward Elgar Publishing Inc.

Lee, R., Anderson, M., & Tuljapurkar, S. (2003). Stochastic forecasts of the social security trust funded. CEDA chapters http://escholarship.org/uc/item/3mw1m56d

Pollard, J. (1973). *Mathematical models of the growth of human population*. Cambridge: Cambridge University Press.

Schmertmann, C. P. (1992). Immigrants' ages and the structure of stationary populations with below-replacement fertility. *Demography, 29*, 595–612.

Schoen, R., & Kim, Y. J. (1996). Stabilization, birth waves, and the surge in the elderly. *Mathematical Population Studies, 6*, 35–53.

United Nations. (2013). *World population prospects: The 2012 revision*. New York: United Nations.

World Bank. (1994). *Averting the old age crisis*. Oxford: Oxford University Press.

Chapter 15
Age-Specific Mortality and Fertility Rates for Probabilistic Population Projections

Hana Ševčíková, Nan Li, Vladimíra Kantorová, Patrick Gerland, and Adrian E. Raftery

15.1 Introduction

Projections of countries' future populations, broken down by age and sex, are used by governments for social, economic and infrastructure planning, by international organizations for development planning and monitoring and global modeling, by the private sector for strategic and marketing decisions, and by academic and other researchers as inputs to social and health research.

Most population projections have until recently been done deterministically, using the cohort component method (Whelpton 1928, 1936). This is an age- and sex-structured version of the basic demographic identity that the population of a country at the next time point is equal to the population at the current time point, plus the number of births, minus the number of deaths, plus the number of immigrants minus the number of emigrants. It was formulated in matrix form by Leslie (1945) and is described in detail in Preston et al. (2001, Chap. 6).

Population projections are currently produced by many organizations, including national and local governments and private companies. The main organizations that have produced population projections for all or most of the world's countries are the United Nations (UN) (United Nations 2011), the World Bank (Bos et al. 1994),

H. Ševčíková
Center for Statistics and the Social Sciences, University of Washington, Seattle, WA 98195-4320, USA
e-mail: hanas@u.washington.edu

N. Li • V. Kantorová • P. Gerland
United Nations Population Division, United Nations, New York, NY 10017, USA
e-mail: li32@un.org; kantorova@un.org; gerland@un.org

A.E. Raftery (✉)
Departments of Statistics and Sociology, University of Washington, Seattle, WA 98195-4322, USA
e-mail: raftery@u.washington.edu

R. Schoen (ed.), *Dynamic Demographic Analysis*,
The Springer Series on Demographic Methods and Population Analysis 39,
DOI 10.1007/978-3-319-26603-9_15

and the United States Census Bureau (U. S. Census Bureau 2009), all of which have until recently used the standard deterministic approach. Among these, the UN produces updated projections for all the world's countries every 2 years, published as the *World Population Prospects*, and these are the de facto standard (Lutz and Samir 2010).

Standard population projection methods are deterministic, meaning that they yield a single projected value for each quantity of interest. However, probabilistic projections that give a probability distribution of each quantity of interest, and hence convey uncertainty about the projections, are widely desired. They are needed for planning purposes. For example, those planning school construction may wish to be reasonably sure of building enough capacity to accommodate all students in the future. For this the relevant projection is an upper quantile of the predictive distribution of the future school population that is relatively unlikely to be exceeded, rather than a "best guess." Probabilistic projections are also useful for assessing change and deviations of population outcomes from expectations, as well as for providing a general assessment of uncertainty about future population.

The most common approach to communicating uncertainty in population projections has been the scenario, or High-Medium-Low, approach. In this approach, a central or main projection is first produced. Then high and low values of the main inputs to the projection model, such as fertility or mortality, are postulated, and a projection is produced with the high values, and another one with the low values. These high and low trajectories are viewed as bracketing the likely future values. This approach has been criticized as having no probabilistic basis and leading to inconsistencies (Lee and Tuljapurkar 1994; National Research Council 2000).

Previous approaches to producing probabilistic population projections include ex-post analysis, time series methods and expert-based approaches (National Research Council 2000; Booth et al. 2006). Ex-post analysis is based on the errors in past forecasts (Keyfitz 1981; Stoto 1983; Alho et al. 2006, 2008; Alders et al. 2007). The time-series analysis approach uses past time series of forecast inputs, such as fertility and mortality, to estimate a statistical time series model, which is then used to simulate a large number of random possible future trajectories. Simulated trajectories of forecast inputs are combined via a cohort component projection model to produce predictive distributions of forecast outputs (Lee and Tuljapurkar 1994; Tuljapurkar and Boe 1999). In the expert-based method (Pflaumer 1988; Lutz et al. 1996, 1998, 2004) experts are asked to provide distributions for each forecast input. These are then used to construct predictive distributions of forecast outputs using a stochastic method similar to the time series method.

The United Nations released official probabilistic population projections for all countries for the first time in July 2014 (Gerland et al. 2014). They were produced by probabilistically projecting the period total fertility rates (TFR) and life expectancies (e_0) for all countries using Bayesian hierarchical models (Alkema et al. 2011; Raftery et al. 2013). These probabilistic projections took the form of a large set of trajectories, each of which was sampled from the joint predictive distribution of TFR and female and male e_0 for all countries and all future time periods to 2100 using Markov chain Monte Carlo (MCMC) methods.[1] Among previous probabilistic methods, the UN approach is most closely related to the time series approach.

For each trajectory, the life expectancies were converted to age- and sex-specific mortality rates, and the total fertility rates were converted to age-specific fertility rates. The population was then projected forward using the cohort-component method. This yielded a large set of trajectories of population by age and sex, and age-specific fertility and mortality rates, for all countries and future time periods jointly. These were summarized by predictive medians and 80 % and 95 % prediction intervals for a wide range of population quantities of interest, for all countries and a wide range of regional and other aggregates. They were published as the UN's Probabilistic Population Projections (PPP), and are available at http://esa.un.org/unpd/wpp.

This chapter focuses on the methods used to convert probabilistic projections of e_0 and TFR to probabilistic projections of age-specific mortality and fertility rates. Some limitations of the methods used for the 2014 PPP are identified, and several improvements are proposed to overcome them. The methods presented in this chapter are implemented in an open source R package called `bayesPop` (Ševčíková and Raftery 2014; Ševčíková et al. 2014).

The chapter is organized as follows. In Sect. 15.2 we describe the current method in PPP for projecting age-specific mortality rates, and our proposed improvements. In Sect. 15.2.1 we outline the Probabilistic Lee-Carter method used in the 2014 PPP. In the rest of Sect. 15.2 we propose several improvements to overcome limitations of this method. These include a new Coherent Kannisto Method for joint projection of future age-specific mortality rates at very high ages that avoids unrealistic crossovers between the sexes (Sect. 15.2.2), application of the Coherent Lee-Carter method to avoid crossovers at lower ages (Sect. 15.2.3), new methods for avoiding jump-off bias (Sect. 15.2.4), and application of the Rotated Lee-Carter method to reflect the fact that when mortality rates are low, they tend to decline faster at older than at younger ages (Sect. 15.2.5). In Sect. 15.3, we describe the current method in PPP for projecting age-specific fertility rates and our proposed improvements. We conclude with a discussion in Sect. 15.4.

[1]This general approach applies to countries experiencing normal mortality trends. For countries having ever experienced 2 % or more adult HIV prevalence during the period 1980–2010, all projected trajectories of life expectancy by sex for each of these countries were adjusted in such a way as to ensure that the median trajectory for each country was consistent with the 2012 Revision of the World Population Prospects deterministic projection that incorporates the impact of HIV/AIDS on mortality, as well as assumptions about future potential improvements both in the reduction of the epidemic and survival due to treatment.

15.2 Age-Specific Mortality Rates for Probabilistic Population Projections

15.2.1 Probabilistic Lee-Carter Method

Our methodology is based on the Lee-Carter model (Lee and Carter 1992). This was originally developed for a single country, and is defined as follows:

$$\log[m_x(t)] = a_x + b_x k(t) + \varepsilon_x(t), \quad \varepsilon_x(t) \sim N(0, \sigma_\varepsilon^2),$$

where $m_x(t)$ is the mortality rate for age x and time period t, the quantity a_x represents the baseline pattern of mortality by age over time, and b_x is the average rate of change in mortality rate by age group for a unit change in the mortality index $k(t)$. The parameter $k(t)$ is a time-varying index of the overall level of mortality, and $\varepsilon_x(t)$ is the residual at age x and time t. Throughout this chapter, log denotes the natural logarithm.

For a given matrix of rates $m_x(t)$, the model is estimated by a least squares method. The baseline mortality pattern a_x is estimated as the average of $\log[m_x(t)]$ over the past time periods with observed data. Since the model is underdetermined, b_x is identified by setting $\sum_x b_x = 1$, where the sum is over all ages or age groups x. Also, $k(t)$ is identified by setting $\sum_{t=1}^{T} k(t) = 0$, where T is the number of past time periods for which data are available. The estimates are then

$$\hat{a}_x = \frac{\sum_{t=1}^{T} \log[m_x(t)]}{T}, \tag{15.1}$$

$$\hat{k}(t) = \sum_x \{\log[m_x(t)] - \hat{a}_x\}, \tag{15.2}$$

$$\hat{b}_x = \frac{\sum_{t=1}^{T} \{\log[m_x(t)] - \hat{a}_x\} \hat{k}(t)}{\sum_{t=1}^{T} \hat{k}(t)^2}. \tag{15.3}$$

To forecast $m_x(t)$, one needs to project $k(t)$ into the future. To project $k(t)$, the Lee-Carter method uses a random walk with a constant drift d per time period, leading to the deterministic projections

$$\hat{k}(t+1) = \hat{k}(t) + \hat{d}, \quad \text{where } \hat{d} = \frac{1}{T-1}[\hat{k}(T) - \hat{k}(1)].$$

Lee and Miller (2001) proposed replacing the step of projecting $k(t)$ by itself by matching future $k(t)$ to future projected $e_0(t)$.

Current calculations are done using a highest age or open interval of 85+. For projections one needs to extend mortality rates to higher ages x, usually beyond 100+, because mortality rates are expected broadly to decline over time in the future, so there will be larger numbers of people at higher ages. For extending the force of mortality at older age groups, the Kannisto model provides a simple but well-fitting way to approximate available mortality rates from age 80 to 100, and to extrapolate mortality rates up to age 130 in a way that is consistent with empirical observations on oldest-old mortality (Thatcher et al. 1998).

The Bayesian probabilistic projections of life expectancy at birth (Raftery et al. 2013, 2014b) provide us with a set of future trajectories of female and male e_0, representing a sample from the joint predictive distribution of future female and male e_0 for all countries and all future time periods. The 2014 PPP used methods for turning a trajectory of future e_0 values into a set of future age-specific mortality based on the ideas of Lee and Miller (2001) and Li and Gerland (2011); see Raftery et al. (2012). They were based on the following algorithm:

Algorithm 15.1

Let $t \in \{1, \ldots, T\}$ and $\tau \in \{T+1, \ldots, T_p\}$ denote the observed and projected time periods, respectively.

1. Using the Kannisto method extend $m_x(t)$ to higher age groups so that $\max(x) = 130+$ for all t.
2. Estimate a_x, $k(t)$ and b_x using the extended historical $m_x(t)$ (Eqs. (15.1), (15.2) and (15.3)), yielding estimates $\hat{a}_x$ and $\hat{b}_x$ for all age groups x.
3. For a given $e_0(\tau)$ in each trajectory and the estimates $\hat{a}_x$ and $\hat{b}_x$ from the previous step, solve for future $k(\tau)$ numerically using life tables. This yields a nonlinear equation which can be solved using the bisection method. More details are given in Sect. 15.2.6.
   This gives a sample of values of $k(\tau)$ for each future time period τ, with one value for each trajectory.
4. Compute age-specific mortality rates by $\log[m_x(\tau)] = \hat{a}_x + \hat{b}_x k(\tau)$ for each trajectory and future time period τ.

Applying these steps to all trajectories of e_0 yields a posterior predictive distribution of $m_x(t)$. Note that in the above algorithm, $k(\tau)$ is trajectory-specific, while $\hat{a}_x$ and $\hat{b}_x$ are the same for all trajectories.

However, this procedure has a number of drawbacks. There is no assurance that the extension of $m_x(t)$ to higher ages yields mortality rates that are coherent between males and females. Similarly, the predicted $m_x(\tau)$ can lead to unwanted crossovers between female and male mortality rates, since they are obtained independently for each sex. In the following sections, we present solutions to these and other limitations of the simple algorithm above, and give more details about Step 3.

15.2.2 Coherent Kannisto Method

A sex-independent extension of the observed mortality rates to higher age categories can lead to unrealistic crossovers at higher ages. We propose a modification of the Kannisto method that treats male and female mortality rates jointly. For simplicity we omit the time index t in this section.

The original Kannisto model has the form

$$m_x = \frac{ce^{dx}}{1 + ce^{dx}} e^{\varepsilon_x}, \quad \text{or}$$
$$\text{logit}(m_x) = \log c + dx + \varepsilon_x,$$

where ε_x is a random perturbation with mean zero. The model is usually estimated independently for each sex, assuming independence across ages and normality of the ε_x, using a maximum likelihood method (Thatcher et al. 1998; Wilmoth et al. 2007). This yields sex-specific parameter estimates $\hat{d}_M, \hat{d}_F, \hat{c}_M, \hat{c}_F$.

We suggest modifying this by forcing the sex-specific parameters d_M and d_F to be equal (i.e. $d_M = d_F = d$), but still allowing the parameters c_M and c_F to differ between the sexes:

$$\text{logit}(m_x^g) = \log c_g + dx + \varepsilon_x^g, \qquad \text{for } g = M, F.$$

This leads to the following model:

$$\text{logit}(m_x^g) = \beta_0 + \beta_1 1_{(g=M)} + \beta_2 x + \varepsilon_x^g,$$

where $1_{(g=M)} = 1$ if $g = M$ and 0 otherwise.

To estimate the β parameters, we fit the model to the observed m_x for ages 80–99 by ordinary least-squares regression, which corresponds to maximum likelihood under the assumptions that the ε_x^g are independent and normally distributed and have the same variance. There are four age groups in the data used for fitting the model, and thus eight points in total for both sexes. Then,

$$\hat{c}_F = e^{\hat{\beta}_0},$$
$$\hat{c}_M = e^{\hat{\beta}_0 + \hat{\beta}_1},$$
$$\hat{d} = \hat{\beta}_2.$$

Figure 15.1 shows the resulting m_x for old ages for Brazil and Lithuania in the last observed time period. From the left panels we see that there are crossovers using the classic Kannisto method, which is unrealistic. However, male mortality stays above female mortality in the coherent version, as can be seen in the right panels; this is more realistic.

15.2.3 Coherent Lee-Carter Method

We adopt an extension of the Lee-Carter method suggested by Li and Lee (2005), the so-called *coherent* Lee-Carter method. It takes into account the fact that mortality patterns for closely related populations are expected to be similar. In our application,

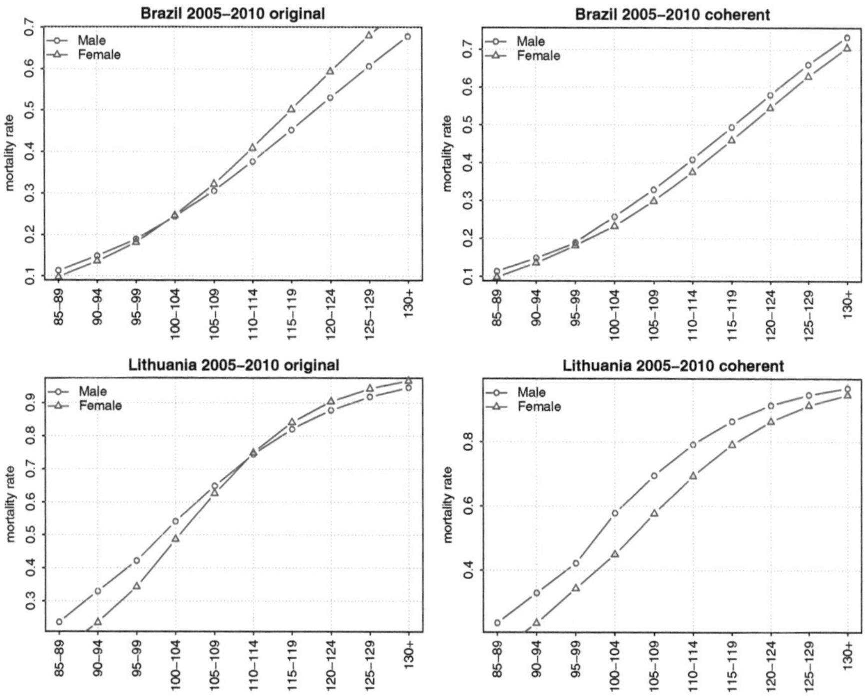

Fig. 15.1 Mortality rates for males and females extended using the coherent Kannisto method (*right panels*), compared to the original Kannisto method (*left panels*) for Brazil and Lithuania in the last observed time period (2005–2010)

these related populations will be males and females in the same country, since there is no expectation that the life expectancy will diverge between these groups. Thus, the Lee-Carter method is extended by two requirements:

$$b_x^M = b_x^F,$$
$$k_M(\tau) = k_F(\tau), \tag{15.4}$$

where M and F denotes male and female sex, respectively. This ensures that the rates of change of the future mortality rates are the same for the two sexes, and thus avoids crossovers.

15.2.4 Avoiding Jump-Off Bias

Mortality rates in the last period of the historical data used for estimation (or jump-off period) are commonly referred to as jump-off rates (Booth et al. 2006). Often there is a mismatch between fitted rates for the last period T and the actual rates (jump-off bias). As a result, a discontinuity between the actual rates in the jump-off period and the rates projected in the first projection period may occur.

A possible solution to avoid jump-off bias is to constrain the model in such a way that $k(t)$ passes through zero in the jump-off period T, and to use m_x only from the last fitting period to obtain a_x (Lee and Miller 2001):

$$a_x = \log[m_x(T)] \implies k(T) = 0. \tag{15.5}$$

A disadvantage of this solution is that in cases where the mortality rates are bumpy in the jump-off period (i.e. not smooth across ages), this "bumpiness" propagates into the future. In general for projections, we suggest using the age-specific mortality rates from the last fitting period and smoothing them over age if necessary (e.g. for small populations with few deaths in some age groups) while preserving the value for the youngest age group:

$$a_x = \text{smooth}_x\{\log[m_x(T)]\}, \text{ with } a_{0-1} = \log[m_{0-1}(T)]. \tag{15.6}$$

Figure 15.2 shows the resulting difference in $\log[m_x(\tau)]$ projected to τ corresponding to 2095–2100 for two countries using the three different methods of computing a_x, namely Eqs. (15.1), (15.5) and (15.6). As can be seen in the case of Bangladesh, the smoothing step removes bumps whereas the averaging method does not.

Figure 15.3 shows the impact of the methods on m_x as time series for Bangladesh for three different age groups. Using the average m_x results in jump-offs for the 5–9 and 95–99 age groups. If the latest raw m_x are used, the jump-offs are eliminated. A smoothed version creates a new jump-off for the age group 75–79.

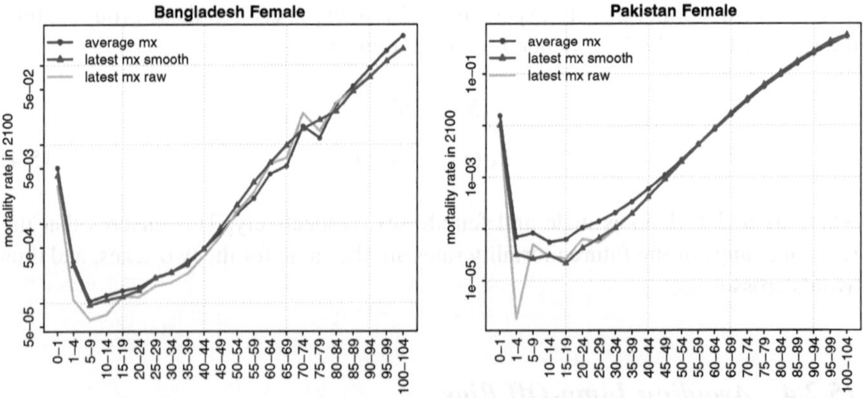

Fig. 15.2 Female age-specific mortality rates for Bangladesh and Pakistan in 2095–2100 projected using three different methods for computing a_x: (1) using an average m_x over time; (2) using the latest smoothed m_x; and (3) using the latest m_x as it is (raw). The *y*-axis is on the logarithmic scale

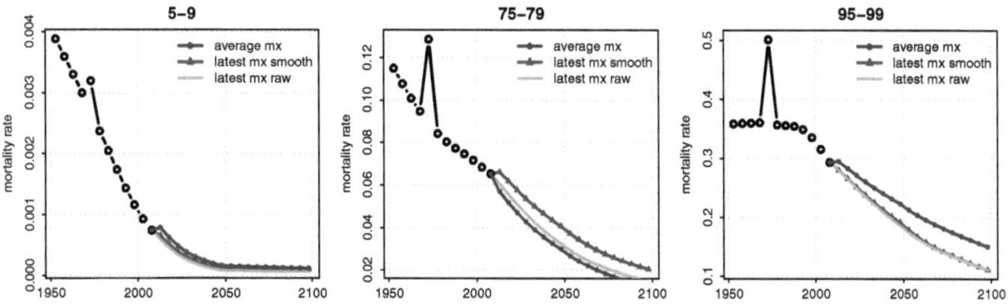

Fig. 15.3 Mortality rates over time for Bangladesh for three different age groups. Line styles correspond to the same methods as in Fig. 15.2

This shows that there is a trade-off between bumpy mortality rates over ages in later projection years and no jump-offs, and smooth mortality rates with no jump-offs. Our solution is to decide on a country-specific basis which method is more appropriate.

15.2.5 Rotated Lee-Carter Method

Li et al. (2013) focused on the fact that in more developed regions, once countries have already reached a high level of life expectancy at birth, the proportional mortality decline decelerates at younger ages and accelerates at old ages. This change in the pace of log mortality decline by age cannot be captured by the original Lee-Carter method, since this constrains the rate of change b_x to be constant over time. They proposed instead rotating the b_x over time to a so-called *ultimate* b_x, denoted by $b_{u,x}$, which is computed as follows.

Let

$$\bar{b}_{15-64} = \frac{1}{10} \sum_{x=15-19}^{60-64} b_x \,.$$

Then

$$b_{u,x} = \begin{cases} \bar{b}_{15-64} & \text{for } x \in \{0-1, 1-4, 5-9, \ldots, 60-64\}, \\ b_x \cdot b_{u,60-64}/b_{65-70} & \text{for } x \in \{65-70, \ldots, 130+\}, \end{cases} \tag{15.7}$$

where $b_{u,x}$ is scaled to sum to unity over all ages.

The rotation is dependent on $e_0(\tau)$, and so the resulting b_x also becomes time-dependent. The rotation finishes at a certain level of life expectancy, denoted by e_0^u. Li et al. (2013) recommended using $e_0^u = 102$. Using the smooth weight function

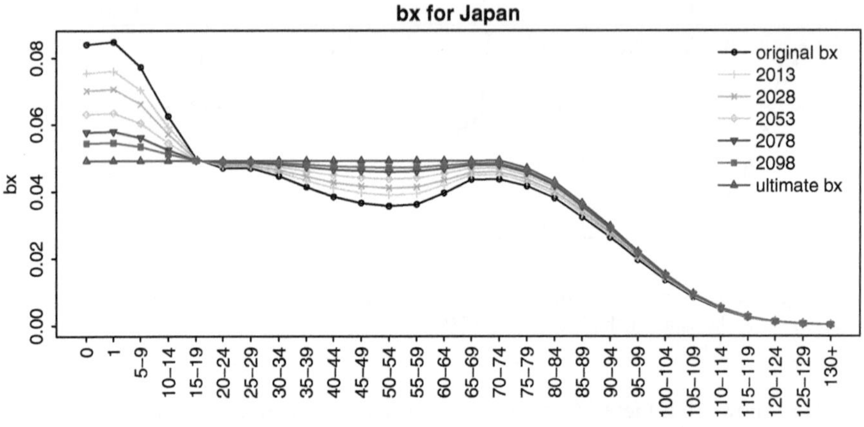

Fig. 15.4 Rotating the parameter b_x over time: data from Japan. The original b_x (*outer curve with circles*) approaches the ultimate $b_{u,x}$ (*outer curve with triangles*) over time starting with the *lightest color* and continuing towards the *darker colors*

$$w(\tau) = \left\{\frac{1}{2}\left[1 + \sin[\frac{\pi}{2}(2w'(\tau) - 1)]\right]\right\}^{\frac{1}{2}} \quad \text{with} \quad w'(\tau) = \frac{e_0(\tau) - 80}{e_0^u - 80},$$

the rotated b_x at time τ, denoted by $B_x(\tau)$, is derived as:

$$B_x(\tau) = \begin{cases} b_x, & e_0(\tau) < 80, \\ [1 - w(\tau)]\, b_x + w(\tau) b_{u,x}, & 80 \leq e_0(\tau) < e_0^u, \\ b_{u,x}, & e_0(\tau) \geq e_0^u. \end{cases} \tag{15.8}$$

Figure 15.4 shows the results for Japan as an example. The original b_x is shown by the black curve with circles. The ultimate $b_{u,x}$, to be reached at life expectancy 102 years, marked by triangles. The remaining curves show the change over time starting with the lightest color and continuing through darker colors towards the ultimate curve.

15.2.6 Computing Life Tables

Step 3 in Algorithm 15.1 calls for matching future $k(\tau)$ to projected $e_0(\tau)$. This is a nonlinear equation in $k(\tau)$. It is solved by an iterative nonlinear procedure in which, for given values of a_x, b_x and $k(\tau)$, a life table is produced, and the resulting life expectancy is computed and compared with the projected $e_0(\tau)$. We used a bisection method to solve the nonlinear equation. This is simple and robust and involves relatively few iterations. It would be possible to use a nonlinear solution method that is more efficient computationally, but the computational gains would be modest and this could make the method much more complex.

In the process of computing life tables, the conversion of mortality rates $m_x(\tau)$ to probabilities of dying $q_x(\tau)$ follows the approach used by the United Nations to compute abridged life tables. This is computed by the LIFTB function in Mortpak (United Nations 1988, 2013a), where at a given time point the probability of dying for an individual between age x and $x + n$ is:

$$ {}_nq_x = \frac{n * {}_nm_x}{1 + (n - {}_nA_x) * {}_nm_x}, \tag{15.9} $$

with n being the length of the age interval and ${}_nA_x$ being the average number of years lived between ages x and $x + n$ by those dying in the interval. With l_x being the number of survivors at age x, we have

$$ l_{x+n} = l_x(1 - {}_nq_x), \tag{15.10} $$

$$ {}_nd_x = l_x - l_{x+n}, \tag{15.11} $$

$$ {}_nL_x = {}_nA_x l_x - (n - {}_nA_x) l_{x+n}, \tag{15.12} $$

where ${}_nd_x$ denotes the number of deaths between ages x and $x + n$ and ${}_nL_x$ denotes the number of person-years lived between ages x and $x + n$. The expectation of life at age x (in years) e_x is given by

$$ e_x = \frac{T_x}{l_x} \quad \text{with} \quad T_x = \sum_{a=x}^{\infty} {}_nL_a, $$

where T_x is the number of person-years lived at age x and older.

For ages 15 and over, the expression for ${}_nA_x$ is derived from the Greville (1943) approach to calculating age-specific separation factors based on the age pattern of the mortality rates themselves with:

$$ {}_nA_x = 2.5 - \frac{25}{12}({}_nm_x - k), \quad \text{where } k = \frac{1}{10}\log\left(\frac{{}_nm_{x+5}}{{}_nm_{x-5}}\right). $$

For ages 5 and 10, ${}_nA_x = 2.5$ and for ages under 5, values from the Coale and Demeny West region relationships are used for ${}_nA_x$ (Coale and Demeny 1966).[2]

[2]The Coale and Demeny West region formulae are used as follows. When ${}_0m_1 \geqslant 0.107$, then ${}_1A_0 = 0.33$ for males and 0.35 for females; ${}_4A_1 = 1.352$ for males and 1.361 for females. When ${}_1m_0 < 0.107$, ${}_1A_0 = 0.045 + (2.684 \cdot {}_1m_0)$ for males and ${}_1A_0 = 0.053 + (2.800 \cdot {}_1m_0)$ for females; ${}_4A_1 = 1.651 - (2.816 \cdot {}_1m_0)$ for males and ${}_4A_1 = 1.522 - (1.518 \cdot {}_1m_0)$ for females.

15.2.7 Summary of Improved Algorithm

We now summarize the modifications described in the previous sections by proposing an improved algorithm for deriving the age-specific mortality rates m_x for potential use in future probabilistic population projections.

Algorithm 15.2

As before, let $t \in \{1, \ldots, T\}$ and $\tau \in \{T+1, \ldots T_p\}$ denote the observed and projected time periods, respectively. Also, let $g \in \{F, M\}$ be an index to distinguish sex-specific measures.

1. Using the Coherent Kannisto Method from Sect. 15.2.2, extend $m_x(t)$ to higher age categories with $\max(x) = 130+$ for all t.
2. Choose a method to estimate a_x, i.e. one of Eqs. (15.1), (15.5) or (15.6), depending on country specifics.[3] Do the estimation for each sex g, obtaining $\hat{a}_x^g$.
3. Estimate $k(t)$ and b_x using the extended historical $m_x(t)$ and Eqs. (15.2) and (15.3) for $g = M, F$ independently, yielding $\hat{b}_x^g$.
4. Given $\hat{b}_x^M$ and $\hat{b}_x^F$ from Step 3, set $\hat{b}_x = \frac{\hat{b}_x^M + \hat{b}_x^F}{2}$.
5. Compute the ultimate $b_{u,x}$ as in Eq. (15.7).
6. For a combined $e_0(\tau) = \left[e_0^M(\tau) + e_0^F(\tau)\right]/2$ in each trajectory, compute $B_x(\tau)$ as in Eq. (15.8).
7. For a given sex-specific $e_0^g(\tau)$ in each trajectory and the estimated $\hat{a}_x^g$ and $B_x(\tau)$ from the previous steps, solve for future $k_g(\tau)$ numerically using life tables. This yields a nonlinear equation which is solved using the bisection method, as described in Sect. 15.2.6.
   As in Algorithm 15.1, this gives a sample of values of $k_g(\tau)$ for each future time period τ, with one value for each trajectory.
8. For each trajectory, time τ and sex g, compute mortality rates by
$$\log[m_x^g(\tau)] = \hat{a}_x^g + B_x(\tau)k_g(\tau).$$
9. Since the previous step does not comply with Eq. (15.4) and thus can lead to crossovers in high ages, an additional constraint is added:
   If $e_0^M(\tau) < e_0^F(\tau)$ then
$$m_x^M(\tau) = \max[m_x^M(\tau), m_x^F(\tau)] \text{ for } x \geq 100.$$

Figure 15.5 shows the resulting probabilistic projection of $m_x(\tau)$ for the period 2095–2100 for both sexes in two selected countries. In addition to the marginal distribution for Kazakhstan in the right panel of Fig. 15.5, its joint distribution for males and females is shown in Fig. 15.6 on a logarithmic scale. Points below the $x = y$ solid line indicate crossovers in the individual trajectories. It can be seen that only a few trajectories experience crossovers when mortality is low, i.e. in young ages, suggesting a low (but non-zero) probability for such an event. There are no

[3]In the bayesPop package this country-specific set of options is controlled through two dummy variables in the `vwBaseYear2012` dataset: (1) whether the most recent estimate of age mortality pattern should be used (LatestAgeMortalityPattern) and (2) whether it should be smoothed (SmoothLatestAgeMortalityPattern). See `help(vwBaseYear2012)` in R.

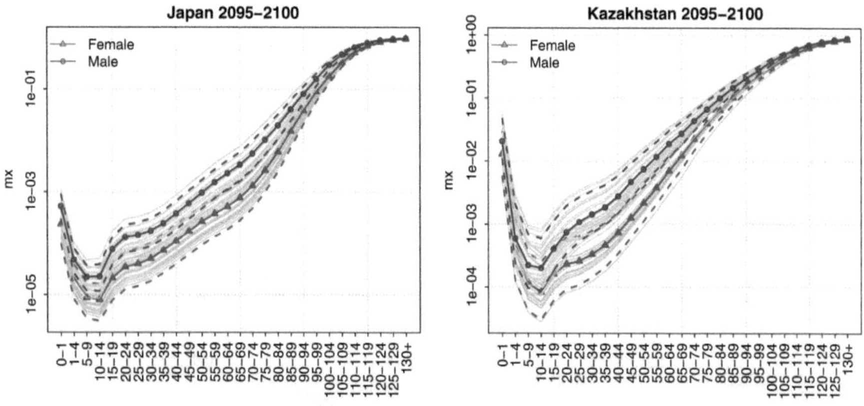

Fig. 15.5 Probabilistic projection of age-specific mortality rates for Japan (*left panel*) and Kazakhstan (*right panel*) in the time period 2095–2100. Both plots show the marginal distributions for males and females where the *dashed lines* mark the 80 % probability intervals and the *solid lines* are 20 randomly sampled trajectories for each sex. The y-axis is on the logarithmic scale

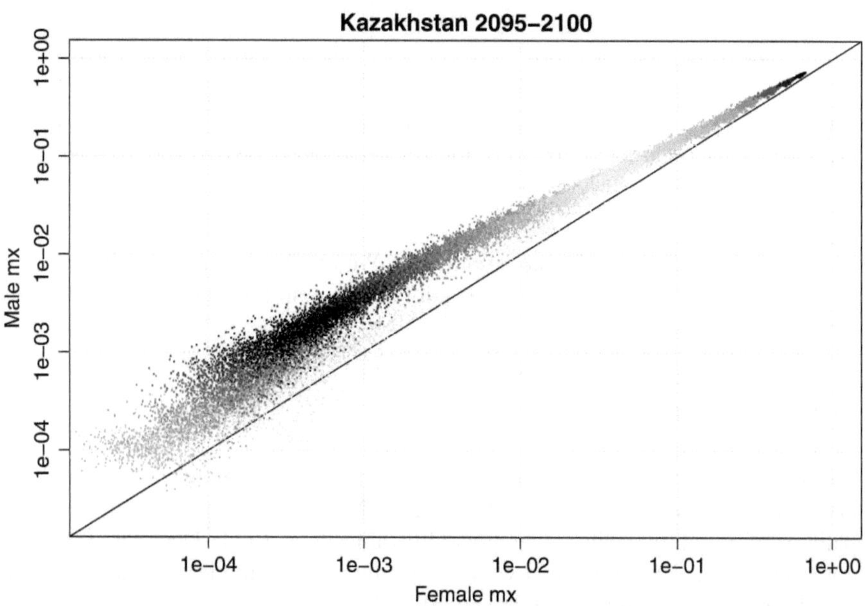

Fig. 15.6 The joint predictive distribution of mortality rates for females and males for Kazakhstan in 2095–2100. It shows mortality rates from all age groups. Age groups are distinguished by color in the online version. Both axes are on the logarithmic scale. There are 1000 points per age group

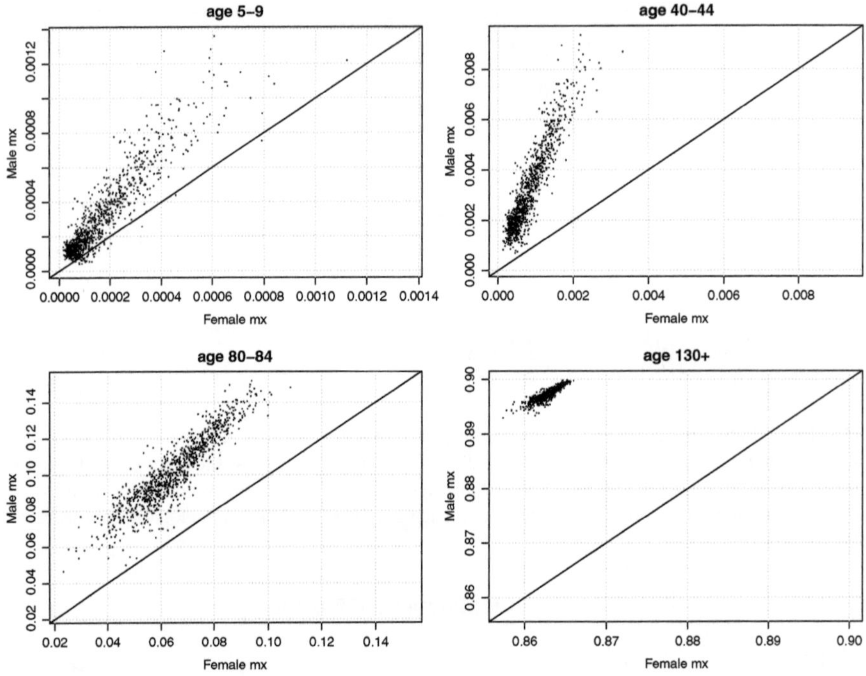

Fig. 15.7 The joint predictive distribution of future age-specific mortality rates for females and males for Kazakhstan in 2095–2100 for four individual age groups. In all panels, 1000 points are shown and all axes are on a normal scale

crossovers for high mortality, i.e. at old ages. We observed similar results for most countries. Figure 15.7 shows the same joint distribution for selected age groups on a normal scale.

15.2.7.1 Exceptions

For about 50 countries, insufficient detailed data about mortality by age and sex are available between 1950 and 2010 (United Nations 2013c). Therefore, the age patterns of mortality are based on model life tables (e.g., Coale-Demeny). For these countries a model b_x associated with one of the regional model life tables is used (see Table 2 page 18 in Li and Gerland 2011).[4]

In addition, for about 40 countries with a generalized HIV/AIDS epidemic, age patterns of mortality since the 1980s have been affected by the impact

[4]In the bayesPop package this country-specific set of options is controlled through two variables in the `vwBaseYear2012` dataset: (1) the type of age mortality pattern used for the estimation period (AgeMortalityType with the option "Model life tables") and (2) the specific mortality pattern used (AgeMortalityPattern with options like "CD West").

of AIDS mortality (especially before the scaling up of antiretroviral treatment starting in 2005). For these countries the application of the conventional Lee-Carter approach is inappropriate.[5] Instead, we introduce a modification where steps 2–6 in Algorithm 15.2 are replaced by the following steps:

1. Start with the most recent a_x (affected by impact of HIV/AIDS on mortality) and smooth it as in Eq. (15.6), obtaining a_x^s.
2. Compute an *ultimate* (or "AIDS-free" target) a_x, denoted by a_x^u, which is a smoothed average of historical $\log(m_x)$ up to 1985 (i.e., prior to the start of the impact caused by HIV/AIDS on mortality), denoted by a_x^v:

$$a_x^v = \frac{\sum_{t=1}^{T_u} \log[m_x(t)]}{T_u} \quad \text{with } T_u \text{ corresponding to 1985}$$
$$a_x^u = \text{smooth}_x\{a_x^v\} \text{ with } a_{0-1}^u = a_{0-1}^v \tag{15.13}$$

3. For each x interpolate from a_x^s to a_x^u assuming that in the long run the excess mortality due to the HIV/AIDS epidemic disappears (or reaches a very low endemic level with negligible mortality impact) both as a result of decreased HIV prevalence, improved access to treatment and survival with treatment.
4. During the projections, pick an $a_x(\tau)$ by moving along the interpolated line of the corresponding x, so that a_x^u is reached by 2100.
5. As above, b_x is associated with one of the regional model life tables.

An example of the resulting projected median age-specific mortality rates for Botswana, a country with a generalized HIV/AIDS epidemic, is shown in Fig. 15.8.

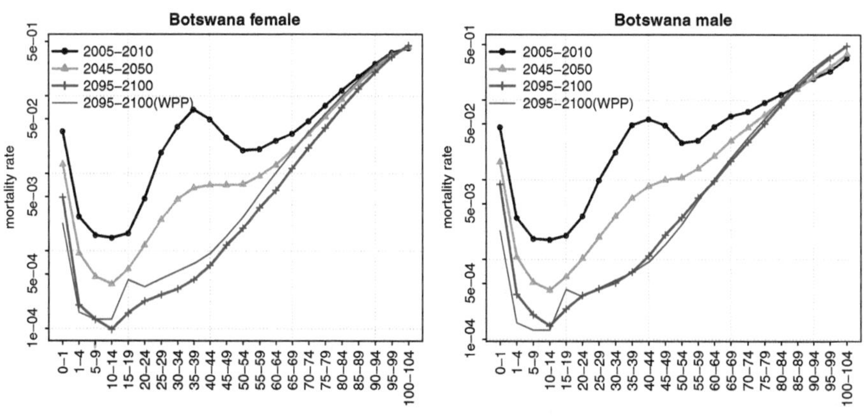

Fig. 15.8 Projected age-specific mortality rates for Botswana, a country with a generalized HIV/AIDS epidemic. The *y*-axis is on the logarithmic scale

[5]In the bayesPop package this specific-set of countries are identified through a dummy variable (WPPAIDS) in the `vwBaseYear2012` dataset.

There has been recent progress in the modelling of age patterns of mortality for countries with generalized HIV/AIDS (Sharrow et al. 2014). This could provide additional options to better incorporate the uncertainty about future HIV prevalence, expanded access to treatment, underlying age mortality patterns, and their interaction on overall mortality by age into probabilistic population projections. Further calibration and validation of these models using empirical estimates from cohort studies (Zaba et al. 2007; Reniers et al. 2014) will be important in this context.

15.3 Age-Specific Fertility Rates for Probabilistic Population Projections

15.3.1 WPP 2012 Method of Projecting Age-Specific Fertility Rates

The United Nations probabilistic population projections released in 2014 (Gerland et al. 2014) used a set of projected age-specific fertility rates for each country obtained by combining probabilistic projections of the total fertility rate with deterministic projections of age patterns of fertility as used in the 2012 revision of the World Population Prospects (United Nations 2014).

For high-fertility and medium-fertility countries, future age patterns of fertility were obtained by interpolating linearly between a starting proportionate age pattern of fertility and a target model pattern. The target model pattern was chosen from among 15 proportionate age patterns of fertility, with mean age at childbearing varying between 24 and 28.5 years. The target pattern was held constant once the country reaches its lowest fertility level, or by 2045–2050 onward.

For low fertility countries, a similar approach was used. It projected future age-specific fertility patterns by assuming that they would reach a target model pattern by 2025–2030. This target was chosen from among five target age patterns of fertility either for the market economies of Europe (with mean age of childbearing varying between 28 and 32 years) or for countries with economies in transition (with mean age of childbearing varying between 26 and 30 years). Once the model pattern was reached, it was assumed to remain constant until the end of the projection period. In some instances, a modified Lee-Carter approach (Li and Gerland 2009) was used to extrapolate the most recent set of proportionate age-specific fertility rates using the rates of change from country-specific historical trends.

All the trajectories making up the probabilistic projection of fertility for a given country used the same age pattern of fertility. The choice of target pattern of fertility for a given country, from among the set of model patterns considered, was driven by country-specific expert opinion about future trends and normative assumptions. No global or regional convergence in age patterns of fertility was imposed.

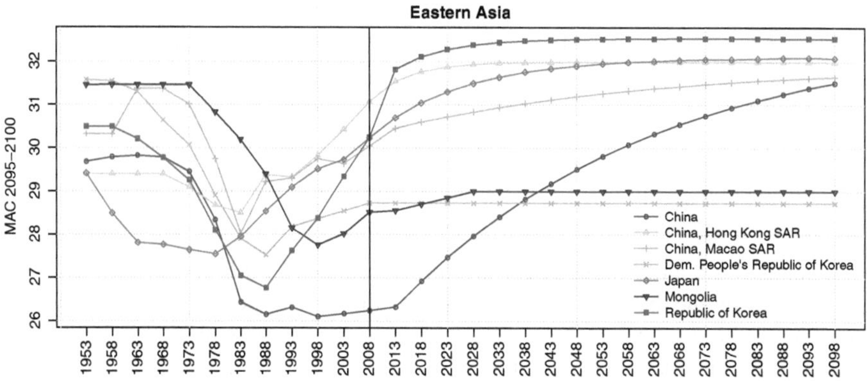

Fig. 15.9 Example of projected Mean Age of Childbearing (MAC) for countries in Eastern Asia in WPP 2012

Figure 15.9 shows the results of the projections for the Mean Age of Childbearing (MAC) for countries in Eastern Asia from the 2012 Revision of the World Population Prospects.

Overall, the method for projecting age-specific patterns of fertility in the 2012 Revision (as well as in previous revisions) has several limitations. First, no global or regional convergence has been imposed despite the overall convergence in total fertility rates observed in the projection period up to 2100. Second, the time point when the target age-specific pattern is reached is not related to the projected total fertility rates. Third, expert assumptions on the target age pattern and method used for individual countries introduce diversity in the age-specific trends that are difficult to explain (see Fig. 15.9—Mongolia and Democratic People's Republic of Korea were done by Analyst 1, all other countries by Analyst 2). Finally, since the analysts have used at least two different methods and 25 target age patterns of fertility, the documentation of the decisions made for individual countries have been challenging.

15.3.2 Convergence Method for Projecting Age-Specific Fertility Rates

We now propose a new method for projecting age-specific fertility rates, to overcome some of the limitations of the existing method used in WPP 2012. This builds on the approach adopted in sets of projections of married or in-union women of reproductive age (MWRA) (United Nations 2013b). Beginning from the most recent observation of the age pattern of fertility in the base period of projection, the projected age patterns of fertility are based on the past national trend combined with the trend towards the global model age pattern of fertility. The projection method is implemented on the proportionate age-specific fertility rates (PASFR) covering seven age groups from 15–19 to 45–49. The final projection of PASFRs for each age group is a weighted average of two preliminary projections:

(a) the first preliminary projection, assuming that the PASFRs converge to the global model pattern, see Sect. 15.3.2.1; and
(b) the second preliminary projection, assuming the observed national trend in PASFRs continues into the indefinite future, see Sect. 15.3.2.2.

The method is applied to all TFR trajectories from 2014 PPP.

We now define the preliminary projections that constitute our overall projection. We use different notation than in Sect. 15.2, so the same symbol may be used to denote different quantities in the two sections.

15.3.2.1 Trend Towards the Global Model Pattern

Let t_r denote the base period of a projection and t_g the year when the global model pattern is reached. For $t_r < t < t_g$, the proportion of the interval $[t_r, t_g]$ that has elapsed at time t is

$$\tau_t = (t - t_r)/(t_g - t_r).$$

Section 15.3.2.4 below gives details about how to estimate t_g.

Let p_r denote PASFR at the base period t_r, and let p_g denote PASFR of the global model pattern.[6] The projections at time t of PASFR towards the global model pattern, denoted by p_t^I, is obtained by:

$$\mathrm{logit}(p_t^I) = \mathrm{logit}(p_r) + \tau_t \left[\mathrm{logit}(p_g) - \mathrm{logit}(p_r)\right] \tag{15.14}$$

Then p_t^I is renormalized so that it sums to unity for all time periods t.

15.3.2.2 Continuing of Observed National Trend

Let T denote the number of 5-year periods over which the model is fitted. Then t_{r-T} is the starting time period of the estimation and $p_{(r-T)}$ is PASFR at t_{r-T}. p_t^{II} is the

[6]In the bayesPop package the global model pattern is created as an average of most recent PASFRs for a set of countries (selected through a dummy variable in the `vwBaseYear2012` dataset). For the purpose of the current analysis, the low fertility countries selected have already reached their Phase III and represent later childbearing patterns with mean age at childbearing close to or above 30 years in 2010–2015: Austria, the Czech Republic, Denmark, France, Germany, Japan, the Netherlands, Norway and the Republic of Korea. The specification of the countries used for the global model pattern can be changed in input file.

projected PASFR at time t, assuming the past trend was to continue into the future under the following rule:

$$\text{logit}(p_t^{II}) = \text{logit}(p_r) + \frac{t - t_r}{t_r - t_{r-T}} [\text{logit}(p_r) - \text{logit}(p_{r-T})] \tag{15.15}$$

As above, p_t^{II} should be scaled to sum to unity for all t. Note that in our implementation we use $T = 3$.

15.3.2.3 Resulting Projection

Projected PASFR at time t, p_t, is calculated as:

$$\text{logit}(p_t) = \tau_t \cdot \text{logit}(p_t^I) + (1 - \tau_t)\,\text{logit}(p_t^{II}) \tag{15.16}$$

Resulting p_t is renormalized to sums to unity for all time periods t.

15.3.2.4 Estimating the Time Period of Reaching Global Pattern

We assume that the transition from the most recent age pattern of fertility to the global model age pattern of fertility is dependent on the timing when TFR enters Phase III, i.e. when the fertility transition is completed and the country reaches low fertility. For the countries in Phase III, a time series model to project TFR was used that assumed that in the long run fertility would approach and fluctuate around country-specific ultimate fertility levels based on a Bayesian hierarchical model (Raftery et al. 2014a). The time series model uses the empirical evidence from low-fertility countries that have experienced fertility increases from a sub-replacement level after a completed fertility transition. At the same time, based on the empirical evidence on the postponement of childbearing in low-fertility countries, profound shifts to later start of childbearing and an increase in the mean age of childbearing are still taking place several periods after the start of Phase III (see Fig. 15.10). The timing and speed of the postponement of childbearing in Phase III is country-specific and in this chapter we implement the assumption that the transition to later childbearing pattern is completed when total fertility approaches country-specific ultimate fertility levels.

To be more specific, we assume that the time t_g of a completion of the transition to a global model pattern corresponds to the time point t_u, when TFR reaches the ultimate fertility level of that country. In probabilistic projections of TFR, we approximate the ultimate fertility level, denoted by f_u, by the median TFR in the last projection period t_e, e.g. $t_e = 2095–2100$, if TFR is in Phase III:

$$\hat{f}_u = \text{median}_i\,[f_i(t_e)] \quad (i \text{ denotes trajectories}) \tag{15.17}$$

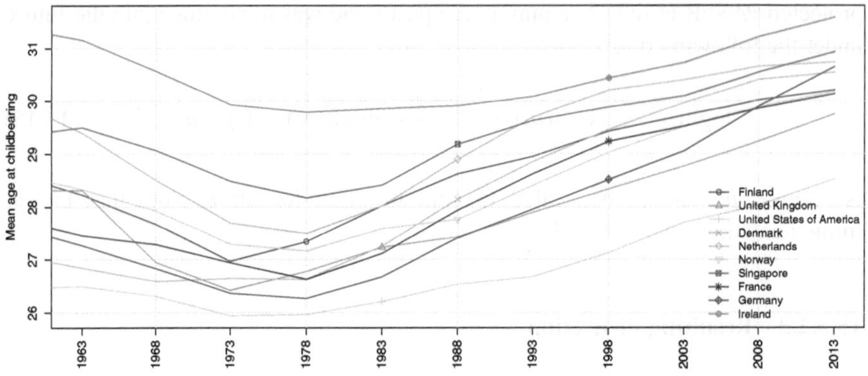

Fig. 15.10 Trends in mean age at childbearing in countries with the start of Phase III of fertility decline before 2000. *Symbols* mark the time period when the country entered Phase III

Then for each TFR trajectory, t_u is the earliest time period, at which the TFR is larger or equal to $\hat{f}_u$:

$$t_u = \min\{t : f(t) \geqslant \hat{f}_u \text{ and } t > t_{P3}\} \tag{15.18}$$

where t_{P3} denotes the start of Phase III. For the estimation of t_g, we will now distinguish two cases, depending if t_{P3} is smaller or larger than the end period t_e.

Case 1: $t_{P3} < t_e$

In this case, t_{P3} is either observed ($t_{P3} \leqslant t_r$) or projected within the projecting period ($t_r < t_{P3} < t_e$). In both cases, if t_u exists,

$$t_g = \max(t_u, t_r + 10)\,. \tag{15.19}$$

This includes a situation where $f(t) \geqslant \hat{f}_u$ for $t \leqslant t_r$. In such a case, the global pattern is reached quickly, namely in two 5-year periods.

If $f(t) < \hat{f}_u$ for all $t_r \leqslant t \leqslant t_e$, then t_u does not exist. In such a case, t_g is set to the end of the projection period, but at least five 5-year periods after t_{P3}:

$$t_g = \max(t_e, t_{P3} + 25) \tag{15.20}$$

Case 2: $t_e \leqslant t_{P3}$

In this case, t_{P3} is unknown, i.e. the TFR trajectory has not reached Phase III at t_e. Thus, we will make an estimate of t_{P3}, denoted by $\hat{t}_{P3}$, and then simply apply

$$t_g = \hat{t}_{P3} + 25\,. \tag{15.21}$$

If the TFR at t_e is low, namely $f(t_e) \leqslant 1.8$, we assume that $\hat{t}_{P3} = t_e$. Otherwise, we approximate t_{P3} by a linear extrapolation of TFR from the last four time periods and determine when such line reaches 1.8, with an upper limit of $\hat{t}_{P3} = t_e + 50$.

15.3.2.5 Exception for Late Childbearing Pattern

Since trajectories for some countries have already observed or – as projected by the algorithm described above – will in near future reach higher MAC than the MAC associated with the global model pattern, we assume that for a given country's trajectory once the maximum MAC is reached in the convergence period the associated PASFR pattern is kept constant for the remaining projection periods. This assumption enables to keep trajectory-specific patterns of late childbearing for trajectories after the Phase III, thus already with low total fertility (see Fig. 15.11 for example of the Czech Republic). Note that this rule is applied only in Case 1 above.

15.3.3 Results of the Convergence Method Applied to Probabilistic Projections

For the 2012 Revision, age-specific fertility estimates are based on empirical data for all countries of the world for the period up to 2010 (or up to 2010–2015 for 37 countries with empirical data up to 2011 or 2012; Gerland et al. 2014). Using the probabilistic projections of TFR, each TFR trajectory has a specific start of Phase III and therefore the timing of convergence to the global model pattern is trajectory-specific. This yields a set of trajectories of PASFR (although not probabilistic) which in turn, when combined with the probabilistic TFR, yield probabilistic projection of age-specific fertility rates.

Figure 15.11 shows an example of the results for PASFR in Niger, Bangladesh and the Czech Republic for selected age groups over time. Figure 15.12 shows an example of the probabilistic results of age-specific fertility rates for Ethiopia, Nepal and Japan at the end of projection period in 2095–2100.

Figure 15.13 shows the development of PASFR for Uganda, India and Germany over time from 2005–2010 to 2095–2100. Here, the methodology was applied to the deterministic projection of TFR from WPP 2012.

In Fig. 15.9 we showed projections of MAC from WPP 2012. This can be compared to Fig. 15.14 where the same measure is shown after applying the new methodology to the TFR of WPP 2012.

Overall, the new method we propose improves on the current methodology in several ways. First, in the very long term (after 2100) the age patterns of fertility are converging to one global pattern, while retaining specific late childbearing patterns

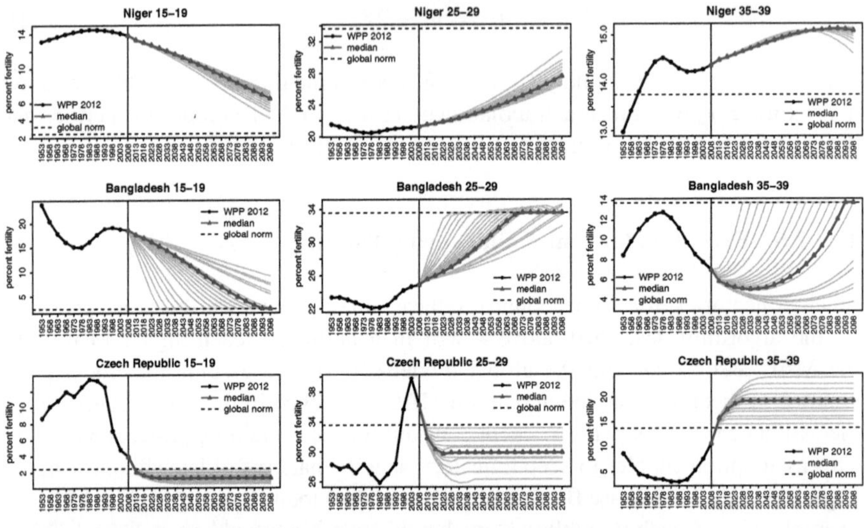

Fig. 15.11 Proportionate age-specific fertility rates (PASFR) by time for age groups 15–19, 25–29 and 35–39 in Niger, Bangladesh and the Czech Republic. Projected median of PASFR approaches global model pattern of PASFR (*dashed line*). The *solid grey lines* are trajectories that correspond to different starting periods for Phase III; they do not represent random samples from a predictive probability distribution

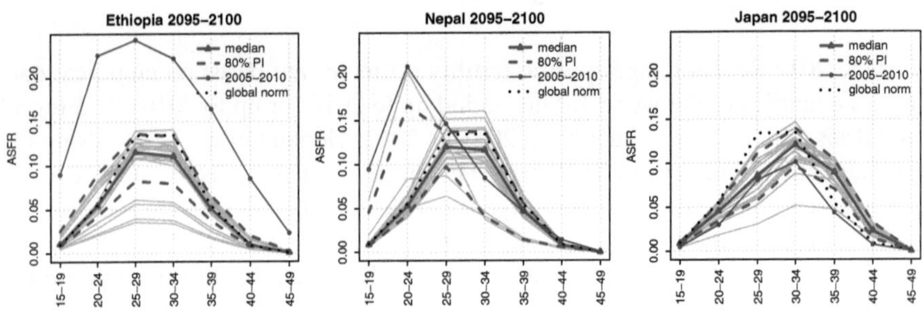

Fig. 15.12 Probabilistic projection of age-specific fertility rates for Ethiopia (*left panel*), Nepal (*middle panel*) and Japan (*right panel*) in the time period 2095–2100. The marginal distribution for age-specific fertility rates (*red lines*) where the *dashed lines* mark the 80 % probability intervals and the *solid grey lines* are randomly sampled trajectories are compared to age-specific fertility rates in the time period 2005–2010 (*blue line*) and to the global model pattern applied to median projection of total fertility for the world in 2095–2100 (*black dotted line*)

for several countries that reach such patterns in the current period or in the near future. Second, the projections of the age pattern of fertility are now linked to projections of the total fertility rate. Finally, for each probabilistic trajectory, the time when the target age pattern is reached depends on the trajectory-specific total fertility rate.

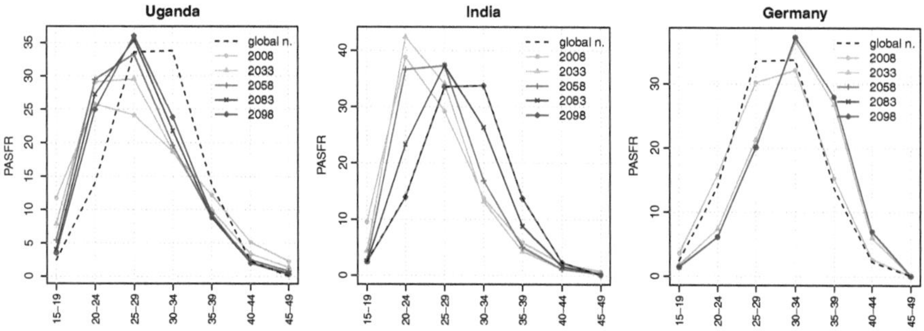

Fig. 15.13 Proportionate age-specific fertility rates (PASFR) by age over time for selected countries

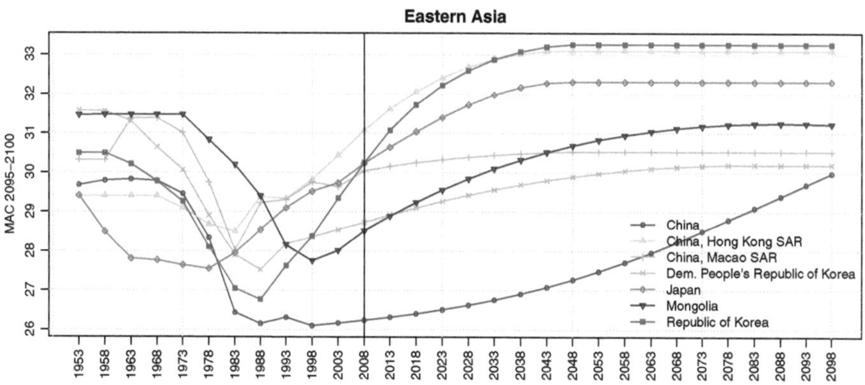

Fig. 15.14 Example of projected MAC for countries in Eastern Asia after applying the proposed methodology

15.4 Discussion

We have described the methods used for converting projected life expectancies at birth and total fertility rates to age-specific mortality and fertility rates in the UN's 2014 probabilistic population projections. We have identified some limitations of these methods and have proposed several improvements to overcome them. These include a new coherent Kannisto method to avoid crossovers in mortality rates between the sexes at very high ages. They also include the application of the coherent Lee-Carter method to avoid crossovers in mortality rates between the sexes at lower ages, methods for avoiding jump-off bias, and the rotated Lee-Carter method to reflect the fact that at high life expectancies, mortality rates tend to decline faster at higher than at lower ages.

It should be noted that the 2014 PPP takes account of uncertainty about the overall level of fertility as measured by the TFR, and also about the overall level of mortality as measured by e_0. Conditional on TFR and e_0, however, the projected vital rates are deterministic. There is thus a missing component of uncertainty, and it would be desirable to extend the methods used to take account of this, particularly of uncertainty about the future mean age at childbearing (Ediev 2013).

Acknowledgements This research was supported by NIH grants R01 HD054511 and R01 HD070936. The views expressed in this article are those of the authors and do not necessarily reflect those of NIH or the United Nations. The authors are grateful to the editor for very helpful comments.

References

Alders, M., Keilman, N., & Cruijsen, H. (2007). Assumptions for long-term stochastic population forecasts in 18 European countries. *European Journal of Population, 23*, 33–69.

Alho, J. M., Alders, M., Cruijsen, H., Keilman, N., Nikander, T., & Pham, D. Q. (2006). New forecast: Population decline postponed in Europe. *Statistical Journal of the United Nations Economic Commission for Europe, 23*, 1–10.

Alho, J. M., Jensen, S. E. H., & Lassila, J. (2008). *Uncertain demographics and fiscal sustainability*. Cambridge/New York: Cambridge University Press.

Alkema, L., Raftery, A. E., Gerland, P., Clark, S. J., Pelletier, F., Buettner, T., & Heilig, G. K. (2011). Probabilistic projections of the total fertility rate for all countries. *Demography, 48*, 815–839.

Booth, H., Hyndman, R. J., Tickle, L., & de Jong, P. (2006). Lee-Carter mortality forecasting: A multi-country comparison of variants and extensions. *Demographic Research, 15*, 289–310.

Bos, E., Vu, M. T., Massiah, E., & Bulatao, R. (1994). *World population projections 1994–95: Estimates and projections with related demographic statistics*. Baltimore: Johns Hopkins University Press for the World Bank.

Coale, A. J., & Demeny, P. G. (1966). *Regional model life tables and stable populations*. Princeton: Princeton University Press.

Ediev, D. M. (2013). *Comparative importance of the fertility model, the total fertility, the mean age and the standard deviation of age at childbearing in population projections*. Presented at the Meeting of the International Union for the Scientific Study of Population, Busan. http://iussp.org/sites/default/files/event_call_for_papers/TF%20MS%20SD_what%20matters_StWr.pdf

Gerland, P., Raftery, A. E., Ševčíková, H., Li, N., Gu, D., Spoorenberg, T., Alkema, L., Fosdick, B. K., Chunn, J. L., Lalic, N., Bay, G., Buettner, T., Heilig, G. K., & Wilmoth, J. (2014). World population stabilization unlikely this century. *Science, 346*, 234–237.

Greville, T. N. (1943). Short methods of constructing abridged life tables. *The Record of the American Institute of Actuaries, XXXII, 1*, 29–42.

Keyfitz, N. (1981). The limits of population forecasting. *Population and Development Review, 7*, 579–593.

Lee, R. D., & Carter, L. (1992). Modeling and forecasting the time series of US mortality. *Journal of the American Statistical Association, 87*, 659–671.

Lee, R. D., & Miller, T. (2001). Evaluating the performance of the Lee-Carter method for forecasting mortality. *Demography, 38*, 537–549.

Lee, R. D., & Tuljapurkar, S. (1994). Stochastic population forecasts for the United States: Beyond high, medium, and low. *Journal of the American Statistical Association, 89*, 1175–1189.

Leslie, P. H. (1945). On the use of matrices in certain population dynamics. *Biometrika, 33*, 183–212.

Li, N., & Gerland, P. (2009). *Modelling and projecting the postponement of childbearing in low-fertility countries*. Presented at the XXVI IUSSP International Population Conference. iussp2009.princeton.edu/papers/90315

Li, N., & Gerland, P. (2011). *Modifying the Lee-Carter method to project mortality changes up to 2100*. Presented at the Annual Meeting of Population Association of America. http://paa2011.princeton.edu/abstracts/110555

Li, N., & Lee, R. D. (2005). Coherent mortality forecasts for a group of populations: An extension of the Lee-Carter method. *Demography, 42*, 575–594.

Li, N., Lee, R. D., & Gerland, P. (2013). Extending the Lee-Carter method to model the rotation of age patterns of mortality decline for long-term projections. *Demography, 50*, 2037–2051.

Lutz, W., & Samir, K. C. (2010). Dimensions of global population projections: What do we know about future population trends and structures? *Philosophical Transactions of the Royal Society B, 365*, 2779–2791.

Lutz, W., Sanderson, W. C., & Scherbov, S. (1996). Probabilistic population projections based on expert opinion. In Lutz, W. (Ed.), *The future population of the world: What can we assume today*? (pp. 397–428). London: Earthscan Publications Ltd. Revised 1996 edition.

Lutz, W., Sanderson, W. C., & Scherbov, S. (1998). Expert-based probabilistic population projections. *Population and Development Review, 24*, 139–155.

Lutz, W., Sanderson, W. C., & Scherbov, S. (2004). *The end of world population growth in the 21st century: New challenges for human capital formation and sustainable development*. Sterling: Earthscan.

National Research Council. (2000). *Beyond six billion: Forecasting the world's population*. Washington, DC: National Academy Press.

Pflaumer, P. (1988). Confidence intervals for population projections based on Monte Carlo methods. *International Journal of Forecasting, 4*, 135–142.

Preston, S. H., Heuveline, P., & Guillot, M. (2001). *Demography: Measuring and modeling population processes*. Malden: Blackwell.

Raftery, A. E., Li, N., Ševčíková, H., Gerland, P., & Heilig, G. K. (2012). Bayesian probabilistic population projections for all countries. *Proceedings of the National Academy of Sciences, 109*, 13915–13921.

Raftery, A. E., Chunn, J. L., Gerland, P., & Ševčíková, H. (2013). Bayesian probabilistic projections of life expectancy for all countries. *Demography, 50*, 777–801.

Raftery, A. E., Alkema, L., & Gerland, P. (2014a). Bayesian population projections for the United Nations. *Statistical Science, 29*, 58–68.

Raftery, A. E., Lalic, N., & Gerland, P. (2014b). Joint probabilistic projection of female and male life expectancy. *Demographic Research, 30*, 795–822.

Reniers, G., Slaymaker, E., Nakiyingi-Miiro, J., Nyamukapa, C., Crampin, A. C., Herbst, K., Urassa, M., Otieno, F., Gregson, S., Sewe, M., Michael, D., Lutalo, T., Hosegood, V., Kasamba, I., Price, A., Nabukalu, D., Mclean, E., Zaba, B., & Network, A. (2014). Mortality trends in the era of antiretroviral therapy: Evidence from the network for analysing longitudinal population based HIV/AIDS data on Africa (ALPHA). *AIDS, 28*, S533–S542.

Ševčíková, H., & Raftery, A. E. (2014). *bayesPop: Probabilistic population projection*. R package version 5.4-0.

Ševčíková, H., Raftery, A. E., & Gerland, P. (2014). *Bayesian probabilistic population projections: Do it yourself*. Presented at the Annual Meeting of Population Association of America. http://paa2014.princeton.edu/abstracts/141301

Sharrow, D. J., Clark, S. J., & Raftery A. E. (2014). Modeling age-specific mortality for countries with generalized HIV epidemics. *PLoS One, 9*, e96447.

Stoto, M. A. (1983). The accuracy of population projections. *Journal of the American Statistical Association, 78*, 13–20.

Thatcher, A. R., Kannisto, V., & Vaupel, J. W. (1998). *The force of mortality at ages 80 to 120* (Volume 5 of odense monographs on population aging series). Odense: Odense University Press.

Tuljapurkar, S., & Boe, C. (1999). Validation, probability-weighted priors, and information in stoachastic forecasts. *International Journal of Forecasting, 15*, 259–271.

United Nations. (1988). *MortPak – The United Nations software package for mortality measurement. Bach-oriented software for the mainframe computer* (ST/ESA/SER.R/78). New York: United Nations. Available at http://www.un.org/esa/population//publications/MortPak_SoftwarePkg/MortPak_SoftwarePkg.htm

United Nations. (2011). *World population prospects: The 2010 revision*. Population Division, Deptartment of Economic and Social Affairs, United Nations, New York.

United Nations. (2013a). MortPak for Windows Version 4.3 – The United Nations software package for demographic mortality measurement.

United Nations. (2013b). *National, regional and global estimates and projections of the number of women aged 15 to 49 who are married or in a union, 1970–2030* (Technical paper 2013/2). Population Division, Deptartment of Economic and Social Affairs, United Nations, New York.

United Nations. (2013c). *World population prospects: The 2012 revision – online and DVD edition – data sources and meta information (POP/DB/WPP/Rev.2012/F0-2)*. Population Division, Deptartment of Economic and Social Affairs, New York.

United Nations. (2014). *World population prospects: The 2012 revision, methodology of the United Nations population estimates and projections. ESA/P/WP.235*. Population Division, Deptartment of Economic and Social Affairs, United Nations, New York.

U. S. Census Bureau (2009). International data base: Population estimates and projections methodology. Available at http://www.census.gov/ipc/www/idb/estandproj.pdf

Whelpton, P. K. (1928). Population of the United States, 1925–1975. *American Journal of Sociology, 31*, 253–270.

Whelpton, P. K. (1936). An empirical method for calculating future population. *Journal of the American Statistical Association, 31*, 457–473.

Wilmoth, J. R., Andreev, K., Jdanov, D., & Glei, D. A. (2007). Methods protocol for the Human Mortality Database. Online publication of the Human Mortality Database. http://www.mortality.org/Public/Docs/MethodsProtocol.pdf

Zaba, B., Marston, M., Crampin, A. C., Isingo, R., Biraro, S., Barnighausen, T., Lopman, B., Lutalo, T., Glynn, J. R., & Todd, J. (2007). Age-specific mortality patterns in HIV-infected individuals: A comparative analysis of African community study data. *Aids, 21*(6), S87–S96.

Part VI
The Age-Period-Cohort Problem

Chapter 16
Modeling the Evolution of Age and Cohort Effects

Sam Schulhofer-Wohl and Y. Claire Yang

16.1 Introduction

Social scientists conceive of many phenomena as depending on age, period, and cohort (APC) effects. For example:

- In demography, vital rates may depend on a person's age, on environmental conditions in the current year (period), and on conditions early in life that created scarring or selection effects (cohort).
- In sociology, behaviors such as going to college or forming a family may depend on individual physiological and social development (age), on major historical

This is a revised version of a paper presented at the 2008 annual meeting of the American Sociological Association and the 2009 annual meeting of the Population Association of America. We are grateful to audiences there and at the University of Cape Town, the Federal Reserve Bank of Minneapolis, Princeton University, and the Massachusetts Institute of Technology, as well as to Stephen Raudenbush, Robert Schoen, and an anonymous reviewer for helpful comments. We thank Kristin Plys and Thu Vu for excellent research assistance. This research was supported by a pilot grant from the Princeton Center for the Demography of Aging, NIH P30 AG024361. The views expressed herein are those of the authors and not necessarily those of the Federal Reserve Bank of Minneapolis or the Federal Reserve System.

S. Schulhofer-Wohl (✉)
Federal Reserve Bank of Minneapolis, 90 Hennepin Avenue, Minneapolis, MN 55408-0291, USA
e-mail: samuel.schulhofer-wohl@mpls.frb.org

Y.C. Yang
Department of Sociology, University of North Carolina, 155 Hamilton Hall CB 3210, Chapel Hill, NC 27599, USA
e-mail: yangy@unc.edu

R. Schoen (ed.), *Dynamic Demographic Analysis*,
The Springer Series on Demographic Methods and Population Analysis 39,
DOI 10.1007/978-3-319-26603-9_16

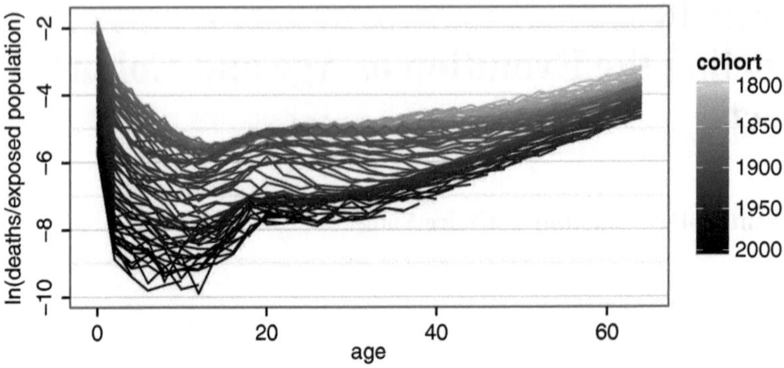

Fig. 16.1 Mortality in Sweden, 1861–2005. Each *line* shows the realized mortality of a particular birth cohort at various ages. Cohorts included are those born in 1797, 1799, ..., 2005. Lines for cohorts born before 1861 or after 1925 omit some ages because the dataset does not cover those ages for those cohorts (Data source: Human Mortality Database 2007)

events and social structural changes that individuals encounter in the current year (period), and on formative experiences of groups of individuals coming of age in different historical and social contexts (cohort).

- In economics, consumption inequality among a group of people born in the same year may depend on stages of the life cycle (age), on economic conditions in the current year (period), and on the group's initial level of inequality (cohort).

Despite the analytic importance of age, period, and cohort effects, how to empirically distinguish them is among the best-known and longest-standing methodological problems in the social sciences. Researchers commonly analyze data consisting of an $(A + 1) \times T$ matrix of outcomes for ages $a = 0, 1, \ldots, A$ and dates $t = 1, 2, \ldots, T$, such as a table of age-specific mortality rates in various years. The diagonals of this matrix correspond to birth cohorts: people born in the same year and aging together. For example, Fig. 16.1 illustrates a typical dataset on mortality in Sweden; we will analyze these data later in the chapter.

Most attempts to date to distinguish age, period, and cohort effects in such data have used linear models. However, it is impossible to distinguish the separate effects of age, period, and birth cohort in a linear model because age, period, and cohort are linearly dependent; for any person, *birth year* = *current year* − *age*. The problem persists even if one specifies age, period, and cohort effects nonparametrically with dummy variables for each possible value, as in the additive model

$$y_{at} = \alpha_a + \beta_t + \gamma_j, \quad j = t - a, \tag{16.1}$$

where y_{at} is some outcome for people of age a in year t (who are therefore members of birth cohort $j = t - a$); α_a is the effect of being age a; β_t is the effect of living in year t; and γ_j is the effect of being a member of cohort j. The separate effects of age, period, and cohort cannot be distinguished in the additive model of Eq. 16.1 because, if Eq. 16.1 is true, then for any constant δ, we also have

$$y_{at} = (\alpha_a + \delta a) + (\beta_t - \delta t) + (\gamma_j + \delta j). \quad (16.2)$$

That is, age, period, and cohort effects are identified only up to an unknown trend δ.

Even though the additive model Eq. 16.1 is not identified, it has been widely adopted to study age, period, and cohort effects. (Examples date to Greenberg et al. 1950; for reviews, see, e.g., Hobcraft et al. 1982 and Robertson et al. 1999). Researchers typically solve the identification problem by imposing one or more constraints on the parameters (e.g., Mason et al. 1973; Mason and Smith 1985; Deaton and Paxson 1994). But such constraints are often unsatisfying because they must depend on potentially unavailable outside information, on the researcher's subjective preferences, or on purely mathematical (as opposed to substantive) considerations. Some researchers have also argued that, empirically, the lack of identification is unlikely to pose practical problems in realistic settings, especially when the assumption of linearity of age, period, and cohort effects is not imposed; see, e.g., Reither et al. (2015) and Yang and Land (2013).

Besides being unidentified, the conventional additive model Eq. 16.1 has serious substantive limitations. The additive model is a quite simple approximation to the process of social change and does not adequately describe many phenomena where age, period, and cohort effects are of interest:

- *The additive model specifies that the marginal effect of age is the same in all time periods and for all cohorts.* In fact, however, the marginal effect of age changes over time and across cohorts. Consider, for instance, the dramatic declines in infant mortality over the past century (United Nations 1997).
- *The additive model specifies that the marginal effect of conditions in the present period is the same for people of all ages.* In reality, period effects are often age-specific. For example, the influenza epidemic of 1918 caused especially high mortality among people in their teens and twenties (Noymer and Garenne 2000).
- *The additive model specifies that cohorts do not change over time.* But cohorts must change, not least because—most obviously in the context of studies of mortality—some members of the cohort die each year, and they are not necessarily identical to those who remain alive (Vaupel et al. 1979).

Cohorts can also change over time for reasons other than composition effects. As Ryder (1965) explained in his seminal article:

> The case for the cohort as a temporal unit in the analysis of social change rests on a set of primitive notions: persons of age a in time t are those who were age $a - 1$ in time $t - 1$; transformations of the social world modify people of different ages in different ways; the effects of these transformations are persistent.

In other words, cohort effects arise because different cohorts live through different social events, or live through the same events at different ages, and these events change the cohort in long-lasting ways. But because new events constantly occur, a model with unchanging cohort effects is appropriate only if all relevant events occur before the initial observation and only if the impact of these events stays fixed as the cohort ages (Hobcraft et al. 1982). One can model the effect of events experienced at earlier ages by including lagged period effects if these events and conditions affect all age groups similarly. However, if, as Ryder argues, cohorts are continuously exposed to events that affect people of different ages in different ways, one needs a more general model—a framework that Hobcraft et al. (1982) labeled "continuously accumulating cohort effects." Despite the widespread theoretical influence of Ryder's paper, the concept of continuous cohort change appears never to have been mathematically formalized or taken to data.

We fill this gap by developing a new model of age, period, and cohort effects that can accommodate the various processes of change described above. Our model improves on the additive model in two ways. First, in our model, age profiles can change over time, and period effects can have different marginal effects on people of different ages. Second, our model operationalizes Ryder's concept of continuously evolving cohort effects by specifying cohort effects as accumulations of age-by-period interactions. These substantive contributions lead to a methodological contribution. We show that our model nests the additive model as a special case. Apart from a set of measure zero of special cases, however, the parameters of our model are identified, unlike those of the additive model. In other words, by broadening the concept of cohort effects, our model avoids the identification problem that has bedeviled the previous literature on the additive model.

Previous researchers, of course, also extended the APC accounting model Eq. 16.1 to include interactions (Fienberg and Mason 1985; James and Segal 1982; Moolgavkar et al. 1979). Our model differs from previous models of interactions both substantively and mathematically. Our model allows outcomes to depend on the accumulation of all the events a group of people experiences over the life course, whereas previous models have assumed that only events in the birth year and in the present year are relevant and that the influence of the birth year never changes. Previous models, further, remain unidentified because the additive part can never be identified without additional constraints.

The chapter proceeds as follows. In Sect. 16.2, we describe the model and discuss how to interpret its parameters. In Sect. 16.3, we analyze conditions under which the parameters are identified when outcomes are measured without error, while Sect. 16.4 extends the analysis to allow measurement error. Section 16.5 applies the model to analyze the evolution of human mortality—a fundamentally important phenomenon in demography—and Sect. 16.6 concludes.

16.2 Model

We model an outcome y_{at} as an accumulation of age-by-period interactions. Specifically, there are $K \geq 1$ sequences of time effects $\mathbf{e}_1, \ldots, \mathbf{e}_K$, where K is assumed to be known a priori. Each sequence $\mathbf{e}_k$ is a list of time effects in various years s: $\mathbf{e}_k = \{e_{k,s}\}_{s=-\infty}^{\infty}$. Time effects that occur in year s affect every cohort alive in that year. However, the impact may depend on the cohort's age and on which sequence contains the time effect: $w_{k,a}e_{k,s}$ is the contribution of time effects from sequence k in year s to the outcomes of people who are age a in year s. We refer to $w_{k,a}$ as the age weight for sequence k at age a. Each sequence of time effects should be thought of as representing a different factor that contributes to the outcome of interest. For example, if the outcome is mortality, one sequence of time effects might represent environmental conditions that affect infant mortality, and another might represent medical technologies that affect the mortality of older people.

The influence of past time effects may increase or fall off over time. Suppose we let $r(k, a, a')$ represent the increase or decay between ages a' and a, where $r(k, a, a) \equiv 1$. Then the impact in year t of time effects from sequence k occurring in year $s \leq t$ for people who were age a' in year s would be

$$r(k, a, a')w_{k,a'}e_{k,s}, \quad a = a' + t - s. \tag{16.3}$$

For example, $r(k, a, a')$ could be a step function if time effects at a young age have no further impact until old age, as in the case of increases in old-age lung cancer mortality that result from increasing popularity of cigarettes when people are young. The general form of increase or decay in Eq. 16.3 may, however, be difficult to analyze. In particular, modeling $r(k, a, a')$ as a step function with steps at unknown ages would lead to nonsmooth likelihood functions, and allowing $r(k, a, a')$ to depend nonparametrically on a and a' would add $A(A + 1)K/2$ parameters to the model. Therefore, in this chapter we restrict attention to the simplest possible smooth approximation—exponential increase or decay:

$$r(k, a, a') = r_k^{a-a'}, \quad r_k \geq 0. \tag{16.4}$$

(We adopt the convention that $0^0 = 1$.) Although the exponential form does not encompass all possible forms of increase or decay, its simplicity makes it easy to analyze, and we will show that it generalizes the additive model. We recognize that other choices of $r(k, a, a')$ may be valuable in particular applications and leave the analysis of such models for future research.

Having defined our building blocks $e_{k,t}$, $w_{k,a}$, and $r(k, a, a')$, we add an intercept and sum up the entire history of time effects to obtain our model for the outcomes for a particular cohort in a particular year:

$$y_{at} = \mu + \sum_{k=1}^{K} \sum_{a'=0}^{a} r_k^{a-a'} w_{k,a'} e_{k,t-a+a'}. \tag{16.5}$$

We now consider how to interpret the parameters of the model. For some parameter values, age and cohort effects in our model evolve over time. For other parameter values, our model generates time-invariant age effects, time-invariant cohort effects, and period effects that have the same influence on people of all ages. We first discuss the parameter values that generate these pure effects before showing how other parameter values can produce effects that evolve over time.

Pure age effects: Suppose that, for the kth sequence of time effects, the same time effects occur every year: $e_{k,s} = \bar{e}_k$ for all years s. Then the contribution of this sequence of time effects to outcomes for people of age a in year t is $\sum_{a'=0}^{a} r_k^{a-a'} w_{k,a'} \bar{e}_k$, which depends only on age a, not on the period t or the cohort $j = t - a$.

Pure period effects: Suppose that, for the kth sequence of time effects, $r_k = 0$ and $w_a^k = 1$ for all a. Then the contribution of the kth sequence to outcomes for age a in year t is simply e_t^k, which depends only on the current year and not on age or birth year.

Pure cohort effects: Suppose that, for the kth sequence of time effects, $r_k = 1$, $w_0^k = 1$, and $w_a^k = 0$ for $a > 0$. Then the contribution of the kth sequence to outcomes for age a in year t is simply e_{t-a}^k, which depends only on the birth year $j = t - a$ and not separately on age or the current year.

Because our model can generate pure age, period, and cohort effects, it nests the additive model Eq. 16.1. Specifically, suppose that $K = 3$, $e_{1,s} = \bar{e}_1$ for all s, $r_2 = 0$, $w_{2,a} = 1$ for all a, $r_3 = 1$, $w_{3,0} = 1$, and $w_{3,a} = 0$ for $a > 0$. Then Eq. 16.5 reduces to

$$y_{at} = \mu + \sum_{a'=0}^{a} r_1^{a-a'} w_{1,a'} \bar{e}_1 + e_{2,t} + e_{3,j}, \quad j = t - a, \tag{16.6}$$

which is equivalent to Eq. 16.1 with $\alpha_a = \mu + \sum_{a'=0}^{a} r_1^{a-a'} w_{1,a'} \bar{e}_1$, $\beta_t = e_{2,t}$, and $\gamma_j = e_{3,j}$.

Three questions are of substantial interest in analyzing age- and period-specific data. First, how do outcomes vary with age, and how has the effect of age changed over time? For example, how does the mortality rate depend on age? Or, how does within-cohort consumption inequality change as the cohort ages? Second, how do outcomes depend on conditions in the current period? And third, does history influence current outcomes in a way that age effects do not fully capture—in other words, is a cohort theory appropriate? The parameters of our model help answer all three of these questions.

We can address the first question by estimating age profiles of outcomes. In our model, we conceive of changes in the age profile over historical time as changes in the time effects that accumulate for different cohorts. Permanent changes in time effects lead to permanent changes in the age profile. Define $m_k(a) = \sum_{a'=0}^{a} r_k^{a-a'} w_{k,a'}$. Then a hypothetical cohort that experienced the same time effects $(\bar{e}_1, \bar{e}_2, \ldots, \bar{e}_K)$ in every year of its life would, at age a, have outcomes

$$y_a(\bar{e}_1, \bar{e}_2, \ldots, \bar{e}_K) = \mu + \sum_{k=1}^{K} \bar{e}_k m_k(a), \tag{16.7}$$

which depends only on the cohort's age a. Two kinds of comparisons are in order. First, by comparing the model profile with the observed outcomes $y_{a,j+a}$ for a particular cohort j, we can see how the cohort's outcomes differ from what we would have predicted if conditions had never changed throughout its life. This comparison applies to actual people—it can help us see whether, and in what way, conditions changed over a particular cohort's life course. Second, consider a different hypothetical cohort that experienced a different set of constant time effects $(\tilde{e}_1, \tilde{e}_2, \ldots, \tilde{e}_K)$ in every year of its life. The second cohort would have outcomes

$$y_a(\tilde{e}_1, \tilde{e}_2, \ldots, \tilde{e}_K) = \mu + \sum_{k=1}^{K} \tilde{e}_k m_k(a). \tag{16.8}$$

The outcomes in Eq. 16.8 again depend only on the cohort's age, but they differ from the outcomes of the first hypothetical cohort in Eq. 16.7. Thus, given any set of time effects, we can calculate the hypothetical age profile that would result if those time effects continued for the entire life of a cohort. So, for example, we can calculate different age profiles corresponding to the time effects of 1900 and the time effects of 2000:

$$\begin{aligned} y_a(e_{1,1900}, e_{2,1900}, \ldots, e_{K,1900}) &= \mu + \sum_{k=1}^{K} e_{k,1900} m_k(a), \\ y_a(e_{1,2000}, e_{2,2000}, \ldots, e_{K,2000}) &= \mu + \sum_{k=1}^{K} e_{k,2000} m_k(a). \end{aligned} \tag{16.9}$$

The profile $y_a(e_{1,1900}, e_{2,1900}, \ldots, e_{K,1900})$ tells us the effect of age on outcome y in 1900. We can interpret $y_a(e_{1,1900}, e_{2,1900}, \ldots, e_{K,1900})$ as a prediction for the outcomes of the 1900 birth cohort if conditions never changed after the cohort's birth. In other words, $y_a(e_{1,1900}, e_{2,1900}, \ldots, e_{K,1900})$ describes the effect of age on outcomes y, holding time effects constant. Similarly, the profile $y_a(e_{1,2000}, e_{2,2000}, \ldots, e_{K,2000})$ tells us the effect of age on outcome y in 2000. By comparing the profiles, we can see how the effect of age on y changed over the twentieth century. This comparison applies not to particular cohorts but to history—it can help us see whether, and in what way, conditions changed between two perhaps widely separated eras.

We can address the second question—how do outcomes depend on current conditions?—by examining both the period effects $e_{k,t}$ and the age weights $w_{k,a}$. Suppose we are examining mortality, and suppose that for some age a, $w_{k,a}$ is positive. Also suppose that for 2 years s and t, $e_{k,s} < e_{k,t}$. The immediate impact of conditions in year t on people of age a is $w_{k,a}e_{k,t}$. Thus, all else equal, mortality for people of age a is predicted to be lower in year s than in year t.

We can address the third question—is a cohort theory appropriate?—by examining the rates of decay r_k. A cohort theory predicts that past events continue to affect a cohort's outcomes at much later ages. Recall that the impact at age a of conditions at age $a' < a$ is $r_k^{a-a'} w_{k,a'} e_{k,t-a+a'}$. A cohort theory says that even if a' is much less than a, this impact is large. But in that case, r_k must be close to one. Therefore, we can tell whether a cohort theory describes the data by examining whether any r_k is close to one.

By combining all of the parameters, we can use our model to see how a particular cohort evolves over the life course. For simplicity, suppose $K = 1$, so there is only one type of time effect, and suppose we are studying mortality. Consider the mortality rate of cohort j. At age a, its mortality is

$$y_{a,j+a} = \mu + \sum_{a'=0}^{a} r_1^{a-a'} w_{1,a'} e_{1,j+a'}. \tag{16.10}$$

The next year, at age $a + 1$, the cohort's mortality is

$$y_{a+1,j+a+1} = \mu + \sum_{a'=0}^{a+1} r_1^{a+1-a'} w_{1,a'} e_{1,j+a'}. \tag{16.11}$$

The change in mortality from age a to age $a + 1$ is thus

$$y_{a+1,j+a+1} - y_{a,j+a} = w_{1,a+1} e_{1,j+a+1} + (r_1 - 1) \sum_{a'=0}^{a} r_1^{a-a'} w_{1,a'} e_{1,j+a'}. \tag{16.12}$$

In words, the change between age a and age $a+1$ is a combination of what happens to the cohort when it reaches age $a + 1$ (the term $w_{1,a+1} e_{1,j+a+1}$) and a decrease in the influence of its past experiences (the term consisting of $(r_1 - 1)$ multiplied by an accumulation of time effects at ages $a' < a + 1$). The cohort evolves because it has new experiences and because the influence of the past diminishes.

16.3 Identification

We have claimed that one advantage of our model over the additive model Eq. 16.1 is that the parameters of our model are identified. We now make this claim precise. Because the additive model is unidentified even when Eq. 16.1 does not contain an error term, we assume for now that the outcomes y_{at} are measured without error; in Sect. 16.4, we show how to handle measurement error.

We say the parameters of our model are identified if there exists a unique set of parameters that can generate any given matrix of outcomes y_{at} for $a = 0, \ldots, A$ and $t = 1, \ldots, T$. That is, the parameters are identified if there is a unique vector

$$\boldsymbol{\theta} = \left[\mu, \left\{ \{e_{k,t}\}_{t=1-A}^{T}, \{w_{k,a}\}_{a=1}^{A}, r_k \right\}_{k=1}^{K} \right]$$

such that Eq. 16.5 holds for all $a = 0, \ldots, A$ and $t = 1, \ldots, T$. It turns out that our model is identified for some values of the true parameters and not for other values. The following definition is therefore helpful:

Definition 16.1. The parameter vector $\boldsymbol{\theta}$, an element of a parameter space $\boldsymbol{\Theta}$, is identified with respect to $\boldsymbol{\Theta}$ if there does not exist any vector $\tilde{\boldsymbol{\theta}} \in \boldsymbol{\Theta}$ distinct from $\boldsymbol{\theta}$ such that, given $\left\{\{y_{at}\}_{a=0}^{A}\right\}_{t=1}^{T}$, Eq. 16.5 holds for both $\boldsymbol{\theta}$ and $\tilde{\boldsymbol{\theta}}$ for all $a = 0, \ldots, A$ and $t = 1, \ldots, T$.

Under normalizations on the parameter space $\boldsymbol{\Theta}$ that do not affect the interpretation of the model, the set of parameter vectors that are not identified with respect to $\boldsymbol{\Theta}$ is of measure zero. The normalizations are:

Normalization 16.1. $r_k \leq r_{k'}$ for all $k < k'$.

Normalization 16.2. $w_{k,0} = 1$ for all k.

Normalization 16.3. If $K > 1$, then $e_{k,s} = 0$ for $s < k - A$.

Normalization 16.1 puts the time effect types in order, which is necessary because switching k with k' would not change the model. (We show below that the unidentified set of measure zero includes the case $r_k = r_{k'}$, so the ordering is strict.) Normalization 16.2 fixes the sign and scale of the age weights $w_{k,a}$ and the time effects $e_{k,s}$; for any $c_k \neq 0$, replacing $w_{k,a}$ by $c_k w_{k,a}$ for all a and $e_{k,s}$ by $e_{k,s}/c_k$ for all s would not change the model. The normalization does not affect the interpretation of results since only the product $w_{k,a} e_{k,s}$ enters the age profiles Eq. 16.7. Finally, we need Normalization 16.3 because the data do not contain adequate information about time effects in the distant past. The normalization is equivalent to dropping all data on the K oldest cohorts. To see why, notice that time effects $e_{k,s}$ at any date $s \leq K{-}A$ influence only the K oldest cohorts; that there are K^2 such time effects e_s^k in the model; and that we have $K(K+1)/2 \leq K^2$ observations (with strict inequality for $K > 1$) on the K oldest cohorts. We therefore have no hope of identifying all the time effects at dates $s \leq K{-}A$. In addition, by appropriately choosing $\{e_{k,s}\}_{s \leq k-A}$, we can perfectly fit the data on the K oldest cohorts regardless of how we choose $\mathbf{r}$, $\mathbf{w}$, μ, and $\{e_{k,s}\}_{s>K-A}$. Since the K oldest cohorts are uninformative, we could drop them and avoid estimating $\{e_{k,s}\}_{s \leq k-A}$. Equivalently, we can normalize some elements of $\{e_{k,s}\}_{s \leq k-A}$ to zero. Since the normalization does not affect $\mathbf{r}, \mathbf{w}, \mu, \{e_{k,s}\}_{s>K-A}$, it does not affect the substantive results.

Proposition 16.1. *Let $K \in \{1,2,3\}$ be known, and let the parameter space $\boldsymbol{\Theta}$ consist of all vectors $\left[\mu, \left\{\{e_{k,t}\}_{t=1-A}^{T}, \{w_{k,a}\}_{a=0}^{A}, r_k\right\}_{k=1}^{K}\right]$ that satisfy Normalizations 16.1 to 16.3. Suppose further that $A \geq K$, that $T \geq A + K$, and that if $K = 1$, then $T \geq 4$; if $K = 2$, then $T \geq 12$; and if $K = 3$, then $T \geq 32$. Then there exists a set $X_K \subset \boldsymbol{\Theta}$ such that X_K is of measure zero and all $\boldsymbol{\theta} \in \boldsymbol{\Theta} \setminus X_K$ are identified with respect to $\boldsymbol{\Theta}$.*

Proof. We prove the result separately for $K = 1$, $K = 2$, and $K = 3$. In each case, the strategy will be to construct a set $X_K \subset \boldsymbol{\Theta}$ such that X_K is of measure zero and such that, unless $\boldsymbol{\theta} = \left[\mu, \left\{\{e_{k,t}\}_{t=1-A}^{T}, \{w_{k,a}\}_{a=1}^{A}, r_k\right\}_{k=1}^{K}\right]$ is in X_K, the equality

$$\mu + \sum_{k=1}^{K} \sum_{a'=0}^{a} r_k^{a-a'} w_{k,a'} e_{k,t-a+a'} = \tilde{\mu} + \sum_{k=1}^{K} \sum_{a'=0}^{a} \tilde{r}_k^{a-a'} \tilde{w}_{k,a'} \tilde{e}_{k,t-a+a'},$$
$$a = 0, \ldots, A,\ t = 1, \ldots, T, \qquad (16.13)$$

implies, under the hypotheses of the proposition, that $\left[\mu, \left\{\{e_{k,t}\}_{t=1-A}^{T}, \{w_{k,a}\}_{a=1}^{A}, r_k\right\}_{k=1}^{K}\right] = \left[\tilde{\mu}, \left\{\{\tilde{e}_{k,t}\}_{t=1-A}^{T}, \{\tilde{w}_{k,a}\}_{a=1}^{A}, \tilde{r}_k\right\}_{k=1}^{K}\right] \equiv \tilde{\boldsymbol{\theta}}$.

Case 1: $K = 1$. Let X_1 be the set of $\boldsymbol{\theta} \in \boldsymbol{\Theta}$ such that either $r_1 + w_{1,1} = 1$ or the vectors $(e_{1,1}, \ldots, e_{1,T-1})$ and $(e_{1,2}, \ldots, e_{1,T})$ are collinear with a constant. X_1 is a set of measure zero. Assume $\boldsymbol{\theta} \in \boldsymbol{\Theta} \setminus X_1$. Specializing Eq. 16.13 to $K = 1$, $a = 0$, and $a = 1$ (by hypothesis, $A \geq 1$) and using Normalization 16.2, we have

$$\mu + e_{1,t} = \tilde{\mu} + \tilde{e}_{1,t}, \quad t = 1, \ldots, T, \qquad (16.14a)$$

$$\mu + r_1 e_{1,t-1} + w_{1,1} e_{1,t} = \tilde{\mu} + \tilde{r}_1 \tilde{e}_{1,t-1} + \tilde{w}_{1,1} \tilde{e}_{1,t}, \quad t = 2, \ldots, T. \qquad (16.14b)$$

Substituting Eq. 16.14a into Eq. 16.14b and collecting terms gives

$$0 = (\mu - \tilde{\mu})(1 - \tilde{r}_1 - \tilde{w}_{1,1}) - (\tilde{r}_1 - r_1) e_{1,t-1} - (\tilde{w}_{1,1} - w_{1,1}) e_{1,t}, \quad t = 2, \ldots, T. \qquad (16.15)$$

By hypothesis, $T \geq 4$, so Eq. 16.15 contains at least three equations. Since (given $\boldsymbol{\theta} \notin X_1$) $e_{1,t-1}$ and $e_{1,t}$ are not collinear with a constant, Eq. 16.15 can hold only if $(\mu - \tilde{\mu})(1 - \tilde{r}_1 - \tilde{w}_{1,1}) = 0$ and the coefficients on $e_{1,t-1}$ and $e_{1,t}$ are both zero. Hence $\tilde{r}_1 = r_1$, $\tilde{w}_{1,1} = w_{1,1}$, and, since $1 - r_1 - w_{1,1} \neq 0$ for $\boldsymbol{\theta} \notin X_1$, we must have $\tilde{\mu} = \mu$. It follows from Eq. 16.14a that $\tilde{e}_{1,t} = e_{1,t}$ for $t = 1, \ldots, T$. Finally, substituting the foregoing results into Eq. 16.13 for $a \geq 2$ shows that $\tilde{e}_{1,t} = e_{1,t}$ for $t \leq 0$ and $\tilde{w}_{1,a} = w_{1,a}$ for $a \geq 2$.

Case 2: $K = 2$. Define the following sets:

$$X_{2,1} = \{\boldsymbol{\theta} \in \boldsymbol{\Theta} : \exists k \text{ s.t. } r_k = 0\}, \quad X_{2,2} = \{\boldsymbol{\theta} \in \boldsymbol{\Theta} : r_1 = r_2\},$$

$$X_{2,3} = \{\boldsymbol{\theta} \in \boldsymbol{\Theta} : w_{1,1} = w_{2,1}\},$$

$$X_{2,4} = \left\{ \boldsymbol{\theta} \in \boldsymbol{\Theta} : \text{rank} \begin{bmatrix} 1 & \begin{pmatrix} e_{k,j} \\ \vdots \\ e_{k,T-4+j} \end{pmatrix}_{\substack{k \in \{1,2\}, \\ j \in \{1,2,3,4\}}} \\ \vdots & \\ 1 & \end{bmatrix} < 9 \right\},$$

$$X_{2,5} = \{\boldsymbol{\theta} \in \boldsymbol{\Theta} : (w_{1,1} - w_{2,1})[-r_2 w_{1,1} + r_1 w_{2,1}] + (w_{1,2} - w_{2,2})(r_2 - r_1) = 0\},$$

$$X_{2,6} = \{\boldsymbol{\theta} \in \boldsymbol{\Theta} : w_{1,1} - w_{2,1} + r_1 - r_2 + (r_2^2 + r_2 w_{2,1} + w_{2,2})(1 - r_1 - w_{1,1})$$
$$- (w_{1,2} + r_1 w_{1,1} + r_1^2)(1 - r_2 - w_{2,1}) = 0\}. \qquad (16.16)$$

Let $X_2 = \cup_{j=1}^{6} X_{2,j}$. X_2 has measure zero. Schulhofer-Wohl and Yang (2011) show that under Normalizations 16.1 to 16.3 and the hypotheses of the proposition, if $\boldsymbol{\theta} \in \boldsymbol{\Theta} \setminus X_2$, then the unique solution to Eq. 16.13 is $\tilde{\boldsymbol{\theta}} = \boldsymbol{\theta}$. The algebra proceeds by using Eq. 16.13 at $a = 0$ and $a = 1$ to eliminate $\tilde{e}_{2,t}$ and obtain a first-order difference equation in $\tilde{e}_{1,t}$; substituting the difference equation into Eq. 16.13 at $a = 2$ to eliminate $\tilde{e}_{1,t}$; and observing that coefficients in a linear combination of a constant with $\{e_{k,t-3}, \ldots, e_{k,t}\}_{k=1}^{2}$ must be zero given $\boldsymbol{\theta} \notin X_{2,4}$. Setting the coefficients to zero yields quadratic equations with two solutions, $(\tilde{r}_1, \tilde{r}_2) = (r_1, r_2)$ and $(\tilde{r}_1, \tilde{r}_2) = (r_2, r_1)$; Normalization 16.1 rules out the latter to give uniqueness.

Case 3: $K = 3$. The approach parallels the $K = 2$ case; see Schulhofer-Wohl and Yang (2011). □

Proposition 16.1 says there may be parameter vectors $\boldsymbol{\theta}$ that are not identified: For each of these $\boldsymbol{\theta}$, there exists some $\boldsymbol{\theta}' \neq \boldsymbol{\theta}$ that would generate the same data as $\boldsymbol{\theta}$. However, the set X of unidentified parameter vectors is of measure zero. For almost all $\boldsymbol{\theta}$, therefore, there does not exist any $\boldsymbol{\theta}' \neq \boldsymbol{\theta}$ that would generate the same data, and by observing y_{at}, we can uniquely determine the true parameter vector $\boldsymbol{\theta}$. We have not proved versions of Proposition 16.1 for $K > 3$, but we conjecture that it holds; a proof would require tedious algebra.

The conditions in Proposition 16.1 are sufficient but not necessary for identification. In particular, the parameters may be identified for T smaller than the values stated, so long as A is sufficiently large. We have not completely characterized the sets X_K of unidentified parameter vectors. In one sense, this is unimportant since almost all parameter vectors lie outside X_K. However, to understand the source of identification, it is helpful to partially characterize X_K. The next proposition gives some necessary conditions for a parameter vector to be identified.

Proposition 16.2. *Under the hypotheses of Proposition 16.1, any parameter vector* $\boldsymbol{\theta} \in \boldsymbol{\Theta}$ *is not identified if either:*

(a) $e_{k,t} = \bar{e}_k$ *for some* k *and all* $t = 1 - A, \ldots, T$, *or*
(b) $K > 1$ *and* $r_k = r_{k'}$ *for some* $k \neq k'$.

Further, $\boldsymbol{\theta}$ *remains unidentified in each of these cases even if* μ *is known.*

Proof. We must show that under each of conditions (a) and (b), Eq. 16.13 has multiple solutions for $(\tilde{\mu}, \tilde{\mathbf{r}}, \tilde{\mathbf{e}}, \tilde{\mathbf{w}})$ in terms of $(\mu, \mathbf{r}, \mathbf{e}, \mathbf{w})$, and that this is so even if $\tilde{\mu} = \mu$.

Condition (a): Without loss of generality, suppose $e_{1,t} = \bar{e}_1$. Choose any $r^* \in [0, 1]$. Let $\{w_a^*\}_{a=1}^{A}$ be the unique solution to the following nonsingular triangular system of linear equations given r^*, r_1 and $\{w_{1,a}\}_{a=1}^{A}$:

$$\sum_{a'=1}^{a} (r^*)^{a-a'} w_{a'}^* = -(r^*)^a + \sum_{a'=0}^{a} r_1^{a-a'} w_{1,a'}, \quad a = 1, \ldots, A. \tag{16.17}$$

Given $e_{1,t} = \bar{e}_1$, the following solves Eq. 16.13: $\tilde{\mu} = \mu$; $\tilde{e}_{j,t} = e_{j,t}\ \forall\ j, t$; $\tilde{r}_1 = r^*$; $\tilde{r}_j = r_j\ \forall\ j > 1$; $\tilde{w}_{1,a} = w_a^*\ \forall\ a$; $\tilde{w}_{j,a} = w_{j,a}\ \forall\ j > 1, a$. Therefore, Eq. 16.13 has a continuum of solutions indexed by $r^* \in [0, 1]$.

Condition (b): Without loss of generality, suppose $r_1 = r_2$. Choose any $x \in (1/2, 1]$. Given $r_1 = r_2$, the following solves Eq. 16.13: $\tilde{\mu} = \mu$; $\tilde{r}_j = r_j\ \forall\ j$; $\tilde{w}_{1,a} = xw_{1,a} + (1-x)w_{2,a}\ \forall\ a$; $\tilde{w}_{2,a} = (1-x)w_{1,a} + xw_{2,a}\ \forall\ a$; $\tilde{e}_{1,t} = \frac{(1-x)e_{2,t} - xe_{1,t}}{1-2x}\ \forall\ t$; $\tilde{e}_{2,t} = \frac{(1-x)e_{1,t} - xe_{2,t}}{1-2x}\ \forall\ t$; $\tilde{e}_{j,t} = e_{j,t}\ \forall\ j > 2, t$; $\tilde{w}_{j,a} = w_{j,a}\ \forall j > 2, a$. Therefore, Eq. 16.13 has a continuum of solutions indexed by $x \in (1/2, 1]$. □

Condition (a) in Proposition 16.2 is the case where the model contains pure age effects. Therefore, although the additive model Eq. 16.1 is a special case of our model, it is an unidentified special case. We emphasize that the potential need to identify the intercept μ has nothing to do with this failure of identification. It is clear that pure age, period, or cohort effects will be unidentified in our model without some normalization on μ for the usual reason that—even without the APC identification problem—one dummy variable in any given category must be omitted in any linear model that contains an intercept. But Proposition 16.2 shows that pure age effects will remain unidentified even with a normalization on μ. The intuition is as follows. Suppose the same time effect happens over and over, i.e., $e_{k,t} = \bar{e}_k$. Then it will be impossible to distinguish whether this time effect has a transitory impact that directly affects people of all ages (a period effect) or a persistent impact that directly affects only the young (so that the effect on the old is indirect, a cohort effect). Pure age effects, in other words, make it impossible to distinguish period from cohort.

16.4 Identification with Measurement Error in y

Suppose that, instead of observing y_{at}, we have data only on a noisy measurement $\bar{y}_{at}$, where

$$\bar{y}_{at} = y_{at} + \epsilon_{at}. \tag{16.18}$$

For example, y_{at} could be the probability of death for individuals age a in year t, and $\bar{y}_{at}$ could be the observed mortality rate, which is a random variable with mean y_{at} when the population is finite. Alternatively, y_{at} could be a measure of consumption inequality among all people age a in year t, and $\bar{y}_{at}$ could be an estimate of inequality calculated from a random sample of the population. We now show conditions on the measurement error ϵ_{at} under which our model remains identified.

Assumption 16.1. $\mathrm{E}[\epsilon_{at}|y_{at}] = 0$ and $\mathrm{E}[\epsilon_{a,t}^2|y_{at}] = \sigma^2$ for all a, t, and $\mathrm{E}[\epsilon_{a,t}\epsilon_{a',t'}|y_{at}, y_{a't'}] = 0$ whenever $a' \neq a$ or $t' \neq t$.

Assumption 16.1 restricts the variance-covariance matrix of the measurement error. We must impose such a restriction because age-period-cohort analysis is, in essence, a decomposition of variance. In Sect. 16.5, we will consider an application in which Assumption 16.1 is plausible.

Proposition 16.3. *Suppose assumption 1 and the hypotheses of Proposition 16.1 hold. Let*

$$\hat{\boldsymbol{\theta}} = \arg\min_{\tilde{\boldsymbol{\theta}}} \sum_{a=0}^{A} \sum_{t=1}^{T} \left(\bar{y}_{at} - \tilde{\mu} - \sum_{k=1}^{K} \sum_{a'=0}^{a} \tilde{r}_k^{a-a'} \tilde{w}_{k,a'} \tilde{e}_{k,t-a+a'} \right)^2 . \tag{16.19}$$

Then, subject to regularity conditions on ϵ_{at}:

(a) $\hat{\boldsymbol{\theta}} \xrightarrow{p} \boldsymbol{\theta}$ in the limit as $\sigma^2 \to 0$ with A and T fixed, and
(b) If $e_{k,t}$ is a stationary and ergodic process, then $\left(\left\{\{\hat{w}_{k,a}\}_{a=1}^{A}, \hat{r}_k\right\}_{k=1}^{K}, \hat{\mu}\right) \xrightarrow{p} \left(\left\{\{w_{k,a}\}_{a=1}^{A}, r_k\right\}_{k=1}^{K}, \mu\right)$ in the limit as $T \to \infty$ with A fixed.

Proof. We assume the distribution of ϵ_{at} satisfies regularity conditions such that a uniform law of large numbers (ULLN) holds. Case (a): If $\sigma^2 = 0$, Eq. 16.19 becomes Eq. 16.13; hence the true parameters uniquely solve Eq. 16.19 when $\sigma^2 = 0$. Since the objective function in Eq. 16.19 is continuous, a ULLN applies, and solutions for $\sigma^2 > 0$ converge to the solution for $\sigma^2 = 0$. Case (b): The predicted values can be written as $\hat{\mathbf{y}}(\tilde{\mathbf{w}}, \tilde{\mathbf{r}}, \tilde{\mu}, \tilde{\mathbf{e}}) = \mathbf{X}(\tilde{\mathbf{w}}, \tilde{\mathbf{r}}, \tilde{\mu})\tilde{\mathbf{e}}$. Hence if we solve Eq. 16.19 for $\hat{\mathbf{e}}$ as a function of the remaining parameters, we obtain $\hat{\mathbf{e}} = [\mathbf{X}(\tilde{\mathbf{w}}, \tilde{\mathbf{r}}, \tilde{\mu})'\mathbf{X}(\tilde{\mathbf{w}}, \tilde{\mathbf{r}}, \tilde{\mu})]^{-1}\mathbf{X}(\tilde{\mathbf{w}}, \tilde{\mathbf{r}}, \tilde{\mu})'(\mathbf{y} + \boldsymbol{\epsilon})$. (Interpret the inverse as a generalized inverse when $\mathbf{X}(\tilde{\mathbf{w}}, \tilde{\mathbf{r}}, \tilde{\mu})$ is not of full rank.) Substituting this solution into Eq. 16.19, we obtain

$$(\hat{\mathbf{w}}, \hat{\mathbf{r}}, \hat{\mu}) = \arg\min_{\tilde{\mathbf{w}}, \tilde{\mathbf{r}}, \tilde{\mu}} \frac{1}{T} [\mathbf{y}'\mathbf{M}(\tilde{\mathbf{w}}, \tilde{\mathbf{r}}, \tilde{\mu})\mathbf{y} + 2\mathbf{y}'\mathbf{M}(\tilde{\mathbf{w}}, \tilde{\mathbf{r}}, \tilde{\mu})\boldsymbol{\epsilon} + \boldsymbol{\epsilon}'\mathbf{M}(\tilde{\mathbf{w}}, \tilde{\mathbf{r}}, \tilde{\mu})\boldsymbol{\epsilon}], \tag{16.20}$$

where $\mathbf{M}(\tilde{\mathbf{w}}, \tilde{\mathbf{r}}, \tilde{\mu}) = \mathbf{I} - \mathbf{X}(\tilde{\mathbf{w}}, \tilde{\mathbf{r}}, \tilde{\mu})[\mathbf{X}(\tilde{\mathbf{w}}, \tilde{\mathbf{r}}, \tilde{\mu})'\mathbf{X}(\tilde{\mathbf{w}}, \tilde{\mathbf{r}}, \tilde{\mu})]^{-1}\mathbf{X}(\tilde{\mathbf{w}}, \tilde{\mathbf{r}}, \tilde{\mu})'$ is a symmetric and idempotent matrix. Since $e_{k,t}$ is stationary and ergodic, so is y_{at}, and so the ergodic theorem and ULLN apply to the new objective function. Hence as $T \to \infty$, the second term in the objective function converges uniformly in probability to zero. Further, since ϵ_{at} is serially uncorrelated and homoskedastic by Assumption 16.1, the third term converges uniformly in probability to $\sigma^2\mathrm{tr}[\mathbf{M}(\tilde{\mathbf{w}}, \tilde{\mathbf{r}}, \tilde{\mu})]$. Since $\mathbf{M}$ is idempotent, its trace equals its rank, which is no smaller than its rank when $\mathbf{X}(\tilde{\mathbf{w}}, \tilde{\mathbf{r}}, \tilde{\mu})$ has full rank. At the true parameters, $\mathbf{X}$ has full rank. Hence, in the limit as $T \to \infty$, the true parameters minimize the third term. Further, in the limit as $T \to \infty$, the first term converges uniformly in probability to a function that is zero at the true parameters and, by Proposition 16.1, strictly positive otherwise. Thus the objective function converges uniformly in probability to a function minimized by the true parameters. Thus $(\hat{\mathbf{w}}, \hat{\mathbf{r}}, \hat{\mu})$ converges in probability to the true parameters. □

Proposition 16.3 says certain parameters can be consistently estimated by nonlinear least squares when outcomes are measured with serially uncorrelated, homoskedastic, mean-zero error. In the limit as the variance of the measurement error goes to zero, all of the parameters can be consistently estimated; this limit applies when $\bar{y}_{at}$ is computed from large populations in each (a,t) cell, as in the case of mortality rates calculated from vital records. In the limit as T goes to infinity with A fixed—as when small samples are collected in each of many years—all parameters except the time effects $e_{k,t}$ can be consistently estimated; parameters indexed by t cannot be consistently estimated because adding data on new time periods does not add information about parameters relevant only to earlier time periods. We do not consider limits as A goes to infinity because the human life span is finite. One can test whether the homoskedasticity requirement ($\mathrm{E}[\epsilon_{a,t}^2|y_{at}] = \sigma^2$) in Assumption 16.1 holds by examining whether the squared residuals $\left(\bar{y}_{at} - \hat{\mu} - \sum_{k=1}^{K}\sum_{a'=0}^{a}\hat{r}_k^{a-a'}\hat{w}_{k,a'}\hat{e}_{k,t-a+a'}\right)^2$ are systematically related to the predicted values $\hat{y}_{at} = \hat{\mu} + \sum_{k=1}^{K}\sum_{a'=0}^{a}\hat{r}_k^{a-a'}\hat{w}_{k,a'}\hat{e}_{k,t-a+a'}$.

16.5 Example: Mortality Rates in Sweden

The demographic transition in Western developed countries over the past 200 years featured gradual mortality declines in response to improvements in features of the environment including water quality, sanitation, nutrition, prevalence of infectious diseases, and medical technology (Elo and Preston 1992; Omran 1982). When did these changes occur? How did they differentially affect people of different ages? And did they have lasting impacts on particular birth cohorts? We answer these questions by estimating our model on the long time series of high-quality mortality data from Sweden in the Human Mortality Database (2007).

We analyze data from 1861 to 2005 on ages 64 and younger. (We drop earlier years and older ages due to data quality concerns described in the Human Mortality Database documentation.) To keep the sample size and number of parameters manageable, we use only data on ages that are multiples of 2 $(0, 2, \ldots, 64)$ and at 2-year intervals $(1861, 1863, \ldots, 2005)$. Figure 16.1 displays the data. Infant mortality has decreased proportionately much more than adult mortality over the past two centuries—exactly the kind of shift in an age profile of outcomes that our model aims to capture.

The dependent variable we analyze is the natural logarithm of the realized mortality rate among people who are age a in year t. We treat Eq. 16.5 as a model of the underlying log probability of death and Eq. 16.18 as a model of log realized mortality, which randomly differs from the log probability of death in a finite population. We estimate the model by nonlinear least squares as in Eq. 16.19, weighting each age-year cell by a consistent estimate of the inverse of the variance of observed log mortality in that cell.[1]

We estimate the additive model Eq. 16.1 as well as the continuously accumulating model Eq. 16.5 for $K = 1$, $K = 2$, and $K = 3$. (We did not attempt models with $K > 3$ because of the large number of parameters involved.) Our purpose in estimating the additive model is not to interpret its parameters but only to test it against the more general $K = 3$ model in which it is nested. For this purpose, the failure of identification in the additive model does not cause problems: We need to obtain only the log likelihood of the additive model, which does not depend on which single identifying constraint we impose on the parameters.

Table 16.1 reports goodness-of-fit statistics for the additive and continuously accumulating models. The continuously accumulating model with $K = 3$ fits best by any criterion: log likelihood, Akaike information criterion (AIC), or Bayesian information criterion (BIC). A likelihood ratio test of the $K = 3$ continuously accumulating model against the additive model nested within it rejects the additive model with a p-value of zero. A possible concern is that our model uses so many parameters that it may overfit the data. However, even when we penalize our model for using more parameters by examining the BIC, we still find that our model is preferred to the additive model.

Figure 16.2 plots the residuals for each model. We observe a great deal of heteroskedasticity in the additive and $K = 1$ models, but relatively little heteroskedasticity in the $K = 3$ model, further evidence that the $K = 3$ model accounts better for the data. (The unusually small residuals at the highest predicted values for $K = 3$ correspond to data points for early years and young ages; one can show that our weights are noisily estimated for these data points and thus that the points may be overweighted in a finite sample, leading to incorrectly small residuals.) None of

[1]This procedure is equivalent to maximum likelihood estimation and generates residuals that satisfy Assumption 16.1. To see this, suppose each individual who is age a at time t has a probability of death p_{at}, and let N_{at} be the population at risk in cell (a, t). If $\bar{p}_{at}$ is the realized mortality rate in the cell, then by the central limit theorem, $\sqrt{N_{at}}(\bar{p}_{at} - p_{at}) \xrightarrow{d} \mathcal{N}[0, p_{at}(1 - p_{at})]$ as $N_{at} \rightarrow \infty$. (The smallest cell in our data has $N_{at} = 19{,}856$, and the median cell has $N_{at} = 92{,}794$, so approximating the distribution by the limit as $N_{at} \rightarrow \infty$ seems reasonable.) By the delta method, $\sqrt{N_{at}}(\ln \bar{p}_{at} - \ln p_{at}) \xrightarrow{d} \mathcal{N}[0, (1 - p_{at})/p_{at}]$. We observe realized log mortality $\bar{y}_{at} \equiv \ln \bar{p}_{at}$ and population N_{at} but not true log mortality $y_{at} \equiv \ln p_{at}$; indeed, the goal is to estimate parameters determining y_{at}. But $\bar{p}_{at} \xrightarrow{p} p_{at}$, so by the continuous mapping theorem, $\sqrt{N_{at}\bar{p}_{at}/(1 - \bar{p}_{at})}(\bar{y}_{at} - y_{at}) \xrightarrow{d} \mathcal{N}(0, 1)$. If p_{at} depends on parameters $\boldsymbol{\theta}$, the log likelihood for data on ages $a = 0, \ldots, A$ and years $t = 1, \ldots, T$ is $\ln L = -\frac{(A+1)T}{2} \ln(2\pi) - \frac{1}{2} \sum_{a=0}^{A} \sum_{t=1}^{T} \frac{N_{at}\bar{p}_{at}}{1-\bar{p}_{at}} [\bar{y}_{at} - y_{at}(\boldsymbol{\theta})]^2$. Maximizing the likelihood is thus equivalent to minimizing the weighted nonlinear least squares objective function for the model $\bar{y}_{at} = \ln(p_{at}(\boldsymbol{\theta})) + \epsilon_{at}$ with weights $\hat{\sigma}_{at}^{-2} = N_{at}\bar{p}_{at}/(1 - \bar{p}_{at})$. The minimized weighted nonlinear least squares objective function, divided by the residual degrees of freedom, is an estimate of dispersion; the dispersion should be 1 if the model fully accounts for variation in mortality. In practice, since we estimate dispersion greater than 1, we compute the log likelihood and standard errors without assuming the dispersion equals 1.

Table 16.1 Goodness-of-fit statistics for six models (Data source: Human Mortality Database 2007)

		Continuously accumulating				3-factor
	Additive	K = 1	K = 2	K = 3	Lee-Carter	nonaccum
Log likelihood	−7728	−9321	−6446	−5163	−7048	−5969
AIC	15,872	18,920	13,444	11,150	14,371	12,567
BIC	17,075	19,724	15,041	13,535	15,164	14,390
Weighted R^2	0.9900	0.9561	0.9965	0.9984	0.9943	0.9976
Dispersion	14.4	52.5	5.13	1.89	7.94	3.52
Cells	2409	2409	2409	2409	2409	2409
Parameters	208	139	276	412	137	315
Residual d.f.	2201	2270	2133	1997	2272	2094

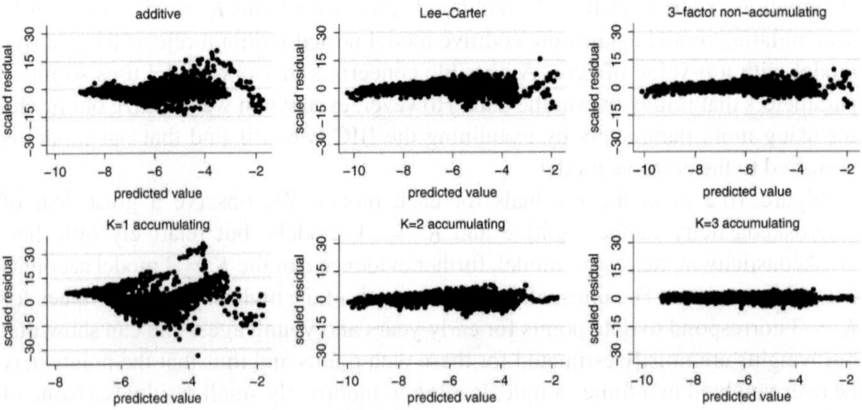

Fig. 16.2 Scaled residuals vs. predicted values. Graphs plot scaled residuals $(\bar{y}_{at} - \hat{y}_{at})/\hat{\sigma}_{at}$ against predicted values $\hat{y}_{at}$ for each model estimated

the models fits perfectly: Even with $K = 3$, the estimated dispersion is 1.89 times what it would be if our model accounted perfectly for the individual probability of death and the residuals were due only to randomness in realized death rates. Still, the fit of the $K = 3$ model is quite good. It accounts for 99.84 % of the variation in the data and has one-eighth the overdispersion of the additive model. Also, as long as the residuals are homoskedastic, the overdispersion does not make our parameter estimates inconsistent.

We investigate the overdispersion by estimating two alternative models besides the additive model. The first alternative is the Lee and Carter (1992) model, which takes the form $y_{at} = w_{1,a} + w_{2,a}e_{2,t}$. The Lee-Carter model is a widely adopted statistical method for forecasting age-specific mortality rates (United Nations 2003); although our model is explicitly *not* intended for forecasting, it is still interesting to compare our model fit with that of another model commonly estimated on mortality data. The second alternative is a three-factor model without lagged effects,

$y_{at} = \sum_{k=1}^{3} w_{k,a} e_{k,t}$. This model is simply our model Eq. 16.5 with $K = 3$ under the restriction that $r_1 = r_2 = r_3 = 0$—that is, under the restriction that lagged effects do not accumulate over time—so estimating the three-factor nonaccumulating model is a way to tell whether the process of accumulation that we model is important. The $K = 3$ continuously accumulating model fits better than either of these two alternatives by all criteria considered. The Lee-Carter model has four times as much overdispersion as the continuously accumulating model, demonstrating that the overdispersion problem is not limited to our model; mortality clearly is a complex phenomenon that any simple statistical model can only imperfectly describe. The rejection of the nonaccumulating model provides evidence that the accumulation of lagged effects is important in mortality data.

One possible reason for the overdispersion may be that we use a simple functional form for $r(k, a, a')$, the impact on outcomes at age a of time effects at age $a' < a$. If the true relationship between time effects in youth and subsequent mortality is not exponential—for example, if early life conditions can cause mortality in early life and old age but not in between (e.g., Horiuchi 1983)—then the form we use for $r(k, a, a')$ will not adequately fit the data. As discussed in Sect. 16.2, more complicated forms for $r(k, a, a')$ would be difficult to implement, and we leave them for future research.

Because the $K = 3$ continuously accumulating model fits the data best, we present parameter estimates only from that model. Table 16.2 reports the estimated rates of decay, and Fig. 16.3 shows the estimated age weights and sequences of time effects.

The first type of time effect, $k = 1$, has short-lasting effects, with an estimated half-life of about 4 months. The estimated age weights $w_{1,a}$ show that these time effects impact mainly the young, not the middle-aged or the old. The estimated time effects $e_{1,t}$ show that mortality related to these time effects first rose and then fell over the years we study. Examining the estimated age weights and time effects together, we conclude that there was a sharp spike in mortality related to these time effects in 1919, when there was a global influenza epidemic.

The second type of time effect, $k = 2$, displays an interesting pattern: It has opposite impacts on the very young as on all others. A time effect of type $k = 2$ that lowers the mortality of 2-year-olds by one log point is predicted to *raise* the mortality of 18-year-olds by about the same amount; impacts on older people are smaller but still raise mortality. In other words, time effects of type $k = 2$ describe a process in which falling infant mortality and rising adult mortality, especially young-adult

Table 16.2 Estimated rates of decay for $K = 3$ continuously accumulating model. Rates of decay are for 1-year intervals. Half-life in years $t_{1/2}$ solves $r_k^{t_{1/2}} = 1/2$ (Data source: Human Mortality Database 2007)

Parameter	Estimate	Standard error	Half-life (years)
r_1	0.138	0.0000	0.35
r_2	0.582	0.0005	1.28
r_3	0.845	0.0005	4.11

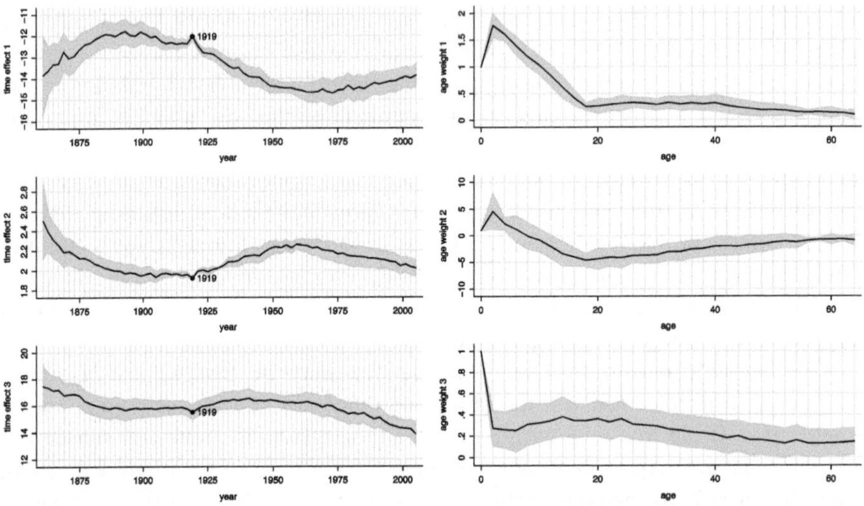

Fig. 16.3 Estimated time effects and age weights. Estimates from $K = 3$ continuously accumulating model. *Black lines* are point estimates; *gray lines* are pointwise 95 % confidence intervals

mortality, are two sides of the same coin. At least two explanations are possible. First is a selection pattern: Reductions in infant mortality may mainly save the lives of unhealthy people who will soon die anyway, so that saving infants inevitably raises adult mortality. Second is a historical explanation: Technological and social changes that reduced infant mortality, such as better sanitation, may have happened in Sweden around the same time as other technological changes, such as the introduction of machinery and motor vehicles, that raise the "accident hump" frequently observed among young adults (Heligman and Pollard 1980). Our model is not designed to discriminate between these or other possible explanations for the estimated pattern, but further investigation via other methods would be worthwhile. There is a downward spike in these time effects in 1919, which raises mortality of young adults and lowers mortality of infants. Combined with the impact of time effects of type 1, events in 1919 then have little impact on infants but increase the mortality of young adults, consistent with the findings of Noymer and Garenne (2000).

The third type of time effect, $k = 3$, has the longest-lasting effects, with a half-life of more than 4 years. It impacts mainly infants, with smaller impacts on the young and virtually no impact on the old. These findings are somewhat consistent with a theory of cohort effects in mortality, since they demonstrate that conditions early in life have lasting consequences. However, given the estimated half-life, the consequences do not necessarily last into old age. Thus, our findings fall in the middle of the debate between Finch and Crimmins (2004), who proposed the cohort morbidity phenotype hypothesis that suggests reductions in early-life mortality due to infections are associated with reductions in mortality at all subsequent ages for the same cohort, and Barbi and Vaupel (2005), who contend that cohort effects are unimportant.

The estimated sequences of time effects $e_{k,t}$ show that mortality related to time effects of type $k = 1$ and $k = 3$ largely fell over the twentieth century. For time effects of type $k = 2$, infant mortality largely fell, but the decrease was disrupted around the time of the Great Depression and World War II.

The high rates of decay for all three types of effects in the $K = 3$ continuously accumulating model suggest that there is little role in the data for traditional cohort effects; early-life influences may have an impact for several years but do not carry through to old age. The $K = 1$ and $K = 2$ continuously accumulating models produced similar results, with half-lives ranging from 0.3 to 4.4 years. It is important to note that the finding of little role for cohort effects does not result mechanically from our assumption of exponential decay, because the rate of decay could have been estimated to be small or zero; the data are telling us that the decay is rapid.

Figure 16.4 shows the observed mortality rates of several birth cohorts and the predicted age profile of mortality based on conditions in each cohort's birth year. Forecasting is not our goal, so differences between the predicted and observed outcomes do not reflect a failure of our model; rather, these differences are interesting because they show us how conditions changed over a cohort's life course. For the 1881 birth cohort, the predicted age profile closely matches the observed mortality rates, showing that conditions changed little over the cohort's life course. The findings for the 1881 cohort are in sharp contrast with those for later cohorts, where observed mortality at most ages is strikingly lower than the predicted age profile based on conditions in the cohort's birth year. The gap between predicted and observed mortality grows larger with each successive cohort, suggesting not only that conditions improved during each cohort's life course but also that the rate of improvement grew over time.

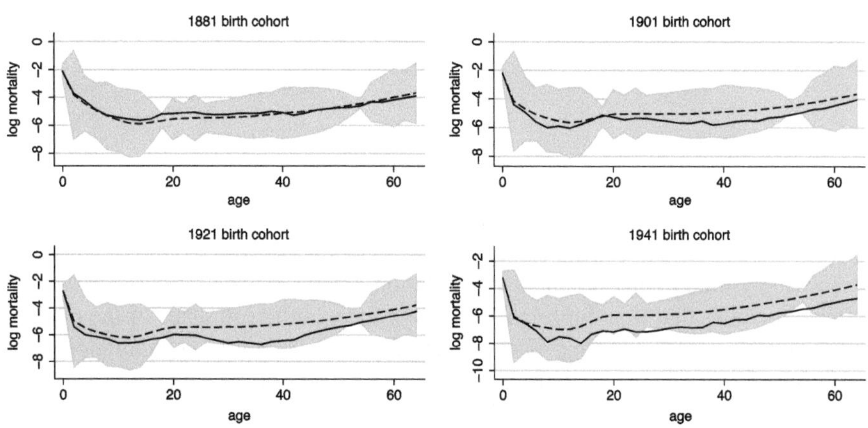

Fig. 16.4 Predicted age profiles vs. actual mortality. Each *solid line* shows the observed mortality of a particular birth cohort at various ages. The *dashed line* represents the mortality that the $K = 3$ continuously accumulating model would predict for that cohort if conditions in its birth year continued throughout its life span

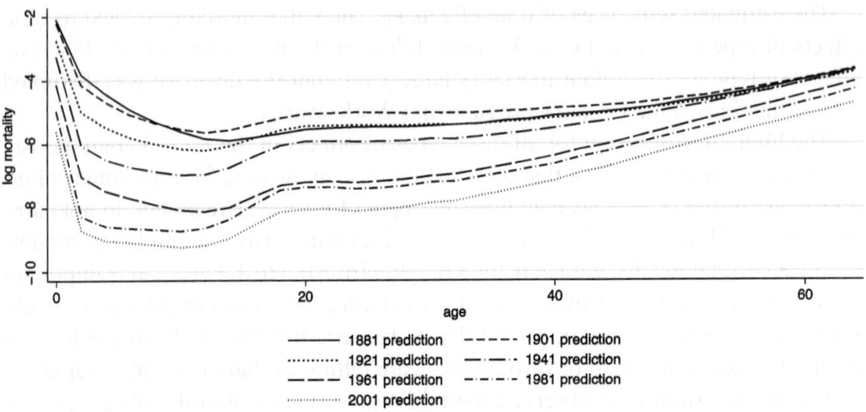

Fig. 16.5 Comparing predicted age profiles. Each *line* represents the mortality that the $K = 3$ continuously accumulating model would predict for the cohort born in a given year if conditions in the birth year continued throughout the cohort's life span

Figure 16.5 shows the predicted age profiles of mortality based on conditions in several birth years. The graph allows us to examine historical declines in mortality through the lens of age profiles—that is, we can analyze how the effect of age on mortality has changed over time. Except for 1881 conditions, we see continuous improvements over time in mortality at all ages. However, the improvements are larger at young adult ages, and improvements in mortality among people in their 50s and 60s are quite small until recent years. Thus, the rate of increase of mortality with age among adults became more steep from 1881 to 1941, but since then the rate of increase of mortality with age among adults has been roughly constant. In the future, it would be worthwhile to extend this analysis to ages beyond 64 when data quality permits.

16.6 Conclusion

The conventional linear model of additive age, period, and cohort effects has been widely used to analyze tabular population-level data. The literature, however, often concludes that it is impossible to obtain meaningful estimates of the distinct contributions to social change of age, time period, and cohort. The methodological problem underlying this conclusion is well recognized: In the additive model, one must resolve the identification problem induced by the exact linear dependency between age, period, and cohort indicators by imposing some identifying constraint, and in many applications, there is no consensus as to what constitutes a satisfactory constraint.

In this chapter, we emphasize that the APC identification problem is inevitable only under the conventional specification of fixed, additive age, period, and cohort effects. But additive effects are merely one approximation to the process of social change. A prominent example of an alternative process is that of continuously accumulating or evolving cohort effects, described decades ago by social demographers who also noted the absence of procedures for empirically investigating such a process (Hobcraft et al. 1982; Ryder 1965). It is this process that we attempt to model in this chapter.

The new model relaxes the assumption of the conventional additive model that the marginal effect of age is the same in all time periods, the marginal effect of present conditions is the same for people of all ages, and cohorts do not change over time. We show that the failure of identification in the conventional model stems precisely from the strong assumptions it makes. When we generalize the model to allow age profiles to change over time, period effects to have different marginal effects on people of different ages, and cohorts to evolve from one period to the next, we obtain a model that *is* identified. More important, we can better capture the essence of social change by directly modeling the process that generates cohort effects: As they age, cohorts are continuously exposed to influences that cumulatively alter their trajectories. Our substantive model of cohort effects is what allows identification in our model, because the model restricts the possible forms that cohort effects can take.

As an example, our data analysis illustrates the utility of our model in studying the evolution of human mortality in Sweden from 1861 to 2005. The model shows how we can measure whether time effects impact the young or the old and whether they have persistent or transitory impacts. The estimates show that all of the time effects have relatively short half-lives, so that the impact of early-life conditions is unlikely to reach to old age. It would be valuable for further research to examine whether the model produces similar findings in other countries and time periods. In addition, although our model is not designed primarily as a forecasting tool, it could be worthwhile for further analyses to investigate the model's success in forecasting, for example by examining out-of-sample performance.

We believe that, beyond demography, the model can find application in economics, sociology, and other social sciences and can potentially provide new stylized facts that are useful for explaining and evaluating theories of social change and structure. For example, the model could be used to investigate the impact of conditions in early adulthood on earnings throughout the life cycle, the impact of conditions throughout life on family formation, or the impact of events at one age on political preferences at later ages. Although carrying out such applications is beyond the scope of this chapter, we look forward to seeing further analyses in future work.

References

Barbi, E., & Vaupel, J. W. (2005). Comment on "Inflammatory exposure and historical changes in human life-spans." *Science, 308*(5729), 1743a.

Deaton, A., & Paxson, C. (1994). Intertemporal choice and inequality. *Journal of Political Economy, 102*(3), 437–467.

Elo, I. T., & Preston, S. H. (1992). Effects of early-life conditions on adult mortality: A review. *Population Index, 58*(2), 186–212.

Fienberg, S. E., & Mason, W. M. (1985). Specification and implementation of age, period and cohort models. In W. M. Mason & S. E. Fienberg (Eds.), *Cohort analysis in social research* (pp. 45–88). New York: Springer.

Finch, C. E., & Crimmins, E. M. (2004). Inflammatory exposure and historical changes in human life-spans. *Science, 305*(5691), 1736–1739.

Greenberg, B. G., Wright, J. J., & Sheps, C. G. (1950). A technique for analyzing some factors affecting the incidence of syphilis. *Journal of the American Statistical Association, 45*(251), 373–399.

Heligman, L., & Pollard, J. H. (1980). The age pattern of mortality. *Journal of the Institute of Actuaries, 107*, 49–80.

Hobcraft, J., Menken, J., & Preston, S. H. (1982). Age, period, and cohort effects in demography: A review. *Population Index, 48*(1), 4–43.

Horiuchi, S. (1983). The long-term impact of war on mortality: Old-age mortality of the first world war survivors in the Federal Republic of Germany. *Population Bulletin of the United Nations, 15*, 80–92.

Human Mortality Database. (2007). University of California, Berkeley (USA), and Max Planck Institute for Demographic Research (Germany). Accessed November 13, 2007. http://www.mortality.org

James, I. R., & Segal, M. R. (1982). On a method of mortality analysis incorporating age-year interaction, with application to prostate cancer mortality. *Biometrics, 38*(2), 433–443.

Lee, R. D., & Carter, L. R. (1992). Modeling and forecasting U.S. mortality. *Journal of the American Statistical Association, 87*(419), 659–671.

Mason, W. M., & Smith, H. L. (1985). Age-period-cohort analysis and the study of deaths from pulmonary tuberculosis. In W. M. Mason & S. E. Fienberg (Eds.), *Cohort analysis in social research* (pp. 151–228). New York: Springer.

Mason, K. O., Mason, W. M., Winsborough, H. H., & Poole, W. K. (1973). Some methodological issues in cohort analysis of archival data. *American Sociological Review, 38*(2), 242–258.

Moolgavkar, S. H., Stevens, R. G., & Lee, J. A. H. (1979). Effect of age on incidence of breast cancer in females. *Journal of the National Cancer Institute, 62*(3), 493–501.

Noymer, A., & Garenne, M. (2000). The 1918 influenza epidemic's effects on sex differentials in mortality in the United States. *Population and Development Review, 26*(3), 565–581.

Omran, A. R. (1982). Epidemiologic transition. In Ross, J. A. (Ed.), *International encyclopedia of population* (pp. 172–183). New York: Free Press.

Reither, E. N., Masters, R. K., Yang, Y. C., Powers, D. A., Zheng, H., & Land, K. C. (2015). Should age-period-cohort studies return to the methodologies of the 1970s? *Social Science & Medicine, 128*, 356–365.

Robertson, C., Gandini, S., & Boyle, P. (1999). Age-period-cohort models: A comparative study of available methodologies. *Journal of Clinical Epidemiology, 52*(6), 569–583.

Ryder, N. B. (1965). The cohort as a concept in the study of social change. *American Sociological Review, 30*(6), 843–861.

Schulhofer-Wohl, S., & Yang, Y. (2011). *Modeling the evolution of age and cohort effects in social research: Technical appendix*. Accessed December 29, 2014. ftp://ftp.mpls.frb.fed.us/pub/research/sas/sr461/ssw_yy_apc_proofs_sr461.pdf

United Nations. (1997). *United Nations demographic yearbook 1997, historical supplement, table 11*. New York: United Nations.

United Nations. (2003). *Long-range population projections: Proceedings of the United Nations technical working group on long-range population projections*. Accessed December 29, 2014. http://www.un.org/esa/population/publications/longrange/long-range_working-paper_final.PDF

Vaupel, J. W., Manton, K. G., & Stallard, E. (1979). The impact of heterogeneity in individual frailty on the dynamics of mortality. *Demography, 16*(3), 439–454.

Yang, Y. C., & Land, K. C. (2013). Misunderstandings, mischaracterizations, and the problematic choice of a specific instance in which the IE should never be applied. *Demography, 50*(6), 1969–1971.

Chapter 17
Bayesian Ridge Estimation of Age-Period-Cohort Models

Minle Xu and Daniel A. Powers

17.1 Introduction

Over the past few decades the age-period-cohort (APC) model has become a core approach for the investigation of trends in numerous social phenomena in demography and sociology. The application and impact of APC models has spread beyond areas in social sciences to epidemiology and biostatistics. Discussions about the use and applicability of APC models to separate cohort effects from age and period effects on time-specific phenomena originated eighty years ago among social scientists (Mason and Wolfinger 2001).

The age-period-cohort accounting model for age by period tabular data arrays involves three temporal components. The first component, age, specifies variation in the outcome of interest pertaining to different age groups due to the biological process of aging, cumulated social experience, and changes in social roles and statuses. The period component represents influences associated with time periods that affect people of all age groups at the same time because of significant social, cultural, economic, political changes. The cohort component reflects variation related to groups of people who experience an initial event, typically birth or marriage at the same year or years, and undergo subsequent social and historical events at the same ages (Keyes et al. 2010; Yang and Land 2013).[1] For example, age,

[1] Suzuki (2012) provides insight into the nature, conceptualization, and meaning of the underlying temporal dimensions.

M. Xu (✉) • D.A. Powers
Department of Sociology and Population Research Center, University of Texas at Austin, 305 E. 23d Street (G-1800), Austin, TX 78712-1699, USA
e-mail: minle_xu@utexas.edu; dpowers@austin.utexas.edu

R. Schoen (ed.), *Dynamic Demographic Analysis*,
The Springer Series on Demographic Methods and Population Analysis 39,
DOI 10.1007/978-3-319-26603-9_17

period, and cohort are all related to human behavior, with age, period, and cohort making distinct contributions. Eliminating one of the three temporal dimensions can leave results subject to spurious effects (Mason et al. 1973). Despite the theoretical and conceptual rationale for incorporating age, period, and cohort simultaneously in one model to study time-specific social phenomena, no consensus has been reached on how to solve the fundamental identification problem of APC models. This methodological challenge arises from the exact linear relationship between age, period, and (birth) cohort: cohort = period − age, which renders it impossible to obtain valid estimates of the distinct effects of age, period, and cohort from standard regression techniques.

In recent decades, a variety of methods have been proposed to solve the identification problem of APC models. These include, constrained generalized linear models (CGLM), the ridge estimator, the intrinsic estimator, and as well as hierarchical APC-cross-classified fixed- and random-effects models (Fienberg and Mason 1979; Fu 2000; O'Brien et al. 2008; Yang et al. 2004; Yang and Land 2008). Here we review the identification problem inherent in APC models, discuss current solutions to the identification problem in detail, and then introduce a Bayesian ridge model as an alternative to handling the identification problem using data on incidence rates of cervical cancer among Ontario women from 1960 to 1994.

17.2 The Identification Problem

Prior to discussing some existing strategies to address the identification problem in APC models, we first review the classical identification problem. As early as the 1970s, Mason and colleagues (1973) specified the APC multiple classification model for cross-classified data. In the age by period two-way table, the rows and columns represent the main effects of age and period respectively, with the diagonals representing the interaction between age and period—the cohort effects. The APC multiple classification model is specified as

$$g\left(Y_{ij}\right) = \mu + \alpha_i + \beta_j + \gamma_k + \varepsilon_{ij} \tag{17.1}$$

where $i = 1, \ldots, a$ for the ith age group; $j = 1, \ldots, p$ for the jth period; and $k = a - i + j$ for a total of $a + p - 1$ birth cohorts. Y_{ij} denotes the outcome of interest for those from the ith age group in the jth period, $g(.)$ is the link function for a generalized linear model (or a suitable transformation of the response), and μ is the scaled grand mean of the response function. The parameter α_i denotes the fixed effect of the ith age group, β_j denotes the fixed effect of the ith period category, and γ_k denotes the fixed cohort effect associated with age category i and period category j. We can interpret the distinctive effects of age, period, and cohort through an analysis of variance (ANOVA) framework by imposing a centered-effects normalization in which

$$\sum_{i=1}^{a}\alpha_i = \sum_{j=1}^{p}\beta_j = \sum_{k=1}^{a+p-1}\gamma = 0. \quad (17.2)$$

Thus, the APC parameters are normalized so that each APC effect represents a deviation from the grand mean. This normalization is important. As noted by Kupper et al. (1985, p. 818), the choice of other constraints, such as cornered-effects (i.e., dummy-variable) coding "can produce misleading patterns in the estimated coefficients." In a linear model specification, ε_{ij} would denote a random error with mean 0 and variance σ^2. Generalized linear models would not necessarily include an error term or the accompanying residual variance parameter. Typical choices for $g(.)$ include the log and logit links, which yield Poisson regression and binomial logit models, respectively. Here we focus exclusively on a linear model for the empirical logged rates.

With a continuous response, model (17.1) can be written in matrix form as:

$$\mathbf{Y} = \mathbf{X}\boldsymbol{\beta} + \boldsymbol{\varepsilon} \quad (17.3)$$

where $\mathbf{Y}$ is a column vector of outcomes, $\mathbf{X}$ is the design matrix composed of a unit vector and the collection of APC factors normalized using centered-effects coding with the last factor level of each as reference. In Eq. (17.3), we let $\boldsymbol{\beta}$ denote this parameter vector,

$$\boldsymbol{\beta} = \left(\boldsymbol{\mu}, \boldsymbol{\alpha}_1, \ldots, \boldsymbol{\alpha}_{a-1}, \boldsymbol{\beta}_1, \ldots, \boldsymbol{\beta}_{p-1}, \boldsymbol{\gamma}_1, \ldots, \boldsymbol{\gamma}_{a+p-2}\right)^T \quad (17.4)$$

and let ε denote a vector of random errors with mean 0 and variance σ^2. In an identified model, ordinary least squares can be used to obtain estimates of the model parameter vector $\boldsymbol{\beta}$ as

$$\mathbf{b} = \left(\mathbf{X}^T\mathbf{X}\right)^{-1}\mathbf{X}^T\mathbf{Y} \quad (17.5)$$

However, a unique estimator $\mathbf{b}$ does not exist due to the perfect linear dependence among the age, period, and cohort factors. In this case, the design matrix $\mathbf{X}$ is one less than full rank, and $\mathbf{X}^T\mathbf{X}$ is singular and cannot be inverted to solve for $\mathbf{b}$ without special numerical methods such as a Moore-Penrose generalized inverse or singular-value decomposition. In the case of the unconstrained APC model, there are infinitely many solutions of $\mathbf{b}$ that fit the data equally well as a result of this perfect linear dependence. This is the fundamental identification issue pertaining to the unconstrained APC model.

17.3 Current Solutions to the Identification Problem

Several decades ago, scholars started to address the identification problems of APC models. One early method proposed by Mason and colleagues (1973) was to impose at least one constraint on the parameter vector $\boldsymbol{\beta}$. For instance, the effects of two age groups, two periods, or two cohorts can be constrained to be equal with a priori reasoning. With such a constraint, APC models become just-identified and unique estimates of model parameters exist. Even though different choices of equality constraints will not affect model fit, the coefficients and significance of age, period, and cohort may vary considerably and the results can be difficult to interpret with arbitrary choices. Thus, in order to use the constrained generalized linear model (CGLM), it is crucial to justify the constraint a-priori based on theoretical reasons (Glenn 1976). However, such theoretical information, or side information, is not always available and differs in every situation.

Ridge regression is another method commonly used to deal with the identification problem caused by perfect multicollinearity. Ridge regression was proposed over 50 years ago as an estimator to accommodate models with highly correlated predictors (Hoerl 1962; Hoerl and Kennard 1970; Marquardt 1970). Modern variants of ridge regression methods exist today in the form of the lasso and lars estimators (Tibshirani 1996; Efron et al. 2004), which are known collectively as regularization methods. These methods are commonly applied to high-dimensional problems where the goal is to select an optimal subset of predictors having coefficients with minimum variance. Kupper and Janis (1980) were perhaps the first to suggest that ridge regression might be applied to APC models. Fu (2000) applied the ridge estimator to the APC multiple classification model and developed a formal link between ridge regression and the intrinsic estimator. This set the stage for further development, exposition, and adoption of the intrinsic estimator as a general tool for APC analysis (Yang et al. 2004; Yang et al. 2008).

The ridge estimator overcomes the identification issue by adding a penalty to the diagonal of $\mathbf{X}^T\mathbf{X}$ (i.e., the "ridge"). Increasing this penalty shrinks the parameter vector toward 0. Let $\mathbf{X}$ be the $n \times m$ $(m < n)$ design matrix and $\mathbf{I}$ the $m \times m$ identity matrix. Letting λ be the shrinkage or ridge penalty parameter $(\lambda \geq 0)$, the ridge estimator is defined as

$$\mathbf{b}_{\mathrm{R}} = \left(\mathbf{X}^T\mathbf{X} + \lambda\mathbf{I}\right)^{-1}\mathbf{X}^T\mathbf{Y} \tag{17.6}$$

Equation (17.6) shows that the ridge estimator yields biased estimates as $\lambda \to \infty$ (i.e., $\mathbf{b}_{\mathrm{R}} \to \mathbf{0}$). Like other shrinkage estimators, increasing the ridge penalty results in estimates that are biased relative to OLS, but with a smaller mean square error. This tradeoff makes the choice of the shrinkage parameter critical. In the unconstrained APC model, any choice of λ will produce the same model fit when gauged by criteria such as the residual sum of squares. A ridge trace plot is typically examined to show the behavior of the coefficient vector under varying values of λ, however it can be difficult to decide which coefficient vector to accept from the visual plot. Alternatively, cross-validation measures can be constructed to find an optimal value that produces a little bias but substantially lowers the variance. In practice, we need to search for a value of λ that minimizes a generalized cross-validation (GCV) measure (Golub et al. 1979). A popular choice is

$$\mathrm{GCV}\,(\lambda) = \frac{1}{n}\sum_{i=1}^{n}\left(\frac{y_i - \widehat{y}_i}{1 - \mathrm{tr}\,(\mathbf{H})\,/n}\right)^2, \tag{17.7}$$

where $\mathbf{H}$ is the hat, or influence, matrix $\mathbf{H} = \mathbf{X}\left(\mathbf{X}^T\mathbf{X} + \lambda\mathbf{I}\right)^{-1}\mathbf{X}^T$, and tr($\mathbf{H}$) is the sum of diagonal elements of $\mathbf{H}$. This approach requires repeated model fitting over a range of values of λ in search of the minimum GCV value.

The ridge estimator can be considered as a specific restriction on the parameters. In particular, the ridge estimator can be regarded as estimation subject to prior knowledge or belief that smaller absolute values of parameters are more likely than larger absolute values. The variance of the distribution of the parameters would, in turn, depend on the parameter σ_β^2, or the prior variance of β. With no prior beliefs about the magnitude of effects (e.g., as in OLS), $\sigma_\beta^2 = \infty$. With a-priori restrictions on the parameter vector, the ridge penalty would be estimated by the ratio σ^2/σ_β^2 (Draper and Smith 1981). This provides a motivation for the Bayesian ridge regression to be discussed later on.Yang et al. (2004) popularized the intrinsic estimator (IE) to cope with the identification problem of the APC model. The IE is more appropriately viewed as an estimable function that determines both the linear and nonlinear trends in the temporal parameters (Fu et al. 2003). Given that the design matrix $\mathbf{X}$ is one less than full column rank, the parameter space $\mathbf{b}$ of the APC model can be decomposed into the sum of two linear subspaces:

$$\mathbf{b} = \mathbf{B} + t\mathbf{B}_0 \tag{17.8}$$

where t is a real value for a specific solution, $\mathbf{B}_0$ refers to the null subspace corresponding to the zero eigenvalue of $\mathbf{X}^T\mathbf{X}$ and only relies on the design matrix (i.e., the number of age, period, and cohort categories), and $\mathbf{B}$ represents the complement non-null subspace orthogonal to the null space and is the intrinsic estimator (Yang et al. 2008). One way to compute intrinsic estimator is to use the Moore-Penrose generalized inverse of $\mathbf{X}^T\mathbf{X}$ denoted by $\left(\mathbf{X}^T\mathbf{X}\right)^+$ (Fu and Hall 2006):

$$\mathbf{b}_{\mathrm{IE}} = \left(\mathbf{X}^T\mathbf{X}\right)^+\mathbf{X}^T\mathbf{Y}. \tag{17.9}$$

This approach is equivalent to a principal component regression (PCR)

$$\mathbf{b}_{\mathrm{IE}} = \left(\mathbf{Q}\mathbf{L}_0^{-1}\mathbf{Q}^T\right)\mathbf{X}^T\mathbf{Y}, \tag{17.10}$$

where $\mathbf{Q}$ is the $m \times m$ orthonormal matrix of eigenvectors of $\mathbf{X}^T\mathbf{X}$, $\mathbf{L}$ is an $m \times m$ diagonal matrix containing the eigenvalues of $\mathbf{X}^T\mathbf{X}$, $\ell_1, \ldots, \ell_m$, and $\mathbf{QLQ}^T = \mathbf{X}^T\mathbf{X}$. To accommodate the singular design, $\mathbf{L}_0$ in Eq. (17.10) is defined as the $m \times m$ diagonal matrix with eigenvalues $\ell_1, \ldots, \ell_{m-1}, 0$ on the diagonal. The intrinsic estimator is obtained as a principal components regression by eliminating the eigenvector corresponding to the 0 eigenvalue (see, e.g., Yang et al. 2008).

The intrinsic estimator has been shown to be a limiting form of the ridge estimator (Fu 2000), with a vanishingly small shrinkage penalty $\lambda \rightarrow 0^+$, where 0^+ is a value close to, but not equal to, 0. When $\lambda > 0$, the variance of the ridge estimator is smaller than that of the intrinsic estimator. Thus, if λ is set to be a very small positive number, the ridge estimator will produce results nearly equal to those of the intrinsic estimator. Therefore, as noted by Kupper and Janis (1980) and Fu (2000), researchers might choose to use the ridge estimator rather than the intrinsic estimator for APC analysis. However, a difficulty of the ridge estimator lies in determining the optimal value of λ for a given dataset, i.e., a value that produces optimal shrinkage with minimum bias of the APC coefficient vector for that data.

Although the ridge estimator is an accessible approach to deal with the identification problem of the APC model, a suitable method to find the optimal λ for a given dataset presents an added step in modeling. Fu (2000) suggested using a generalized cross-validation (GCV) approach to select an optimal ridge penalty. As noted earlier, this approach requires a series of ridge regressions carried out over a grid of λ values in search of the value yielding the smallest GCV. While this is a straightforward procedure, an alternative approach to determine the optimal shrinkage parameter would be to determine it jointly along with other APC parameters using Bayesian methods. In this case, we obtain the posterior distribution of λ and can readily determine the posterior mean, standard deviation, and other quantiles. A general Bayesian interpretation of the ridge estimator has been recognized since the 1970s (Hsiang 1975; Marquardt 1970). Congdon (2006) described the use of Bayesian ridge priors as a possible solution to multicollinearity. Bayesian approaches to APC modeling are common in epidemiology and biostatistics (see, e.g., Baker and Bray 2005; Berzuini et al. 1994; Knorr-Held and Rainer 2001; Schmid and Held 2007). However, as far as we know, a Bayesian ridge regression approach has not been applied to the APC multiple classification model. Our approach is a relatively straightforward extension of the traditional ridge estimator. In this case, we utilize Bayesian ridge priors to deal with the identification problem of APC model using data from Fu (2000) on cervical cancer incidence among Ontario women from 1960 to 1994.[2] We then compare the results to those obtained using the intrinsic estimator and using a conventional ridge estimator.

[2]Identification may be less an issue using a Bayesian approach where inference is carried out using simulation, as opposed to the traditional numerical methods using least squares.

17.4 Methods

Before introducing the Bayesian ridge approach, we will briefly review Bayesian statistical methods. Unlike the frequentist statistical paradigm that treats a parameter θ as an unknown fixed parameter, the Bayesian statistical approach views θ as a random quantity and uses a prior probability distribution to describe its variation. This prior distribution of θ is updated by taking account of information from the data to obtain the posterior distribution of θ. According to Bayes' theorem, the posterior distribution of θ is summarized as

$$p(\theta|y) = \frac{p(y|\theta)\,p(\theta)}{p(y)}, \tag{17.11}$$

where $p(y|\theta)$ is the likelihood function, $p(\theta)$ is the prior distribution of θ before seeing the data, and $p(y)$ is the marginal distribution of the data defined by marginalizing out θ from the product of the likelihood of the data and the prior distribution of θ, $p(y) = \int_\theta p(y|\theta)\,p(\theta)\,d\theta$. This integral can be complicated and is often analytically intractable. However, since θ is integrated out, $p(y)$ is a normalizing constant that guarantees that $p(\theta|y)$ is a proper density. Therefore, Bayes' theorem is usually expressed as $p(\theta|y) \propto p(y|\theta)\,p(\theta)$. One commonly used Bayes estimator is the mean of the posterior distribution of θ given by

$$\widehat{\theta} = \int_\theta \theta p(\theta|y)\,d\theta. \tag{17.12}$$

Other summary statistics include the posterior median, mode, variance, credible interval, and interquartile range. When the posterior distribution $p(\theta|y)$ is from a known density function, such summary statistics can be easily calculated. However, this is usually not the case, especially when dealing with high-dimensional models. Under such circumstances, Bayesian statisticians have resorted to sampling-based estimation methods—Markov Chain Monte Carlo (MCMC)—to draw inferences about θ. Sample summary statistics calculated based on relatively large samples from the posterior distribution using iterative MCMC methods tend to equate to posterior summary statistics. One useful Markov chain algorithm is the Gibbs sampler (Geman and Geman 1984; Casella and George 1992), which samples iteratively from the full conditional posterior distribution of each parameter obtained from the joint density. Each parameter is updated sequentially and conditionally on all the other parameters. When models involve standard distributions, the conditional posterior distributions of the parameters are also likely to be standard densities and sampling from such conditional posterior distributions is straightforward.

17.5 A Bayesian Ridge Model Specification

The ridge estimator proposed for the APC identification problem can be viewed from a Bayesian perspective (Congdon 2006). For the standard regression model $\mathbf{Y} = \mathbf{X}\boldsymbol{\beta} + \boldsymbol{\varepsilon}$, with $\boldsymbol{\varepsilon}$ distributed normally with mean 0 and variance σ^2, the prior on $\boldsymbol{\beta}$ can be assumed to come from a common normal density with mean zero and variance σ^2/λ. Then the mean of the posterior distribution of $\boldsymbol{\beta}$ has the form $\left(\mathbf{X}^T\mathbf{X} + \lambda\mathbf{I}\right)^{-1}\mathbf{X}^T\mathbf{Y}$, which is identical to the ridge estimator. Different ridge priors for age, period, and cohort coefficients can also be specified. The inclusion of different ridge priors extends the model to the form of a generalized ridge estimator and the posterior mean of $\boldsymbol{\beta}$ then becomes $\left(\mathbf{X}^T\mathbf{X} + \boldsymbol{\Lambda}\mathbf{I}\right)^{-1}\mathbf{X}^T\mathbf{Y}$, where $\boldsymbol{\Lambda}$ represents a vector of λ's. In the case of APC models, $\boldsymbol{\Lambda}$ is a 3×1 vector of values specific to each temporal dimension, and thus allows for differential shrinkage of each APC parameter vector. Noninformative priors are usually adopted so that the inferences are predominantly based on information from the data. However, a Bayesian approach has advantages over the other frequentist methods (e.g., the conventional ridge estimator) because the specification of priors can draw upon information from previous research as well as take into account uncertainties associated with estimating the parameters of the present study. Moreover, priors based on past research may facilitate a more meaningful interpretation of inference.

The data used here to demonstrate and compare the Bayesian ridge prior model with models estimated by the intrinsic estimator and the ridge estimator were originally presented by Fu (2000). These data document cervical cancer incidence rates of Ontario women aged 20 and above from 1960 to 1994. As shown in Table 17.1, there are 98 observations (or data cells), with 14 age groups, 7 period groups, and 20 diagonals of birth cohorts.

Table 17.1 Cervical cancer Incidence rates in Ontario women 1960–1994 (per 10^5 person-years)

Age/Year	60–64	65–69	70–74	75–79	80–84	85–89	90–94
20–24	3.89	3.24	2.90	2.05	2.19	1.76	1.73
25–29	16.01	11.18	8.92	9.74	8.48	7.43	7.54
30–34	26.02	21.14	16.23	15.84	14.54	13.67	12.71
35–39	38.84	25.09	21.07	18.74	18.80	18.04	18.18
40–44	47.65	32.50	22.71	20.01	18.78	16.19	18.12
45–49	51.48	36.69	22.15	19.20	17.74	17.29	18.31
50–54	49.12	37.26	25.51	18.41	16.66	15.41	14.07
55–59	51.48	40.87	34.70	21.83	16.97	17.69	13.73
60–64	47.68	42.80	29.76	22.71	20.16	17.69	16.94
65–69	40.44	39.17	31.44	28.79	23.35	19.26	19.16
70–74	42.40	35.32	27.78	24.31	20.27	20.19	14.95
75–79	42.44	36.68	28.75	25.22	21.17	21.08	19.43
80–84	41.50	29.74	31.54	22.31	20.04	15.25	21.28
85+	30.79	32.43	37.10	19.81	16.42	14.87	12.06

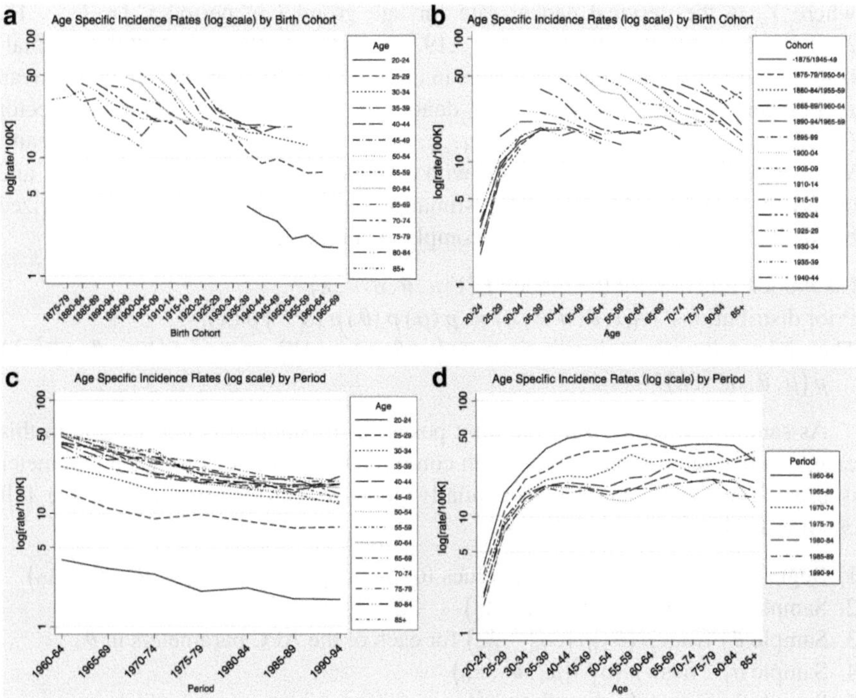

Fig. 17.1 Descriptive statistics (**a**) Age-cohort specific incidence rates (across cohorts) (**b**) Age-cohort specific incidence rates (across ages) (**c**) Age-period specific incidence rates (across periods); (**d**) Age-period specific incidence rates (across ages)

Figure 17.1 provides descriptive information on the trends in these data through various two-way plots of logged age-specific incidence rates (see e.g., Yang and Land 2013). Figure 17.1a shows incidence rates for each age group unfolding by birth cohort, with later cohorts exhibiting uniformly lower incidence at any specific age. Figure 17.1b shows each 5-year cohort's age-specific incidence and reflects the contribution of each birth cohort to the relevant age-specific rates. Figure 17.1c shows period trends across the age groups represented in the data and Fig. 17.1d shows the age specific rates for the 5-year historical periods. As might be expected, incidence rates are generally increasing up to middle age—likely reflecting biological processes—and are moderately decreasing over time—likely reflecting technological improvements in screening and treatment.

While descriptive statistics are useful for exploring empirical patterns, it is generally not possible to gain a concise summary of the temporal trends using graphical depictions alone. This requires a full accounting model. A log transformation is applied to the incidence rates of cervical cancer, yielding the following APC model specification

$$\log\left(Y_{ij}\right) = \mu + \alpha_i + \beta_j + \gamma_k + \varepsilon_{ij}, \tag{17.13}$$

where Y_{ij} is the cervical cancer rate for age group i in period j, $i=1,\ldots,13$, $j=1,\ldots,6$, and birth cohort $k=1,\ldots,19$.[3] A centered-effects or ANOVA normalization is used to center the parameters in model Eq. (17.13) around the grand mean μ. For purposes of exposition, let $\boldsymbol{\theta}$ denote the complete APC parameter vector (i.e., $\boldsymbol{\theta} = (\alpha_1,\ldots,\alpha_{13},\beta_1,\ldots,\beta_6,\gamma_1,\ldots,\gamma_{19})$, which excludes the grand mean). We will denote the Bayesian model with a single ridge prior for age, period, and cohort coefficients as Model 1. The estimation steps for this model are summarized below, but extend generally to more complicated models.

Likelihood function for the model: $f\left(Y|\mu,\boldsymbol{\theta},\sigma^{-2},\lambda\right)$.[4]
Prior distributions: $p\left(\mu,\boldsymbol{\theta},\sigma^{-2},\lambda\right) = p\left(\mu\right)p\left(\boldsymbol{\theta}\right)p\left(\sigma^{-2}\right)p\left(\lambda\right)$.
The joint posterior distribution: $p\left(\mu,\boldsymbol{\theta},\sigma^{-2},\lambda|Y\right) \propto f\left(Y|\mu,\boldsymbol{\theta},\sigma^{-2},\lambda\right) p\left(\mu,\boldsymbol{\theta},\sigma^{-2},\lambda\right)$.

As sampling directly from the joint posterior distribution is not feasible in this case, a Gibbs sampler that works with conditional distributions for each parameter is used. The Gibbs sampler sequentially updates each parameter from the full conditional distribution as follows:

1. Begin with a vector of starting values for all the parameters: $\left(\mu_0,\boldsymbol{\theta}_0,\sigma_0^{-2},\lambda_0\right)$
2. Sample μ_1 from $p\left(\mu_1|\boldsymbol{\theta}_0,\sigma_0^{-2},\lambda_0\right)$
3. Sample $\boldsymbol{\theta}_1$ from $p\left(\boldsymbol{\theta}_1|\mu_1,\sigma_0^{-2},\lambda_0\right)$ for each of the APC parameters in $\boldsymbol{\theta}_1$
4. Sample σ_1^{-2} from $p\left(\sigma_1^{-2}|\mu_1,\boldsymbol{\theta}_1,\lambda_0\right)$
5. Sample λ_1 from $p\left(\lambda_1|\mu_1,\boldsymbol{\theta}_1,\sigma_1^{-2}\right)$
6. Repeat steps 2 through 5: e.g., sample μ_2 from $p\left(\mu_2|\boldsymbol{\theta}_1,\sigma_1^{-2},\lambda_1\right)$

Conditionally conjugate priors are used for all the parameters in the APC model. First, a normal density with $N(0,\sigma^2/\lambda)$ is used as the common prior distribution for all the age, period, and cohort coefficients (i.e., $\boldsymbol{\theta} \sim N\left(\mathbf{0},\sigma^2/\lambda\right)$. A noninformative prior distribution of μ is N(0,1.0E6) and a vague gamma prior is used for the precision of the error term (Gelman, et al. 2013):

$$\sigma^{-2} \sim gamma\left(0.001, 0.001\right). \tag{17.14}$$

The Bayesian ridge penalty for Model 1 may be assigned a noninformative prior

$$\lambda \sim gamma\left(1, 1\right). \tag{17.15}$$

[3]This implies that last category effects are: $\alpha_a = -\sum_{i=1}^{a-1}\alpha_i$, $\beta_p = -\sum_{j=1}^{p-1}\beta_j$, and $\gamma_{a+p-1} = -\sum_{k=1}^{a+p-2}\gamma_k$. The Bayesian approach adopted here makes it straightforward to monitor these quantities along with the other parameters of interest.

[4]Variance components in Bayesian models are typically parameterized in terms of precision, i.e., σ^{-2} rather than variance σ^2.

For the data considered in this example, the posterior means of the age, period, and cohort effects are very similar regardless of the choice of the prior distribution of λ. The choice of a *gamma*(1,1) prior ensures sampling over a wide range of values with a lower bound at 0. Key aspects of the distribution of model parameters can be gained once the Markov chain has run for a large number of iterations. Multiple chains are specified from different starting points to assess convergence and mixing of the chain. The posterior means and standard deviations based on a sequence of M draws of $s = \left(\mu, \boldsymbol{\theta}, \sigma^{-2}, \lambda\right)$ are obtained as summary statistics:

$$\hat{s} = \frac{1}{M}\sum_{i=1}^{M} s_i \tag{17.16}$$

$$\text{std. dev. } (\hat{s}) = \sqrt{\frac{1}{M}\sum_{i=1}^{M}(s_i - \hat{s})^2} \tag{17.17}$$

17.6 APC-Specific Ridge Priors

One idea that fits substantively with the APC theory would be to define different priors for age, period, and cohort effects rather than to use a common ridge prior. Suppose that λ_A, λ_P, and λ_C correspond to the ratio of the error variance to the variances of the age, period, cohort coefficients, respectively. For example, let $\lambda_A = \sigma^2/\sigma_A^2$, with similar expressions applying to the period and cohort effect shrinkage parameters (i.e., λ_P, and λ_C). In other words, the age, period, cohort coefficients are permitted to have distinct variances σ_A^2, σ_P^2, and σ_C^2. In this case, exchangeable ridge priors for the age, period, cohort coefficients can be specified as

$$\alpha_i \sim N\left(0, \sigma^2/\lambda_A\right) \tag{17.18}$$

$$\beta_j \sim N\left(0, \sigma^2/\lambda_P\right) \tag{17.19}$$

$$\gamma_k \sim N\left(0, \sigma^2/\lambda_C\right) \tag{17.20}$$

Under this assumption, the priors used for λ_A, λ_P, and λ_C in Model 2a are

$$\lambda_j \sim gamma\,(1, 1) \qquad j \in \{A, P\} \tag{17.21}$$

and

$$\lambda_C \sim gamma\,(1, 100)\,. \tag{17.22}$$

The choice of an informative *gamma*(1,100) prior ensures that sampling of the cohort penalty takes place over a narrow range. In particular, the prior distribution for the cohort penalty is centered at 0.01 with a variance of 0.0001, yielding sampling bounds of [0,1). We denote this specification (i.e., an informative prior for the cohort penalty) as Model 2a. The prior distributions for the age and period penalties are non-informative (or less informative), and are centered at 1 with a variance of 1, yielding wider sampling bounds over [0,10). The prior distributions of μ and the precision of the error term remain unchanged from those of Model 1. To test the influence of priors on model performance, the priors used for APC-specific shrinkage parameters in Model 2b are defined as:

$$\lambda_j \sim gamma\,(1,1) \qquad j \in \{A, P, C\}, \tag{17.23}$$

which ensures sampling over the range [0,10) for the penalty parameters for all three temporal effects. Table 17.2 provides a summary of the models considered thus far.

In the present study, all analyses are conducted using the R statistical software (R Core Team 2013), with Bayesian inferences conducted using JAGS (Plummer 2003) via the rjags R package (Plummer 2014). The first 500 iterations are used as burn-in and all parameter estimation is based on 20,000 to 50,000 posterior draws and 4 chains.

17.7 Results

Table 17.3 presents estimates of the APC model parameters using the intrinsic estimator, the conventional ridge estimator, a Bayesian model with a common prior on the ridge penalties for the age, period, and cohort effects (Model 1), as well as a Bayesian model with APC-specific shrinkage parameters (Model 2a). The four approaches generate very similar patterns for the age, period, and cohort trends as shown by the estimates and levels of significance. The 95 % credible interval indicates that the significance of age, period, and cohort effects from the Bayesian ridge prior model is consistent with results from the intrinsic and ridge estimators. For instance, the 95 % credible interval for the age effect of the 30–34 age group is (-0.102, 0.190). The inclusion of zero in this interval implies that the age effect for this group is 0. The results from the intrinsic estimator and the ridge estimator also indicate that the risk of cervical cancer among women aged 30 to 34 is insignificant, given that the ratio of the age coefficient to its standard error is less than 1.96. Generalized cross-validation (GCV) is used for selection of the optimal λ for the conventional ridge estimator. The GCV plot shown in Fig. 17.2a illustrates that the minimum value of GCV is about 0.017 corresponding to $\lambda = 0.050$. The posterior distribution λ from the Bayesian implementation is shown in Fig. 17.2b along with posterior mean ($\lambda = 0.080$), which is similar in magnitude to that of the conventional approach. The 95% credible interval indicates the true mean of λ is within the

Table 17.2 Alternative models for estimating age-period-cohort effects

Conventional Models
Intrinsic Estimator: $\mathbf{b}_{\text{IE}} = \left(\mathbf{X}^T\mathbf{X}\right)^{+}\mathbf{X}^T\mathbf{Y}$, where "+" indicates Moore-Penrose (generalized) inverse.
Ridge Estimator: $\mathbf{b}_{\text{R}} = \left(\mathbf{X}^T\mathbf{X} + \lambda\mathbf{I}\right)^{-1}\mathbf{X}^T\mathbf{Y}$, where λ is the ridge penalty parameter.
Bayesian Models
Model 1: Bayesian ridge regression model with common ridge penalty
$\mathbf{Y} \sim N\left(\mu + \boldsymbol{\theta}\mathbf{X}, \sigma^2\right), \quad \boldsymbol{\theta} = \left(\alpha_1, \ldots, \alpha_{a-1}, \beta_1, \ldots, \beta_{p-1}, \gamma_1, \ldots, \gamma_{a+p-2}\right)$ $\mathbf{X}$ centered-effects coded design matrix $\mu \sim N(0, 1.0\text{E}6)$ $\boldsymbol{\theta} \sim N\left(\mathbf{0}, \sigma^2/\lambda\right)$ $\sigma^{-2} \sim gamma\,(1.0\text{E-}3, 1.0\text{E-}3)$ $\lambda \sim gamma\,(1, 1)$
Model 2a: Bayesian ridge regression model with APC-specific ridge penalties
$\mathbf{Y} \sim N\left(\mu + \boldsymbol{\theta}\mathbf{X}, \sigma^2\right), \quad \boldsymbol{\theta} = \left(\alpha_1, \ldots, \alpha_{a-1}, \beta_1, \ldots, \beta_{p-1}, \gamma_1, \ldots, \gamma_{a+p-2}\right)$ $\mathbf{X}$ centered-effects coded design matrix $\mu \sim N(0, 1.0\text{E}6)$ $\alpha_i \sim N\left(0, \sigma^2/\lambda_A\right)$ $\beta_j \sim N\left(0, \sigma^2/\lambda_P\right)$ $\gamma_k \sim N\left(0, \sigma^2/\lambda_C\right)$ $\sigma^{-2} \sim gamma\,(1.0\text{E-}3, 1.0\text{E-}3)$ $\lambda_A \sim gamma\,(1, 1), \; \lambda_P \sim gamma\,(1, 1), \; \lambda_C \sim gamma\,(1, 100)$
Model 2b: Bayesian ridge with APC-specific ridge penalties
$\mathbf{Y} \sim N\left(\mu + \boldsymbol{\theta}\mathbf{X}, \sigma^2\right), \quad \boldsymbol{\theta} = \left(\alpha_1, \ldots, \alpha_{a-1}, \beta_1, \ldots, \beta_{p-1}, \gamma_1, \ldots, \gamma_{a+p-2}\right)$ $\mathbf{X}$ centered-effects coded design matrix $\mu \sim N(0, 1.0\text{E}6)$ $\alpha_i \sim N\left(0, \sigma^2/\lambda_A\right)$ $\beta_j \sim N\left(0, \sigma^2/\lambda_P\right)$ $\gamma_k \sim N\left(0, \sigma^2/\lambda_C\right)$ $\sigma^{-2} \sim gamma\,(1.0\text{E-}3, 1.0\text{E-}3)$ $\lambda_A \sim gamma\,(1, 1), \; \lambda_P \sim gamma\,(1, 1), \; \lambda_C \sim gamma\,(1, 1)$

interval (0.042, 0.138) with 95% probability. In this case, the conventional ridge parameter ($\lambda = 0.050$) is within the 95% credible interval.

Figure 17.3 presents the graphical convergence diagnosis of the MCMC algorithm for selected parameters from Model 1. For each selected parameter, the trace plot shows the posterior sample values of that parameter during the runtime of the chain. The marginal density plot is the smooth histogram of the parameter values from the trace plot. The first three parameters represent the effects of the first age group (20–24), the first period (1960–1964), the first cohort group (–1875). The trace plots provide evidence of satisfactory convergence of the MCMC algorithms for these three parameters. The last three parameters represent the error variance, ridge parameter, and the variance of the APC effects. The trace plots

Table 17.3 Alternative estimates of age-period-cohort effects

					Model 1		95 % credible interval		Model 2a		95 % credible interval	
Parameter	Intrinsic estimator	Std. error	Ridge estimator	Std. error	Posterior mean	Std. dev.	Lower .025	Upper .025	Posterior mean	Std. dev.	Lower .025	Upper .025
Intercept	2.945	−0.014	2.930	0.014	2.941	0.014	2.912	2.969	2.943	0.013	2.918	2.968
Age 20–24	−1.879	0.042	−1.818	0.116	−1.852	0.101	−2.053	−1.656	−1.904	0.147	−2.192	−1.615
Age 25–29	−0.509	0.039	−0.490	0.099	−0.502	0.086	−0.674	−0.334	−0.536	0.125	−0.781	−0.292
Age 30–34	0.047	0.039	0.047	0.084	0.045	0.074	−0.102	0.190	0.022	0.104	−0.183	0.226
Age 35–39	0.316	0.039	0.304	0.070	0.309	0.063	0.184	0.431	0.294	0.084	0.128	0.459
Age 40–44	0.368	0.039	0.351	0.057	0.359	0.053	0.253	0.463	0.350	0.065	0.222	0.478
Age 45–49	0.354	0.040	0.335	0.047	0.344	0.046	0.254	0.434	0.341	0.049	0.245	0.437
Age 50–54	0.243	0.040	0.226	0.041	0.235	0.041	0.153	0.316	0.237	0.038	0.162	0.312
Age 55–59	0.298	0.040	0.280	0.041	0.290	0.041	0.208	0.371	0.297	0.038	0.222	0.373
Age 60–64	0.273	0.040	0.258	0.047	0.267	0.046	0.176	0.358	0.279	0.049	0.183	0.375
Age 65–69	0.278	0.039	0.266	0.057	0.274	0.053	0.169	0.379	0.290	0.066	0.162	0.418
Age 70–74	0.122	0.039	0.118	0.070	0.122	0.063	−0.002	0.248	0.141	0.085	−0.024	0.307
Age 75–79	0.138	0.039	0.140	0.084	0.141	0.074	−0.005	0.287	0.164	0.104	−0.040	0.369
Age 80–84	0.036	0.039	0.047	0.099	0.043	0.086	−0.124	0.213	0.070	0.125	−0.175	0.314
Age 85+	−0.084	0.041	−0.066	0.115	−0.074	0.101	−0.271	0.128	−0.045	0.146	−0.330	0.241
Period 1960–1964	0.476	0.026	0.475	0.056	0.476	0.049	0.380	0.574	0.488	0.070	0.351	0.626
Period 1965–1969	0.270	0.026	0.268	0.042	0.270	0.039	0.194	0.346	0.279	0.050	0.181	0.377
Period 1970–1974	0.081	0.026	0.079	0.031	0.081	0.030	0.021	0.141	0.086	0.032	0.023	0.149
Period 1975–1979	−0.103	0.026	−0.105	0.026	−0.104	0.027	−0.156	−0.050	−0.102	0.024	−0.149	−0.055
Period 1980–1984	−0.190	0.026	−0.190	0.031	−0.190	0.030	−0.249	−0.131	−0.193	0.033	−0.258	−0.129
Period 1985–1989	−0.263	0.026	−0.260	0.042	−0.262	0.038	−0.338	−0.187	−0.271	0.050	−0.368	−0.173
Period 1990–1994	−0.272	0.027	−0.266	0.057	−0.271	0.051	−0.373	−0.172	0.363	0.163	0.043	0.683

Cohort −1875	0.090	0.098	0.060	0.184	0.078	0.164	−0.244	0.399	0.040	0.221	−0.392	0.473
Cohort 1875–1879	0.308	0.070	0.280	0.157	0.293	0.138	0.019	0.562	0.255	0.195	−0.128	0.638
Cohort 1880–1884	0.334	0.058	0.318	0.137	0.324	0.120	0.086	0.558	0.287	0.172	−0.050	0.625
Cohort 1885–1889	0.268	0.052	0.263	0.119	0.262	0.105	0.055	0.466	0.229	0.150	−0.064	0.523
Cohort 1890–1894	0.156	0.047	0.160	0.103	0.155	0.091	−0.024	0.331	0.125	0.127	−0.123	0.374
Cohort 1900–1904	0.180	0.044	0.187	0.086	0.180	0.077	0.027	0.330	0.154	0.106	−0.053	0.362
Cohort 1905–1909	0.133	0.041	0.144	0.071	0.135	0.064	0.008	0.261	0.114	0.085	−0.052	0.281
Cohort 1910–1914	0.210	0.042	0.225	0.059	0.214	0.055	0.105	0.323	0.197	0.067	0.066	0.329
Cohort 1915–1919	0.148	0.043	0.167	0.049	0.155	0.049	0.059	0.251	0.141	0.051	0.041	0.242
Cohort 1920–1924	−0.013	0.043	0.012	0.044	−0.003	0.045	−0.091	0.086	−0.011	0.041	−0.092	0.069
Cohort 1925–1929	−0.133	0.043	−0.104	0.044	−0.120	0.045	−0.207	−0.032	−0.124	0.041	−0.205	−0.044
Cohort 1930–1934	−0.205	0.042	−0.176	0.049	−0.192	0.048	−0.286	−0.097	−0.190	0.051	−0.290	−0.091
Cohort 1930–1939	−0.233	0.041	−0.208	0.058	−0.221	0.055	−0.329	−0.112	−0.213	0.066	−0.343	−0.083
Cohort 1940–1944	−0.233	0.040	−0.219	0.070	−0.226	0.064	−0.351	−0.101	−0.210	0.084	−0.375	−0.045
Cohort 1945–1949	−0.189	0.042	−0.179	0.086	−0.184	0.076	−0.332	−0.034	−0.162	0.105	−0.368	0.044
Cohort 1950–1954	−0.102	0.045	−0.100	0.102	−0.100	0.090	−0.275	0.078	−0.072	0.126	−0.320	0.176
Cohort 1955–1959	−0.138	0.050	−0.142	0.119	−0.138	0.104	−0.340	0.068	−0.104	0.148	−0.395	0.187
Cohort 1960–1964	−0.145	0.057	−0.159	0.137	−0.148	0.119	−0.382	0.088	−0.107	0.171	−0.442	0.227
Cohort 1965–1969	−0.190	0.069	−0.214	0.157	−0.196	0.138	−0.465	0.077	−0.147	0.194	−0.527	0.234
Cohort 1970–1974	−0.245	0.109	−0.317	0.192	−0.268	0.177	−0.615	0.083	−0.203	0.231	−0.035	0.248
σ^2	0.011		0.011		0.011	0.002	0.008	0.016	0.009	0.002	0.006	0.012
λ			0.057		0.080	0.025	0.042	0.138	NA	NA	NA	NA
λ_A									0.030	0.013	0.004	0.056
λ_P									0.175	0.178	−0.174	0.525
λ_C									0.078	0.029	0.021	0.135
σ_A^2									−0.287	0.070	−0.424	−0.151
σ_P^2									0.079	0.060	−0.039	0.197
σ_C^2									0.141	0.090	−0.035	0.317

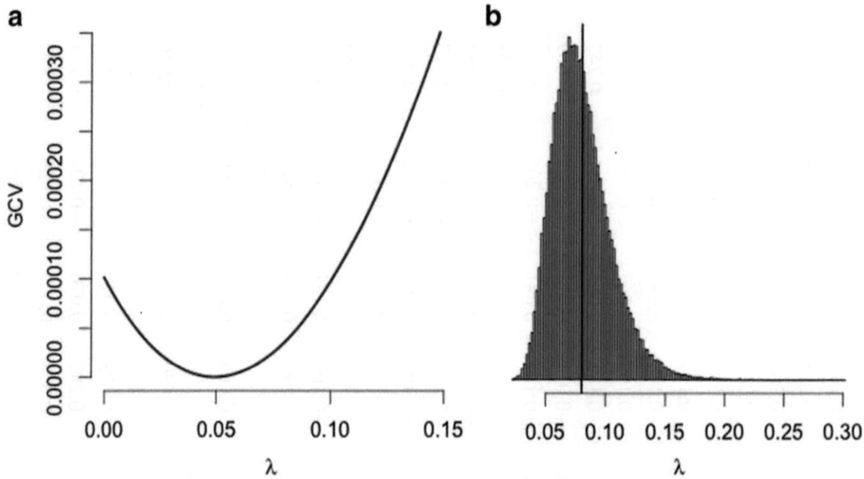

Fig. 17.2 (**a**) Selection of λ for conventional ridge estimator via GCV (**b**) Posterior distribution of λ from Bayesian ridge model (Model 1)

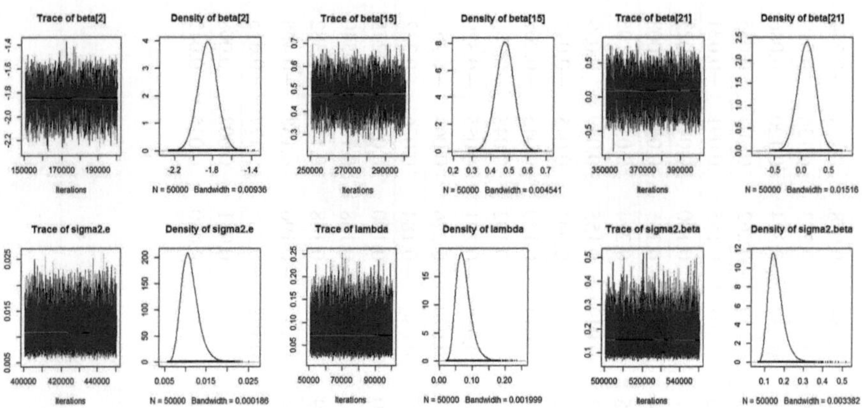

Fig. 17.3 Trace and density plots for the posterior samples for selected parameters (Model 1)

indicate that each chain is mixing well. The Gelman-Rubin (GR) convergence diagnostic is used as a formal test for convergence and assesses whether parallel chains with dispersed initial values converge to the same target distribution. The GR diagnostic shows that the scale reduction factor (SRF) for each parameter is equal to one indicating no difference between the chains for a particular parameter. The multivariate potential SRF is also one, suggesting joint convergence of the chains over all the parameters. Figure 17.4 shows the GR diagnostic plots for selected parameters. For each parameter, the GR plot shows the development of Gelman and Rubin's shrink factor as the number of iterations increases and the shrink factor of each parameter eventually stabilizes around one.

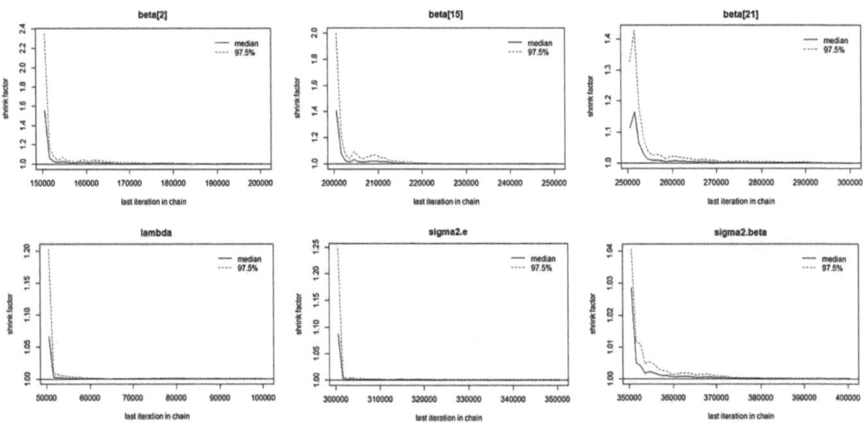

Fig. 17.4 Plots of Gelman-Rubin's diagnostic for selected parameters (Model 1)

Results from Bayesian Model 2a with different ridge priors for age, period, and cohort effects are shown in Table 17.3. The estimated posterior means of the age, period, cohort effects are similar to those from the model with a common prior for the APC effect penalties. However, we see that each coefficient's vector is now subject to differential shrinkage toward zero, with the cohort effects being most affected.

17.8 Model Comparisons

It is recommended to validate results according to side information when possible and to keep conclusions tentative (Glenn 2005). Alternative model specifications may provide additional clues about the plausibility of results. Thus far, we have achieved similar results from Model 1 and Model 2a. We would like to validate these results against Model 2b, as well as with results from alternative approaches. For comparative purposes, we consider two distinct approaches. First, we cast the model in the form of a mixed-effects age-period-cohort model that treats age categories as fixed effects and period and cohort parameters as cross-classified random effects (see, e.g., O'Brien et al. 2008). This approach differs from the hierarchical age-period-cohort cross-classified random effects model (HAPC-CCREM) of Yang and Land (2013), which is applied to repeated cross-sectional micro-level data. In this specification, individual ages—along with additional covariates of interest—are modeled as fixed effects nested within cells created by the cross-classification of birth cohorts and survey years, which are modeled as cross-classified random effects. For tabular rate data, a mixed-effects APC model can be specified as

$$\log(Y_{ij}) = \mu + \alpha_i + u_j + v_k + \varepsilon_{ij}, \tag{17.24}$$

with age parameterized as set of fixed effects $(\alpha_i, \ldots, \alpha_{a-1})$ and period and cohort treated as crossed random effects, u_j and v_k, with the following assumptions

$$u_j \sim N\left(0, \sigma_u^2\right), \quad j = 1, \ldots, p \text{ and } v_k \sim N\left(0, \sigma_v^2\right), \quad k = 1, \ldots, a+p-1, \tag{17.25}$$

with index $k = a - i + j$. We adopt a Bayesian approach with noninformative uniform [0,100] priors for σ_u^2 σ_v^2, and σ_ε^2, and noninformative normal priors for the grand mean μ and the fixed effects of age α_i, specifically $\mu \sim N(0, 1.0\text{E}4)$, and $\alpha_i \sim N(0, 1.0\text{E}4)$, $i = 1, \ldots, a-1$. Centered-effects normalization is used for the age effects, so that $\alpha_a = -\sum_{i=1}^{a-1} \alpha_i$.

Our second approach for comparison uses a table-raking technique known as the median polish, which applies a data-smoothing algorithm to two-way tables (Tukey 1977; Mosteller and Tukey 1977).[5] For the age by period table of log rates, the algorithm works by performing alternating row and column sweeps, which consist of subtracting the age and period median log rates from each cell and replacing each cell entry with a residual. After several iterations, this yields stable row (age) and column (period) effects and an $a \times p$ matrix of residuals, in which both the row and column medians are approximately 0. Cohort effects are the partial interactions of the age (row) and period (column) effects and are recovered from a linear regression of the residuals on the $a+p-1$ cohort dummies using a centered-effects normalization.

Figure 17.5 shows the age, period, and cohort trends from the Bayesian models with different specifications for the ridge penalty priors, along with the mixed-effects APC model and median polish results, which are labeled CCrem and MedPol, respectively. Model 1 refers to the Bayesian model with a common *gamma*(1,1) prior for the ridge parameter. Model 2a, specifies *gamma*(1,1) priors for the ridge penalties λ_A and λ_P, whereas λ_C is distributed as *gamma*(1,100). Model 2b uses a *gamma*(1,1) prior for λ_A, λ_P and λ_C. Figure 17.5 clearly shows that the patterns of age, period, and cohort trends from Model 2a resemble those from the Bayesian model with a common prior (Model 1), and by extension, those based on the IE and conventional ridge regression. However, we have dramatically constrained the sampling range for the cohort ridge penalty in Model 2a. In particular, Model 2a shows significant differences in incidence rates of cervical cancer between the early cohorts (born in the late nineteenth century) and later cohorts (born in the late twentieth century). Model 2b produces age and period patterns similar to those from Model 2a, but with attenuated cohort effects. The implication from Model 2b is that incidence rates of cervical cancer for the early

[5] Keyes et al. (2010) provide an application to APC modeling and a useful comparison with other methods.

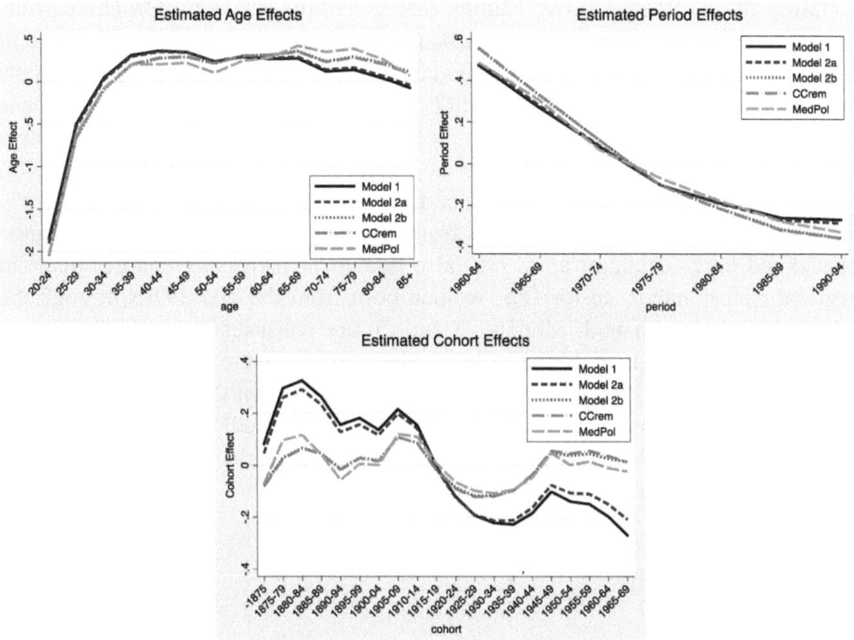

Fig. 17.5 Model comparison of alternative estimates of age, period, and cohort effects of cervical cancer incidence rates in Ontario women

cohorts do not significantly differ from those of later cohorts. This is due to greater shrinkage towards 0 of the cohort parameter vector due to the higher posterior mean of the cohort ridge penalty ($\lambda_C = 1.67$ in Model 2b vs. $\lambda_C = 0.08$ in Model 2a). The mixed-effects (CCrem) specification yields very similar results, with age and period patterns that are similar to the other models and cohort effects that coincide with those from Model 2b. Finally, the median polish (MedPol) fit to the age by period table yields results similar to both Model 2b and the mixed-effects APC model.

It is important to note that the pattern of cohort effects remains consistent across all models, showing a cyclical pattern of increasing and decreasing effects for women born before 1915, decreasing effects for women born from 1915–1935, increasing effects among the 1939–1949 birth cohorts, and moderately decreasing (or flat) effects for cohorts born after 1950. However, the scaling of these effects differs markedly between Models 2a and 2b. The sensitivity of cohort effects to the choice the ridge penalty prior presents a challenge in this case. If we sample over a more restricted range for λ_C using an informative prior as in Model 2a, cohort effects are indistinguishable from those produced by both the conventional ridge regression and Bayesian Model 1. Choosing a noninformative prior (Model 2b) tends to produce considerable shrinkage in the cohort estimates. It should be noted that the cohort effects span a much narrower range compared to the age and period effects, and the absolute difference in the cohort effects from Models 2a and 2b range from 0.0015 to 0.065.

Given these differences, we caution that specifying *distinct* informative priors may require strong justification. In particular, prior knowledge would be needed to impose a less constrained range for the cohort shrinkage parameter for these data, and validation on additional data might be warranted given these results. Evidence from past research indicates increasing cervical cancer incidence among younger Canadian women over time (Arraiz et al. 1990). Similarly, Vizcaino et al. (1998) find increased cervical cancer incidence for U.S. and Canadian women aged 25–49 in the 1935 and later birth cohorts from international data covering the cohorts considered here. Zheng et al. (1996) also document increasing cohort effects in cervical cancer incidence for U.S. women born from the mid-1920s through the mid-1940s. Although both Models 2a and 2b are consistent with this evidence, we generally prefer models that remain agnostic about the prior distribution of the cohort penalty (i.e., Model 2b). Moreover, results from the mixed-effect APC model (CCrem) as well as the median polish technique (MedPol) appear to validate Model 2b. In general, we recommend undertaking comparisons among alternative approaches as a part of the APC model choice process.

17.9 Discussion

The age-period-cohort accounting model serves as a critical framework to study temporal change in phenomena such as mortality, fertility, and disease rates. The importance of separating age, period, and cohort effects for time-specific phenomena poses a challenge in producing unique estimates of age, period, and cohort effects simultaneously due to the perfect linear relationship between age, period and cohort. The last few decades have witnessed a proliferation of methods proposed to deal with the identification problem caused by this particular form of multicollinearity, such as the intrinsic estimator, the ridge estimator, the partial least squares approach of Tu et al. (2011, 2013), the maximum-entropy approach of Browning et al. (2012), and the APC mixed-effects model (O'Brien et al. 2008). In our experience, these approaches tend to agree on overall age-period-cohort trends more often than not.

This paper builds upon the traditional ridge estimator but approaches the identification problem from the Bayesian interpretation of ridge estimation. In so doing, it avoids the inherent limitations related to solving systems of equations in favor of iterated conditional sampling, and offers the added advantages of obtaining the posterior distribution of the ridge penalty parameters along with the parameters of substantive interest. For the data used here, the results from a Bayesian model with a common ridge prior for age, period, and cohort effects are almost identical to those from a traditional ridge estimator and the intrinsic estimator, suggesting that Bayesian ridge regression provides a sensible alternative method in this setting. The Bayesian ridge regression approach has an advantage over the intrinsic estimator in that it does not impose an a-priori 0 shrinkage penalty. It offers advantages over the conventional ridge estimator insofar as it obtains the ridge parameter jointly with the substantive parameters, rather than through a cross-validation exercise. Additionally, the resulting posterior distribution of the shrinkage parameter provides a gauge on both its magnitude and its uncertainty.

Although the optimal ridge parameter enables the model to be identified, it does not have a substantively meaningful interpretation except as a penalty on the magnitude of the APC coefficients. For the Bayesian ridge model, there is no need to assign a single value to the ridge parameter because it is considered a random variable and, as such, researchers are able to incorporate uncertainties about the age, period, and cohort effects for a specific study and simultaneously take advantage of information from existing research. Summary statistics from the posterior samples of the ridge parameter provide more information about this parameter than would be available from cross-validation. In particular, the random property of the ridge parameter in the Bayesian model allows construction of credible intervals to quantify uncertainty.

A natural extension of the Bayesian model with a common prior for the ridge parameter is to define distinct priors for the corresponding ridge parameters for age, period, and cohort effects. This approach accords with the theory of APC modeling in essence and has a potential advantage if prior information on the age, period, and cohort effects is available from meta-analysis based on previous findings. In this case, information from the relevant literature might be incorporated into model estimation by specifying informative priors for age, period, and cohort ridge parameters and the posterior estimation of age-period-cohort effects may more accurately reflect the true trends. However, our study demonstrates that without such information, the choice of alternative prior distributions for the ridge penalty parameters can have important consequences for the posterior means of the temporal effects. In our example data, this is particularly evident with respect to the cohort effects. Based on our limited experience we have found consistent patterning of APC effects on several data sets using alternative estimation approaches. The data used here, present an important departure in light of the distinct cohort effects produced by different methods.

Although this study touches on the sensitivity associated with choices of prior distributions, further work is needed to thoroughly examine the influences of different prior distributions on the APC model performance. As indicated here, one should be cautious when adopting distinct priors for ridge parameters, as the choices of informative priors may have a large influence on posterior inferences, especially in the case of the typical samples sizes associated with tabular APC data. If prior information is unavailable, the use of a noninformative or diffuse prior distribution for the ridge parameters is recommended, as noninformative priors are more objective compared to subjective elicited priors and often lead to Bayesian posterior means close to the maximum likelihood estimates (Congdon 2006). For the results presented here on Ontario cervical cancer incidence, imposing a single common penalty or distinct penalties under strong prior information, produce results consistent with at least two popular approaches. Imposing distinct noniformative priors on the APC ridge penalties produces estimates of cohort effects that are consistent with the mixed-effects specification of the APC model and the median polish procedure. The close similarity of results obtained from different models and procedures leads us to favor Bayesian ridge approaches that make fewer assumptions about the prior distributions of the distinct APC ridge penalties.

References

Arraiz, G. A., Wigle, D. T., & Mao, Y. (1990). Is cervical cancer increasing among young women in Canada? *Canadian Journal of Public Health, 81*, 396–397.

Baker, A., & Bray, I. (2005). Bayesian projections: What are the effects of excluding data from younger age groups? *American Journal of Epidemiology, 162*, 798–805.

Berzuini, C., Clayton, D., & Bernardinelli, L. (1994). Bayesian inference on the Lexis diagram. *Bulletin of the International Statistical Institute, 55*, 149–164.

Browning, M., Crawford, I., & Knoef, M. (2012). *The age-period cohort problem: Set identification and point identification* (CEMMAP working paper CWP02/12). Retrieved from http://dx.doi.org/10.1920/wp.cem.2012.0212

Casella, G., & George, E. I. (1992). Explaining the Gibbs sampler. *The American Statistician, 46*, 167–174.

Congdon, P. (2006). *Bayesian statistical modelling* (Wiley series in probability and statistics). doi:10.1002/9780470035948.

Draper, N. R., & Smith, H. (1981). *Applied regression analysis* (2nd ed.). New York: Wiley.

Efron, B., Hastie, T., Johnstone, I., & Tibshirani, R. (2004). Least angle regression. *The Annals of Statistics, 32*(2), 407–499.

Fienberg, S. E., & Mason, W. M. (1979). Identification and estimation of Age-Period-Cohort models in the analysis of discrete archival data. *Sociological Methodology, 10*, 1–67. doi:10.2307/270764.

Fu, W. J. (2000). Ridge estimator in singular design with application to age-period-cohort analysis of disease rates. *Communications in Statistics Theory and Methods, 29*, 263–278. doi:10.1080/03610920008832483.

Fu, W. J., & Hall, P. (2006). Asymptotic properties of estimators in age-period-cohort analysis. *Statistics and Probability Letters, 76*, 1925–1929. doi:10.1016/j.spl.2006.04.051.

Fu, W. J., Hall, P., & Rohan, T. (2003). *Age-period-cohort analysis: Structure of estimators, estimability, sensitivity and asymptotics*. Technical Report, Department of Epidemiology, Michigan State University, East Lansing.

Gelman, A., Carlin, J. B., Stern, H. S., Runson, D. B., Vehtari, A., & Rubin, D. B. (2013). *Bayesian data analysis* (3rd ed.). Boca Raton: Chapman and Hall/CRC.

Geman, S., & Geman, D. (1984). Stochastic relaxation, Gibbs distributions, and the Bayesian restoration of images. *IEEE Transactions on Pattern Analysis and Machine Intelligence, 6*, 721–741.

Glenn, N. D. (1976). Cohort analysts' futile quest: Statistical attempts to separate age, period and cohort effects. *American Sociological Review, 41*, 900–904.

Glenn, N. D. (2005). *Cohort analysis* (2nd ed.). Thousand Oaks: Sage.

Golub, G. H., Heath, M., & Wahba, G. (1979). Generalized cross-validation as a method for choosing a good ridge parameter. *Technometrics, 21*, 215–223.

Hoerl, A. E. (1962). Application of ridge analysis to regression problems. *Chemical Engineering Progress, 58*, 54–59.

Hoerl, A. E., & Kennard, R. W. (1970). Ridge regression: Biased estimation for non-orthogonal problems. *Technometrics, 12*, 55–67.

Hsiang, T. C. (1975). A Bayesian view on ridge regression. *The Statistician, 24*, 267–268. doi:10.2307/2987923.

Keyes, K. M., Utz, R. L., Robinson, W., & Li, G. (2010). What is a cohort effect? Comparison of three statistical methods for modeling cohort effects in obesity prevalence in the United States, 1971–2006. *Social Science and Medicine, 70*, 1100–1108.

Knorr-Held, L., & Rainer, E. (2001). Projections of lung cancer mortality in West Germany: A case study in Bayesian prediction. *Biostatistics, 2*, 109–129.

Kupper, J. J., & Janis, J. M. (1980). *The multiple classification model in age, period, and cohort analysis: Theoretical considerations* (Institute of Statistics Mimeo No. 1311). Chapel Hill: Department of Biostatistics University of North Carolina.

Kupper, J. J., Janis, J. M., Karmous, A., & Greenberg, B. G. (1985). Statistical age-period-cohort analysis: A review and critique. *Journal of Chronic Disease, 38*, 811–830.

Marquardt, D. W. (1970). Generalized inverses, ridge regression, biased linear estimation, and nonlinear estimation. *Technometrics, 12*, 591–612. doi:10.2307/1267205.

Mason, W. M., & Wolfinger, N. H. (2001). Cohort analysis. *International Encyclopedia of the Social and Behavioral Sciences*, 2189–2194. doi:10.1016/b0-08-043076-7/00401-0.

Mason, K. O., Mason, W. M., Winsborough, H. H., & Poole, W. K. (1973). Some methodological issues in cohort analysis of archival data. *American Sociological Review, 38*, 242–258. doi:10.2307/2094398.

Mosteller, F., & Tukey, J. W. (1977). *Data analysis and regression*. Reading: Addison-Wesley.

O'Brien, R. M., Hudson, K., & Stockard, J. (2008). A mixed model estimation of age, period, and cohort effects. *Sociological Methods & Research, 36*, 402–428.

Plummer, M. (2003). JAGS: A program for analysis of Bayesian graphical models using Gibbs sampling. *Proceedings of the 3rd International Workshop on Distributed Statistical Computing.* Retrieved from http://www.ci.tuwien.ac.at/Conferences/DSC-2003/

Plummer, M. (2014). *rjags: Bayesian graphical models using MCMC*. R package version 3-13. http://CRAN.R-project.org/package=rjags

R Core Team. (2013). *R: A language and environment for statistical computing*. R Foundation for Statistical Computing, Vienna, Austria. URL http://www.R-project.org/

Schmid, V. J., & Held, L. (2007). Bayesian age-period-cohort modeling and prediction – BAMP. *Journal of Statistical Software, 21*(8), 1–15.

Suzuki, E. (2012). Time changes, so do people. *Social Science and Medicine, 75*, 452–456.

Tibshirani, R. (1996). Regression shrinkage and selection via the Lasso. *Journal of the Royal Statistical Society, Series B, 58*(1), 267–288.

Tu, Y. K., Smith, G. D., & Gilthorpe, M. S. (2011). A new approach to age-period-cohort analysis using partial least squares: The trend in blood pressure in Glasgow Alumni Cohort. *Plos One, 6*(4), e19401. 1371/journal.pone. 001901.

Tu, Y. K., Kramer, N., & Lee, W. (2013). Addressing the identification problem in age-period-cohort analysis: A tutorial on the use of partial least squares and principle components analysis. *Epidemiology, 23*, 583–593.

Tukey, J. W. (1977). *Exploratory data analysis*. Reading: Addison-Wesley.

Vizcaino, A. P., Moreno, V., Bosch, F. X., Munoz, N., Barros-Dios, X. M., & Parkin, D. M. (1998). International trends in the incidence of cervical cancer I: Adenocarcinoma and adenosquamous cell carcinomas. *International Journal of Cancer, 75*, 536–545.

Yang, Y., & Land, K. C. (2008). Age-period-cohort analysis of repeated cross-section surveys: Fixed or random effects? *Sociological Methods and Research, 36*, 297–326. doi:10.1177/0049124106292360.

Yang, Y., & Land, K. C. (2013). *Age-period-cohort analysis*. Chapman & Hall/CRC Interdisciplinary Statistics Series. doi:10.1201/b13902.

Yang, Y., Fu, W. J., & Land, K. C. (2004). A methodological comparison of age-period-cohort models: The intrinsic estimator and conventional generalized linear models. *Sociological Methodology, 34*, 75–110. doi:10.1111/j.0081-1750.2004.00148.x.

Yang, Y., Schulehoffer-Wohl, S., Fu, W. J., & Land, K. C. (2008). The intrinsic estimator for age-period-cohort analysis: What it is and how to use it. *American Journal of Sociology, 113*, 1697–1736.

Zheng, T., Hofford, T. R., Ma, Z., Chen, Y., Liu, W., Ward, B. A., & Boyle, P. (1996). The continuing increase in adenocarcinoma of the uterine cervix: A birth cohort phenomenon. *International Journal of Epidemiology, 25*, 252–258.

MIX
Papier aus verantwortungsvollen Quellen
Paper from responsible sources
FSC® C105338

If you have any concerns about our products,
you can contact us on
ProductSafety@springernature.com

In case Publisher is established outside the EU,
the EU authorized representative is:
Springer Nature Customer Service Center GmbH
Europaplatz 3, 69115 Heidelberg, Germany

Printed by Libri Plureos GmbH
in Hamburg, Germany